Fast Solution of Discretized Optimization Problems

Workshop held at the Weierstrass Institute for Applied Analysis and Stochastics, Berlin, May 8–12, 2000

Karl-Heinz Hoffmann
Ronald H.W. Hoppe
Volker Schulz
Editors

Springer Basel AG

Editors:

Karl-Heinz Hoffmann
Center of Advanced European Studies
and Research (caesar)
Friedensplatz 16
53111 Bonn
Germany
e-mail: hoffmann@caesar.de

Ronald H.W. Hoppe
Lehrstuhl für Angewandte Analysis
mit Schwerpunkt Numerik
Universität Augsburg
Universitätsstr. 14
86159 Augsburg
Germany
e-mail: hoppe@math.uni-augsburg.de

Volker Schulz
Weierstrass Institute for Applied Analysis
and Stochastics
Mohrenstrasse 39
10117 Berlin
Germany
e-mail: schulzv@wias-berlin.de

2000 Mathematics Subject Classification 49K20, 49M30, 65K05, 65M32, 65N22, 90C06, 90C20, 90C55

A CIP catalogue record for this book is available from the Library of Congress, Washington D.C., USA

Deutsche Bibliothek Cataloging-in-Publication Data

Fast solution of discretized optimization problems : workshop held at the Weierstrass Institute for Applied Analysis and Stochastics, Berlin, May 8 - 12, 2000 / Karl-Heinz Hoffmann ..., ed.. - Basel ; Boston ; Berlin : Birkhäuser, 2001
(International series of numerical mathematics ; Vol. 138)
ISBN 978-3-0348-9484-5 ISBN 978-3-0348-8233-0 (eBook)
DOI 10.1007/978-3-0348-8233-0

Originally published by Birkhäuser Verlag in 2000
Softcover reprint of the hardcover 1st edition 2000
Printed on acid-free paper produced of chlorine-free pulp. TCF ∞

http://www.birkhauser-science.ch

Table of contents

Preface

Differential equations – partial as well as ordinary – are one of the main tools for the modeling of real world application problems. Pursuing the ultimate aim of influencing these systems in a desired way, one is confronted with the task of optimizing discretized models.

This volume contains selected papers presented at the International Workshop on "Fast Solution of Discretized Optimization Problems", which took place at the Weierstrass Institute for Applied Analysis and Stochastics in Berlin from May 08 until May 12, 2000. The conference was attended by 59 scientists from 10 countries. The scientific program consisted of 8 invited lectures presented by

H. G. Bock (IWR Heidelberg)
M. Heinkenschloss (Rice University, Houston)
K. Kunisch (University of Graz)
U. Langer (University Linz)
B. Mohammadi (University of Montpellier)
J. Petersson (University of Linköping)
E. Sachs (University of Trier)
F. Tröltzsch (Technical University of Chemnitz)

and 28 contributed talks. The aim of this workshop was to foster the exchange of ideas between the still comparatively separated disciplines of nonlinear optimization on the one side and numerical methods for differential equations on the other side. This is necessary for the successful solution of various current optimization problems in practical applications (shape optimization, topology optimization, process optimization ...). Therefore the organizing committee as well as the speakers have come from both these communities.

The various topics discussed at the workshop included

- discretizations for optimization problems,
- large scale optimization,
- iterative linear and quadratic solvers,
- QP-preconditioning,
- sparse QP- and linear solvers,
- suboptimal solution approaches,
- real time optimization,
- interior point methods,
- adaptivity.

Particular emphasis has been laid on numerical methods which are motivated by specific application problems.

The organizers thank the German research foundation (DFG) and the Weierstrass Institute for supporting this workshop. Special thanks go also to Mrs. Giese for her professional assistence in the preparation of the final manuscript of the proceedings volume.

Karl-Heinz Hoffmann, Ronald W. Hoppe, Volker Schulz
March 2001

Block Preconditioners for KKT Systems in PDE–Governed Optimal Control Problems

Astrid Battermann and Ekkehard W. Sachs

Abstract. This paper is concerned with block preconditioners for linear KKT systems that arise in optimization problems governed by partial differential equations. The preconditioners exhibit, like the KKT system, a specific block structure and are composed of preconditioners for submatrices in the general case. Implementation issues are discussed. Numerical results are given for an elliptic boundary control problem from ground water modeling.

1. Introduction

This paper is concerned with a number of block preconditioners for linear systems arising in problems of optimal control that are governed by partial differential equations. A sequential quadratic programming approach is often favored for such problems. This approach requires the solution of a linear quadratic optimal control problem in each step. Upon discretization, linear systems

$$K\,x = r \tag{1}$$

have to be solved with a system matrix K of a specific structure. This matrix is derived from the Karush–Kuhn–Tucker optimality conditions, see Section 1.1. For a short introduction to the different areas where such systems arise, see, e.g., [10].

Since, in the course of optimization, linear systems (1) have to be solved repeatedly, efficient solution procedures are crucial for the overall performance. This makes preconditioning important. It is reviewed in Section 1.2. The systems under consideration exhibit a specific structure that should be exploited in order to develop an efficient algorithm. We assume that the considered linear systems are sparse and of large dimension. Due to these features, we will not consider standard preconditioners for the full system K like, e.g., incomplete decompositions although these can be used efficiently, see, e.g., [28], or like those preconditioners derived from basic iterative schemes, see, e.g., [19]. Instead, we are interested in preconditioners for such matrices K that maintain the block structure of the system and that are composed of subblocks themselves. For the submatrices of K, standard preconditioners can be incorporated into our approach in the same way as preconditioners derived in a multigrid or domain decomposition framework or designed specifically for the underlying partial differential equation. In maintaining the block structure, this work is in one line of thinking with, e.g., [5], [6], [7], [16], [21], [26], [29], [30], [31]. In this work, the indefinite preconditioner analyzed in [6] is compared to the symmetric positive definite preconditioners analyzed in [5].

The preconditioners of the papers [5] and [6] are composed of blocks, e.g.

$$P_1 = \begin{pmatrix} P_\phi^{1/2} & 0 & 0 \\ 0 & P_u^{1/2} & 0 \\ 0 & 0 & P_A P_\phi^{-1/2} \end{pmatrix}, \quad P_4 = \begin{pmatrix} 0 & 0 & P_A^T \\ 0 & P_H & B^T \\ P_A & B & 0 \end{pmatrix}.$$

The subpreconditioners P_ϕ, P_u, P_A, and P_H are – in what we call the "ideal" case – exact inverses for submatrices of K or for products of submatrices, or – in the "general" case – preconditioners for submatrices. The analysis of the preconditioners is given in Sections 2 and 3. We address the eigenvalue distribution of the preconditioned systems. We are also concerned with the applicability of the preconditioners. These two issues – effect and cost – in general represent opposing tasks.

The preconditioners are tested on an example problem arising from an application in ground water modeling. The linear systems under consideration are solved iteratively with the Krylov subspace method MINRES [25] and GMRES [27]. Numerical results for the elliptic model problem are shown in Section 4 for the symmetric positive definite preconditioners of [5] and the indefinite preconditioner of [6].

1.1. Review of the KKT System

In various areas of applied mathematics, linear equations have to be solved where the coefficient matrix has the form

$$\begin{pmatrix} L_{zz} & C^T \\ C & 0 \end{pmatrix}, \tag{2}$$

see, e.g., [10], [33]. In optimization, the matrix for such a Karush–Kuhn–Tucker system is usually called a KKT matrix. We now briefly review the derivation of the KKT matrix for the setting that we have in mind. This will motivate our assumptions on the submatrices.

Upon discretization of a general nonlinear, PDE–constrained problem of optimal control, a nonlinear finite–dimensional equality–constrained optimization problem

$$\min_{(\phi,u)} f(\phi,u) \text{ s.t. } h(\phi,u) = 0 \tag{3}$$

has to be solved, where $f : \mathbb{R}^{m+k} \to \mathbb{R}$ and $h : \mathbb{R}^{m+k} \to \mathbb{R}^m$ for $m, k \in \mathbb{N}$. Here, the decision variables are partitioned into the state variables, $\phi \in \mathbb{R}^m$, and the control variables, $u \in \mathbb{R}^k$. Usually, the dimension m of the state, ϕ, will be considerably larger than the dimension k of the control, u. This condition is especially pronounced for boundary control problems like the one that we consider in our numerical experiments, see Section 4. The partitioning into state and control variables is typical for problems arising from the discretization of an optimal control problem. A popular method for such problems (3) is a sequential quadratic programming (SQP) approach, see, e.g., [7], [20]. SQP is an iterative approach in which successively quadratic subproblems are solved. It also can be viewed as Newton's method applied to the necessary optimality conditions. These are formulated using the Lagrangian function with a multiplier $\lambda \in \mathbb{R}^m$, given by

$$L(\phi, u, \lambda) = f(\phi, u) + \lambda^T h(\phi, u). \tag{4}$$

If z^* is a minimum and if the matrix $\nabla_\phi h(z_*)$ is invertible, then there exists a Lagrange multiplier λ_* such that

$$\nabla_z L(z_*, \lambda_*) = \nabla_z f(z_*) + \lambda_*^T \nabla_z h(z_*) = 0.$$

Hence the necessary optimality conditions are

$$L_\phi = 0, \; L_u = 0, \; h(\phi, u) = 0.$$

An SQP method or Newton's method on the previous nonlinear system requires the solution of the following linear system, the so–called KKT system,

$$K \begin{pmatrix} \phi \\ u \\ \lambda \end{pmatrix} = \begin{pmatrix} L_{\phi\phi} & L_{\phi u} & A^T \\ L_{u\phi} & L_{uu} & B^T \\ A & B & 0 \end{pmatrix} \begin{pmatrix} \phi \\ u \\ \lambda \end{pmatrix} = \begin{pmatrix} -L_\phi \\ -L_u \\ -h \end{pmatrix}. \tag{5}$$

We set $A = h_\phi$, $B = h_u$, and call K the system matrix. The dimensions of the quantities involved are

$$L_{\phi\phi} \in \mathbb{R}^{m\times m}, \; L_{uu} \in \mathbb{R}^{k\times k}, \; L_{\phi u} \in \mathbb{R}^{m\times k}, \; A \in \mathbb{R}^{m\times m}, \; B \in \mathbb{R}^{m\times k}. \tag{6}$$

We also use

$$C = (A\; B), \quad D = -A^{-1}B, \quad L_{zz} = \begin{pmatrix} L_{\phi\phi} & L_{\phi u} \\ L_{u\phi} & L_{uu} \end{pmatrix}, \quad W = \begin{pmatrix} D \\ I_k \end{pmatrix}. \tag{7}$$

Here, W is the representation for the null space of the constraints that is canonical for optimal control problems, see, e.g., [24]. See also the discussion at the end of Section 1.2. The reduced Hessian for the optimization problem is given by

$$H = W^T L_{zz}\, W = B^T A^{-T} L_{\phi\phi} A^{-1} B - L_{u\phi} A^{-1} B - B^T A^{-T} L_{\phi u} + L_{uu}\,. \tag{8}$$

Under second order sufficiency conditions for optimality, the reduced Hessian H for the optimization problem is symmetric positive definite. If A is invertible, the system matrix K is nonsingular if and only if H is nonsingular, see, e.g., [17, Cor. 3.1]. For the analysis of the preconditioners, only the nonsingularity of K (not the symmetric positive definiteness of H) is effectively needed. The assumptions on the system are summarized in Assumption 1.1.

Assumption 1.1. *Let for the system in (5) and its entries the following hold.*

1. *The system K is invertible.*
2. *The matrix A is invertible.*
3. *The matrices $L_{\phi\phi}$ and L_{uu} are symmetric positive semidefinite and $L_{\phi u} = L_{u\phi}^T$.*

Furthermore, note that the matrices A, although nonsingular, are often ill–conditioned. This is the case for the numerical examples given in [1], [5], [7], [26]. In PDE–constrained optimization and optimal control problems, the solution with the matrix A corresponds to solving the discretized governing PDE, and the solution with A^T corresponds to solving the adjoint equation. Typically, this constitutes the main work in solving with any of the subblocks of K.

1.2. Review of Preconditioning

The general issue in preconditioning is to construct a system that is equivalent to the original system and that is easier to solve. With positive definite two–sided preconditioning one solves instead of the original system (1) the equivalent system

$$P^{-1}KP^{-T}\,P^T x = P^{-1}\,r\,. \tag{9}$$

In (9), the symmetry of the original system is maintained. Thus, the widely used Krylov subspace methods for indefinite symmetric matrices MINRES and SYMMLQ [25] can be employed to solve the preconditioned system (9) as well as the original problem (1). Since the performance of iterative solution methods heavily depends on the eigenvalue distribution, one tries to effect favorable changes herein. Preferably, the eigenvalues should be clustered around 1. However, clustering on only one side of the origin cannot be achieved with the usage of a symmetric positive definite preconditioner for K. It is a well known fact that K is indefinite, e.g., [17, Cor. 3.1]. Moreover, if the constraints C have full rank and if L_{zz} is symmetric positive definite on the nullspace of the constraints, then K has m negative and $m+k$ positive eigenvalues.

Although the system matrix K is indefinite, the search for preconditioners of (5) is often restricted to symmetric positive definite preconditioners, e.g., in [16], [26], [30], [31]. The preconditioners of [5], revisited in Section 2, are also symmetric positive definite. The goal to approximate the inverse of the original system matrix is one intuitive reason to admit indefinite preconditioners. For a discussion of definite versus indefinite preconditioning see, e.g., [12], [13], [15]. With left–sided indefinite preconditioning we consider

$$P_L^{-1}K\,x = P_L^{-1}\,r \tag{10}$$

instead of (1). The preconditioned system matrix $P_L^{-1}K$ is in general non–symmetric. To solve the resulting linear systems, one can use extensions of the conjugate gradient method, e.g., symmetric QMR [15] as a special case of the quasi–minimal residual method [14] that can handle symmetric indefinite preconditioners. We chose the most general Krylov subspace method, GMRES [27], for use with the indefinite preconditioner.

If used in conjunction with a symmetric positive definite preconditioner for the symmetric indefinite K, GMRES behaves like MINRES [25], and the solver's performance can be completely described by the eigenvalues alone, see, e.g., [5], [19]. Descriptiveness of the eigenvalues is valid also for GMRES on normal matrices. When the system is non–normal, the issues are considerably more involved, see, e.g., [11], [19]. Different tools have been proposed for the description of convergence behavior, the most prominent being eigenvalues, field of values, and pseudospectra. Each one of them offers specific insight, but can on its own not necessarily provide the full picture. See the works [8], [11], [19], [32] for discussions of these issues, and also [23] for numerical experience with nonsymmetric matrix iterations. Eigenvalue information proves very useful for the system preconditioned with the indefinite preconditioner of [6]. See Section 3.

The null space representation of the constraints that we employ, compare Equations (7) and (8), allows an explicit representation of the reduced Hessian H in terms of the submatrices of K. Note, however, that we do not assume that H can be efficiently assembled in the computations. In fact, the assembly of H will often not be possible, e.g., if only routines are provided to solve with the submatrices of K or to apply them. Sometimes it is convenient to use an orthogonal null space representation $C^T = QR$, see, e.g., [24]. However, the computation of such a decomposition is very expensive and impractical when dealing with large scale problems, impossible even when H cannot be assembled.

2. Three Positive Definite Preconditioners

In [5], three symmetric positive definite preconditioners have been analyzed for KKT systems K with block diagonal upper left part, i.e., $L_{\phi u} = 0$, $L_{u\phi} = 0$. In this paper we are interested in system K with nonzero off–diagonal entries $L_{\phi u}$, $L_{u\phi}$. The preconditioners are considered here in an extended version to account for that. Also, it was assumed in [5] that $L_{\phi\phi}$ and L_{uu} are symmetric positive definite, while in this work we do not want to assume more than positive semidefiniteness unless these are used in the preconditioners themselves. We only address the general form of P_1 in this work, but, for sake of a brief presentation, not the general forms of P_2 and P_3. See [5], also [3], [4] for details.

2.1. First Preconditioner

The first preconditioner is given by P_1 or, equivalently, by its inverse P_1^{-1}, in its general version as

$$\tilde{P}_1 = \begin{pmatrix} P_\phi^{1/2} & 0 & 0 \\ 0 & P_u^{1/2} & 0 \\ 0 & 0 & P_A P_\phi^{-1/2} \end{pmatrix}, \quad \tilde{P}_1^{-1} = \begin{pmatrix} P_\phi^{-1/2} & 0 & 0 \\ 0 & P_u^{-1/2} & 0 \\ 0 & 0 & P_\phi^{1/2} P_A^{-1} \end{pmatrix}.$$

Although $\tilde{P}_1$ has been designed for block diagonal upper part of K, it can be applied for nonzero $L_{\phi u}$, $L_{u\phi}$ as well, leading to the preconditioned system

$$\tilde{P}_1^{-1} K \tilde{P}_1^{-T} = \begin{pmatrix} \tilde{I}_\phi & P_\phi^{-1/2} L_{\phi u} P_u^{-1/2} & P_\phi^{-1/2} \tilde{I}_A P_\phi^{1/2} \\ P_u^{-1/2} L_{u\phi} P_\phi^{-1/2} & \tilde{I}_u & P_u^{-1/2} B^T P_A^{-T} P_\phi^{1/2} \\ P_\phi^{1/2} \tilde{I}_A^T P_\phi^{1/2} & P_\phi^{1/2} P_A^{-1} B P_u^{-1/2} & 0 \end{pmatrix}.$$

Here we used the following notation.

Notation 2.1. *By $\tilde{I}_\phi$ and $\tilde{I}_u$ we denote specifically the symmetrically preconditioned*

$$\tilde{I}_\phi := P_\phi^{-1/2} L_{\phi\phi} P_\phi^{-1/2} \; \tilde{I}_u := P_u^{-1/2} L_{uu} P_u^{-1/2}.$$

The $(1,2)$– and $(2,1)$–entries are obviously zero when the case $L_{\phi u} = 0$, $L_{u\phi} = 0$ is considered. In case the singular values of $P_\phi^{-1/2} L_{\phi u} P_u^{-1/2}$ are small, and the singular values of

$$\tilde{G} = P_\phi^{1/2} P_A^{-1} B P_u^{-1/2} \tag{11}$$

are of moderate size, then iterative methods can be expected to perform better on the preconditioned system than on the original system.

If the exact preconditioner P_1 is used and if $L_{\phi u} = 0$, $L_{u\phi} = 0$, we obtain

$$P_1^{-1} = \begin{pmatrix} L_{\phi\phi}^{-1/2} & 0 & 0 \\ 0 & L_{uu}^{-1/2} & 0 \\ 0 & 0 & L_{\phi\phi}^{1/2} A^{-1} \end{pmatrix}, \quad K_1 = \begin{pmatrix} I_m & 0 & I_m \\ 0 & I_k & G^T \\ I_m & G & 0 \end{pmatrix}.$$

In the ideally preconditioned $K_1 = P_1^{-1} K P_1^{-T}$, G is the ideal version of $\tilde{G}$ in (11),

$$G = L_{\phi\phi}^{1/2} A^{-1} B \, L_{uu}^{-1/2}. \tag{12}$$

In this and the next subsections we often are dealing with the matrix $GG^T + I_m$ and its eigenvalues. Therefore we summarize the results in the following lemma.

Lemma 2.2. *Let Assumption 1.1 hold and let $L_{\phi\phi}$, L_{uu} be symmetric positive definite. Let G be given by (12) and $L_{\phi u} = 0$, $L_{u\phi} = 0$. Then the matrix $GG^T + I_m$ has an eigenvalue 1 with geometric multiplicity of at least $m - k$ and k eigenvalues λ_i $(i = 1, ..., k)$, where each λ_i is also an eigenvalue of the matrix $L_{uu}^{-1} H$.*

Proof. The matrix $GG^T \in \mathbb{R}^{m\times m}$ is rank-deficient and has rank k because G is of rank k. Therefore $GG^T + I_m$ has an eigenvalue 1 with multiplicity $m - k$.
In order to compute the other possibly non-vanishing eigenvalues of GG^T note that all the nonvanishing eigenvalues of $GG^T \in \mathbb{R}^{m\times m}$ are also eigenvalues of $G^T G \in \mathbb{R}^{k\times k}$, because $GG^T x = \lambda x$ implies $G^T G G^T x = \lambda G^T x$. In order to examine the eigenvalues of $G^T G + I_k$ recall that in the case $L_{\phi u} = 0$, $L_{u\phi} = 0$ the reduced Hessian (8) simplifies so that

$$\begin{aligned} L_{uu}^{-1} H &= L_{uu}^{-1} B^T A^{-T} L_{\phi\phi} A^{-1} B + I_k \\ &= L_{uu}^{-1/2} \left(L_{uu}^{-1/2} B^T A^{-T} L_{\phi\phi} A^{-1} B L_{uu}^{-1/2} + I_k \right) L_{uu}^{1/2} \\ &= L_{uu}^{-1/2} \left(G^T G + I_k \right) L_{uu}^{1/2}. \end{aligned}$$

Thus the k eigenvalues of $L_{uu}^{-1} H$ and $G^T G + I_k$ coincide and as previously shown, the k eigenvalues of $G^T G + I_k$ and the k non-unit eigenvalues of $GG^T + I_m$ coincide also. This completes the proof of the lemma. □

We return to the analysis of the eigenvalue distribution of the system K_1.

Theorem 2.3. *Let Assumption 1.1 hold. Let in addition $L_{\phi\phi}$, L_{uu} be symmetric positive definite, let $L_{\phi u} = 0$, $L_{u\phi} = 0$. Let λ_j be the k eigenvalues of $L_{uu}^{-1}H$ ($j = 1, \dots, k$). For the eigenvalues μ_i of the ideally preconditioned system K_1 we obtain*

$$\begin{aligned} \mu_i &= 1 && (i = 1, \dots, k)\,, \\ \mu_i &= 1/2\,(1 \pm \sqrt{5}) && (i = k+1, \dots, 2m-k)\,, \\ \mu_i &= 1/2\,(1 \pm \sqrt{1 + 4\,\lambda_j}) && (i = 2m-k+1, \dots, 2m+k)\,. \end{aligned}$$

Proof. The eigenvalues of K_1 are given as the solution to the set of equations

$$x_1 \quad + \quad x_3 = \mu\, x_1\,, \tag{13}$$

$$x_2 + G^T x_3 = \mu\, x_2\,, \tag{14}$$

$$x_1 + G\, x_2 = \mu\, x_3\,, \tag{15}$$

with $x_1 \in \mathbb{R}^m$, $x_2 \in \mathbb{R}^k$, $x_3 \in \mathbb{R}^m$. We can use the first equation as $x_3 = (\mu - 1)x_1$ to eliminate x_3 and obtain the equations

$$(\mu - 1)G^T x_1 = (\mu - 1)x_2 \tag{16}$$

$$G x_2 = (\mu^2 - \mu - 1)x_1. \tag{17}$$

A first observation is that 1 is an eigenvalue of K_1 with geometric multiplicity k. For $\mu = 1$, we have $x_3 = 0$ and from (17) $G x_2 = -x_1$. There are at most k linearly independent solutions to the latter equation because under the stated assumptions G has rank k.
In the case $\mu \neq 1$, the equation (16) yields $G^T x_1 = x_2$. This is substituted into (17) to obtain $GG^T x_1 = (\mu^2 - \mu - 1)x_1$ or

$$(GG^T + I_m)x_1 = (\mu^2 - \mu)x_1$$

as an equation for the eigenvectors of $GG^T + I_m$. Lemma 2.2 states that its eigenvalues are 1 with multiplicity $m - k$ and λ_i ($i = 1, ..., k$), the eigenvalues of $L_{uu}^{-1}H$. Solving the corresponding equations $\mu^2 - \mu = 1$ and $\mu^2 - \mu = \lambda_i$ yields the statement of the theorem. □

The previous theorem for a symmetric iteration matrix gives an estimate on the number of distinct eigenvalues of K_1. Therefore, under the assumptions of Theorem 2.3 the MINRES iteration in exact arithmetic will terminate after at most $2k + 3$ steps.

Under the assumptions posed in this section, the preconditioner P_1 can be viewed as an approximation to the preconditioner M in [22]. This preconditioner M makes use of the Schur complement S of K, $S = -C\,L_{zz}^{-1}C^T = -(AL_{\phi\phi}^{-1}A^T + BL_{uu}^{-1}B^T)$. We have

$$P_1 P_1^T = \begin{pmatrix} L_{\phi\phi} & 0 & 0 \\ 0 & L_{uu} & 0 \\ 0 & 0 & A\,L_{\phi\phi}^{-1}A^T \end{pmatrix}, \quad \text{while } M = \begin{pmatrix} L_{zz} & 0 \\ 0 & -S \end{pmatrix}.$$

We assume that a solution with A and A^T or their preconditioners P_A and P_A^T, respectively, dominate the computations. Obviously, since P_1 is blockdiagonal, the application of the preconditioner, i.e., of P_1 together with its transpose P_1^T, requires essentially one solution with A and one solution with A^T in each step of an iterative solution method.

2.2. Second Preconditioner

The second preconditioner P_2, given in its ideal version by its inverse

$$P_2^{-1} = \begin{pmatrix} L_{\phi\phi}^{-1/2} & 0 & 0 \\ 0 & L_{uu}^{-1/2} & 0 \\ -L_{\phi\phi}^{-1/2} & -L_{\phi\phi}^{1/2} A^{-1} B L_{uu}^{-1} & L_{\phi\phi}^{1/2} A^{-1} \end{pmatrix}, \tag{18}$$

was designed to transform the original system with $L_{\phi u} = 0$ and $L_{u\phi} = 0$ into the block diagonal system K_2,

$$K_2 = P_2{}^{-1} K P_2{}^{-T} = \begin{pmatrix} I_m & 0 & 0 \\ 0 & I_k & 0 \\ 0 & 0 & -(I_m + GG^T) \end{pmatrix}. \tag{19}$$

The matrix G on the lower diagonal is given in (12).

Theorem 2.4. *Let Assumption 1.1 hold. Let in addition $L_{\phi\phi}$, L_{uu} be symmetric positive definite, let $L_{\phi u} = 0$, $L_{u\phi} = 0$. Let λ_j be the eigenvalues of $L_{uu}^{-1} H$ ($j = 1, \ldots, k$). For the eigenvalues μ_i of the preconditioned system K_2 in (19) we obtain*

$$\begin{aligned} \mu_i &= -\lambda_i & &(i = 1, \ldots, k), \\ \mu_i &= -1 & &(i = k+1, \ldots, m), \\ \mu_i &= 1 & &(i = m+1, \ldots, 2m+k). \end{aligned}$$

Proof. The spectrum of a blockdiagonal matrix is the union of the spectra of the diagonal blocks, see, e.g., [18]. Then the assertion of the Theorem follows directly from Lemma 2.2. □

With the same arguments as in the preceding section we see that MINRES will terminate after at most $k + 2$ steps.

Although the preconditioned system is blockdiagonal for $L_{\phi u} = 0$, $L_{u\phi} = 0$, there are no zero blocks in the preconditioned system when the off–diagonal entries $L_{\phi u}$, $L_{u\phi}$ are nonzero. Applying P_2 for the original K with $L_{u\phi} \neq 0$, $L_{\phi u} \neq 0$, K is transformed into

$$P_2{}^{-1} K P_2{}^{-T} = \begin{pmatrix} I_m & K_{2,12} & -K_{2,12} G^T \\ K_{2,12}^T & I_k & -K_{2,12}^T \\ -G K_{2,12}^T & -K_{2,12} & K_{2,33} \end{pmatrix}, \tag{20}$$

where the entries are given by

$$K_{2,12} = L_{\phi\phi}^{-1/2} L_{\phi u} L_{uu}^{-1/2}, \quad K_{2,33} = -(I_m + (GG^T)) + G\, K_{2,12}^T + K_{2,12}\, G^T.$$

This constitutes a drawback of this preconditioner. Note also that in the non–ideal case, i.e., with preconditioners P_A, P_ϕ, and P_u instead of the exact inverses, the preconditioned system is full even if $L_{\phi u} = 0$, $L_{u\phi} = 0$. See [3]. This renders the eigenvalue analysis very difficult.

The application of the preconditioner P_2 can be done as follows. Let $z = (z_1, z_2, z_3)^T$ with $z_1 \in I\!R^m$, $z_2 \in I\!R^k$, $z_3 \in I\!R^m$, let $x = (x_1, x_2, x_3)^T$ with $x_1 \in I\!R^m$, $x_2 \in I\!R^k$, $x_3 \in I\!R^m$. The transformed vector $x = P_2^{-1} z$ can be computed as follows.

$$\begin{aligned} x_1 &= L_{\phi\phi}^{-1/2} z_1, \\ x_2 &= L_{uu}^{-1/2} z_2, \\ x_3 &= L_{\phi\phi}^T A^{-1}(z_3 - B\, L_{uu}^{-1/2} x_2) - x_1. \end{aligned}$$

It remains to compute $w = P_2^{-T} x$, where $w = (w_1, w_2, w_3)^T$ with $w_1 \in I\!R^m$, $w_2 \in I\!R^k$, $w_3 \in I\!R^m$. This can be accomplished by solving the linear systems

$$\begin{aligned} w_1 &= L_{\phi\phi}^{-1/2}(x_1 - x_3), \\ w_3 &= A^{-T} L_{\phi\phi} x_3, \\ w_2 &= L_{uu}^{-1/2}(x_2 - L_{uu}^{-1/2} B^T w_3). \end{aligned}$$

Under the assumption that the solutions with A and its transpose are the computationally expensive parts of the process, the cost of this preconditioner is comparable to that of P_1.

2.3. Third Preconditioner

A third preconditioner has been considered in [5], essentially transforming the spectrum of K into the spectrum of the reduced Hessian H. See Equation (8). This preconditioner was designed for block diagonal upper part as well, compare [5], but can be extended to the case of nonzero $L_{u\phi}$, $L_{\phi u}$. The preconditioner P_3 is, with this additional modification, given by its inverse as

$$P_3^{-1} = \begin{pmatrix} -B^T A^{-T} & I_k & (B^T A^{-T} L_{\phi\phi} - L_{u\phi}) A^{-1} \\ 0 & 0 & A^{-1} \\ I_m & 0 & -1/2\, L_{\phi\phi} A^{-1} \end{pmatrix}.$$

The preconditioner transforms K such as to obtain the block diagonal system

$$P_3^{-1} K P_3^{-T} = \begin{pmatrix} H & 0 & 0 \\ 0 & 0 & I_m \\ 0 & I_m & 0 \end{pmatrix}. \tag{21}$$

Additional preconditioning of K_3 is possible. Obviously, when $P_H = P_H^{1/2} P_H^{1/2}$ is a preconditioner for H, one can transform (21) into the equivalent system K_3,

$$\begin{pmatrix} P_H^{-1/2} & 0 \\ 0 & I_{2m} \end{pmatrix} P_3^{-1} K P_3^{-T} \begin{pmatrix} P_H^{-1/2} & 0 \\ 0 & I_{2m} \end{pmatrix} = \begin{pmatrix} P_H^{-1/2} H P_H^{-1/2} & 0 & 0 \\ 0 & 0 & I_m \\ 0 & I_m & 0 \end{pmatrix}.$$

See [24] for an approach to precondition the reduced system and [9] for an experience based on this approach. Compare also [7], where the additional preconditioning step is inspired by a quasi–Newton framework.

The eigenvalue distribution of the preconditioned system is easily established.

Theorem 2.5. *Let Assumption 1.1 hold. Let P_H be symmetric positive definite. Let λ_j be the eigenvalues of the subsystem $P_H^{-1}H \in \mathbb{R}^{k\times k}$, $j = 1,\dots,k$. For the eigenvalues μ_i of the preconditioned system K_3 we obtain*

$$\begin{aligned} \mu_i &= -1 && (i = 1,\dots,m)\,, \\ \mu_i &= 1 && (i = m+1,\dots,2m)\,, \\ \mu_i &= \lambda_i && (i = 2m+1,\dots,2m+k)\,. \end{aligned}$$

Similar to the situation with P_1 and P_2 an upper bound for the iterates of MINRES is given by $k+2$ due to the previous theorem.

Note that the application of this preconditioner requires essentially the double amount of work of the preconditioners P_1 and P_2. This is due to the fact that the application of P_3^{-1} involves both A^{-1} and A^{-T}, and that the same is true for P_3^{-T}. Assuming that the cost of solving with A dominates the other computations, we can see that the cost of the application of P_3 is essentially twice the cost of applying the other preconditioners that have been considered.

The application of the preconditioner P_3 can be done in the following way. Note that the vector x is partitioned differently from z and w. Let $z = (z_1, z_2, z_3)^T$ with $z_1 \in \mathbb{R}^m$, $z_2 \in \mathbb{R}^k$, $z_3 \in \mathbb{R}^m$, let $x = (x_1, x_2, x_3)^T$ with $x_1 \in \mathbb{R}^m$, $x_2 \in \mathbb{R}^m$, $x_3 \in \mathbb{R}^k$. The transformed vector $x = P_3^{-1}z$ can be computed as follows. First get

$$x_2 = A^{-1}z_3.$$

Then set $t = L_{\phi\phi}x_2$ to obtain

$$x_1 = z_2 - L_{u\phi}\, x_2 + B^T A^{-T}(t - z_1), \quad x_3 = z_1 - \frac{1}{2}t.$$

The application of P_3^T can be done in a similar way. Let $w = (w_1, w_2, w_3)^T$ with $w_1 \in \mathbb{R}^m$, $w_2 \in \mathbb{R}^k$, $w_3 \in \mathbb{R}^m$. The solution $w = P_3^{-T}x$ is furnished by setting $w_2 = x_1$, solving for $t = A^{-1}Bx_1$, and subsequently determining

$$w_1 = x_3 - t, \quad w_3 = A^{-T}(x_2 - L_{\phi u}\, x_1 + L_{\phi\phi}\,(t - \frac{1}{2}x_3)).$$

3. An Indefinite Preconditioner

In this section we review the preconditioner

$$P_4 = \begin{pmatrix} 0 & 0 & P_A^T \\ 0 & P_H & B^T \\ P_A & B & 0 \end{pmatrix} \tag{22}$$

that was proposed in [6] for use in the iterative solution of linear systems $K\,x = r$ with a system matrix K of the structure given in (5). Instead of using the exact C throughout, compare [21], it allows for preconditioning the component acting on the state,

$$\tilde{C} = (P_A\ B).$$

The reason for this choice is that in the case of a constraint in form of a partial differential equation, A contains the solver routine. For these matrices there are various especially designed preconditioners available. It is sensible in the design a preconditioner for the KKT systems to take advantage of this fact. For sake of a consistent presentation, we restrict our analysis to the case $P_A = A$ in this work. We refer to [6] for the general case.

Like with P_1, both the preconditioner and its inverse are known. For P_H invertible and $P_A = A$, P_4 and its inverse are

$$P_4 = \begin{pmatrix} 0 & 0 & A^T \\ 0 & P_H & B^T \\ A & B & 0 \end{pmatrix}, \quad P_4^{-1} = \begin{pmatrix} DP_H^{-1}D^T & DP_H^{-1} & A^{-1} \\ -P_H^{-1}D^T & P_H^{-1} & 0 \\ A^{-T} & 0 & 0 \end{pmatrix}.$$

Hence we can also compute explicitly the iteration matrix $K_4 = P_4^{-1}K$,

$$\begin{pmatrix} DP_H^{-1}(D^TL_{\phi\phi}+L_{u\phi}) + I_m & D(P_H^{-1}(D^TL_{\phi u}+L_{uu}) - I_k) & 0 \\ P_H^{-1}(D^TL_{\phi\phi}+L_{u\phi}) & P_H^{-1}(D^TL_{\phi u}+L_{uu}) & 0 \\ A^{-T}L_{\phi\phi} & A^{-T}L_{\phi u} & I_m \end{pmatrix}.$$

By left preconditioning with P_4, the linear system (1) is modified to (10), $P_4^{-1}K\,x = \tilde{r}$, where $\tilde{r} = P_4^{-1}\,r$. The preconditioned system is not normal; it is easily observed that $(P_4^{-1}K)^T(P_4^{-1}K) \neq (P_4^{-1}K)(P_4^{-1}K)^T$. Despite the non–normality, the eigenvalues are useful in providing an upper limit for the number of possible iterations of GMRES. In case the constraints are maintained accurately in the preconditioner ($P_A = A$), the spectrum of the preconditioned system consists of the eigenvalues of the low–dimensional system $P_H^{-1}H \in \mathbb{R}^{k\times k}$ and of the request eigenvalue 1. The eigenvalue 1 has high geometric multiplicity $2m$ and low algebraic multiplicity 2. Thus, convergence of GMRES is after an initial delay by a single step well described by the eigenvalue distribution. This follows from a result by Saad and Schultz [27, Prop. 2]. GMRES will (in exact arithmetic) take no more steps than the degree of the minimum polynomial of the relevant system.

In [6] the iteration matrix $P_4^{-1}K$ was analyzed and the following result was shown.

Theorem 3.1. *Let $P_A = A$ and P_H be invertible. Let λ_j be the eigenvalues of the subsystem $P_H^{-1}H$ ($j = 1,\ldots,k$). For the eigenvalues μ_i of the preconditioned system $P_4^{-1}K$ we obtain*

$$\begin{aligned} \mu_i &= \lambda_i & (i = 1,\ldots,k)\,, \\ \mu_i &= 1 & (i = k+1,\ldots,2m+k)\,. \end{aligned}$$

This result has immediate consequences for the convergence of GMRES as shown in [6].

Theorem 3.2. *Let $P_A = A$, and let P_H be invertible. GMRES will terminate in exact arithmetic the iteration on the preconditioned system $P_4{}^{-1}K\,x = P_4{}^{-1}\,r$ after at most $k+2$ iterations.*

When both ideal preconditioners $P_A = A$ and $P_H = H$ are used, K_4 has the single eigenvalue 1 with geometric multiplicity 3. Then, three steps are sufficient to build up the solution in the appropriate Krylov subspaces.

The choice $P_H = H$ can be fairly expensive. Although H is small in dimension and a decomposition should not be prohibitive in case H can be computed explicitly, it is just this assembly of the matrix that might not be possible. If it is only possible to solve with the submatrices involved and to apply them, e.g., if the matrices A and B, are not readily available, H cannot be computed explicitly. Instead, a solution with H then requires several solutions with A and is likely to be very expensive. An alternative approach for a preconditioner for the reduced Hessian could be a quasi–Newton approach, compare [7]. Other inexpensive variants are for instance $P_H = I$ and $P_H = L_{uu}$, if applicable.

Of course, the original (2,2)–entry must be invertible to admit its use as a preconditioner. Then one can apply the previous theorem to derive information about the eigenvalue structure. As in the previous subsections, the eigenvalues of the matrix $L_{uu}^{-1}H$ play an important role.

Similar to how the preceding preconditioners are applied, solutions with P_4 can be done in an efficient way. In order to solve with an iterative solution method

$$P_4^{-1} K\, x = \tilde{r}, \tag{23}$$

$\tilde{r} = P_4^{-1}\, r$, in each iteration the action of $P_4^{-1}K$ on a vector x has to be computed, see, e.g., [2, p. 158ff.]. This can be achieved by the successive solution of three linear subsystems.

Provided that P_A and P_H are invertible, the matrix–vector product

$$P_4^{-1} K\, x = x_+ \tag{24}$$

can be carried out at the cost of solving three linear subsystems, with matrices P_A, P_A^T, and P_H, respectively.

Let $z = (z_1, z_2, z_3)^T$ with $z_1 \in \mathbb{R}^m$, $z_2 \in \mathbb{R}^k$, $z_3 \in \mathbb{R}^m$, and let $x = (x_1, x_2, x_3)^T$ with $x_1 \in \mathbb{R}^m$, $x_2 \in \mathbb{R}^k$, $x_3 \in \mathbb{R}^m$.

Since Equation (24) is equivalent to $K\,x = P_4\, x_+$, and since $K\,x$ can be rewritten $K\,x = P_4\, x + (K - P_4)\, x$, we can compute the action of the matrix $P_4^{-1}K$ on a vector x by solving

$$P_4\,(x_+ - x) = (K - P_4)\, x \tag{25}$$

for $z = x_+ - x$ and then setting $x_+ = x + z$. The solution $z = x_+ - x$ of system (25) can be computed by successively solving the linear systems

$$\begin{aligned} P_A^T\, z_3 &= L_{\phi\phi}\, x_1 + L_{\phi u}\, x_2 + (A - P_A)^T\, x_3 \\ P_H\, z_2 &= L_{u\phi}\, x_1 + (L_{uu} - P_H)\, x_2 - B^T z_3 \\ P_A\, z_1 &= (A - P_A)\, x_1 - B\, z_2\,, \end{aligned}$$

for z_3, z_2, and z_1.

We see that in order to have a practical algorithm, is is necessary not only to solve with the preconditioners P_A, P_A^T, and P_H at moderate cost, but also to be able to apply the preconditioners.

Assuming that solving with A, or with its preconditioner P_A, dominates the computations, the application of the preconditioner P_4 requires essentially one solution with A, respectively P_A, and one solution with A^T, respectively P_A^T, in each step of an iterative solution method. Thus, the cost of P_4 is comparable to the cost of P_1 and P_2, while the cost of P_3 is higher.

4. A Ground Water Modeling Problem

We consider an optimal control problem governed by a partial differential equation. The example problem is motivated by problems of ground water modeling. We assume homogeneous and isotropic conditions in the fully saturated porous medium domain for which the continuum approach is employed. We briefly derive the example problem and state the relevant equations. We subsequently sketch the discretization and turn to the finite–dimensional solution of the optimization problem. For details see [4]. The numerical performance of the preconditioners derived in [5] is compared to that of the preconditioner $\tilde{K}$ explored in the work [6].

4.1. The Optimal Control Problem

We consider a boundary control problem defined on a rectangular domain. It occurs as a mixing problem when extracting ground water on Γ_6 which originates from a stream, represented by the constant potential ϕ_4 on Γ_4, and an additional surface water body on Γ_2.

The height of water on Γ_2 is considered as the control. Leakage occurs from the surface water on Γ_2 into the domain, giving rise to mixed boundary conditions. No–flow conditions are imposed on the boundary parts Γ_1, Γ_3, and Γ_5. The resulting state equation is

$$\begin{aligned}
\Delta\phi(x,z) &= 0 && \text{in } \Omega\,,\\
\tfrac{\partial}{\partial n}\phi(x,z) &= 0 && \text{on } \Gamma_1\cup\Gamma_3\cup\Gamma_5\,,\\
\phi(\bar{x},z) &= \phi_4 && \text{on } \Gamma_4\,,\\
\phi(0,z) &= \phi_6 && \text{on } \Gamma_6\,,\\
\tfrac{\partial}{\partial n}\phi(x,\bar{z}) &= \tfrac{1}{d}\left(u(x)-\phi(x,\bar{z})\right) && \text{on } \Gamma_2\,,
\end{aligned} \tag{26}$$

where, as before, ϕ denotes the state, and u is the control variable. The domain is two–dimensional, thus $\phi=\phi(x,z)$, while the control acts on the boundary part Γ_2 only, $u=u(x)$. The domain is $\Omega=(0,\bar{x})\times(0,\bar{z})$, and its boundary parts Γ_i $(i=1,\ldots,6)$ are defined by

$$\begin{aligned}
\Gamma_1 &= \{(x,\bar{z}) : x\in[0,\bar{x}_1]\}\,, & \Gamma_2 &= \{(x,\bar{z}) : x\in(\bar{x}_1,\bar{x}_2]\}\,,\\
\Gamma_3 &= \{(x,\bar{z}) : x\in(\bar{x}_2,\bar{x}]\}\,, & \Gamma_4 &= \{(\bar{x},z) : z\in[0,\bar{z}]\}\,,\\
\Gamma_5 &= \{(x,0) : x\in[0,\bar{x}]\}\,, & \Gamma_6 &= \{(0,z) : z\in[0,\bar{z}]\}\,.
\end{aligned}$$

It is assumed that the water coming in from Γ_2 is of worse quality than that coming in from Γ_4, thus requiring treatment. Treatment cost for the extracted water is assumed to rise quadratically with inflow from Γ_2. In addition, control cost arises if the water level u deviates from a fixed level u_v on Γ_2. Also relevant is the pumping cost for the well, proportional to the amount of water extracted

FIGURE 1. The geography of the domain Ω.

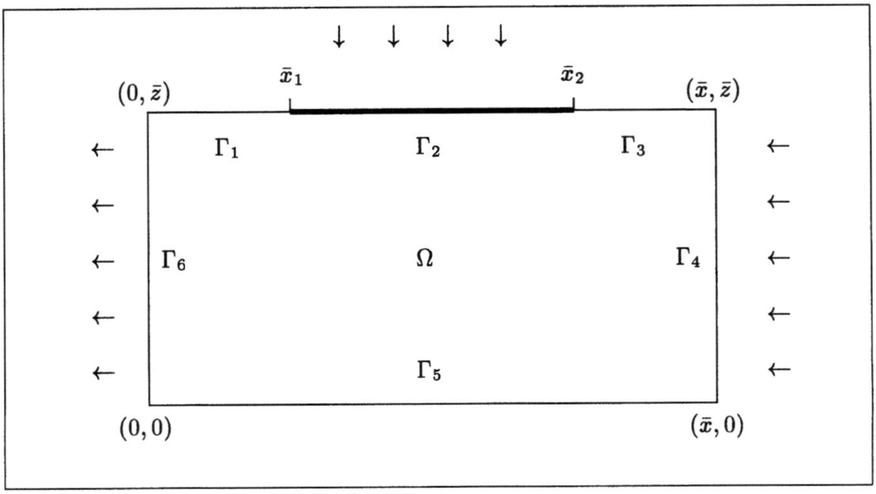

on Γ_6, and the pumping height ϕ_f. For appropriate choices of the cost parameters β_i $(i = 1, 2, 3)$ the goal to minimize the total cost is described by

$$\begin{aligned} \min_{(\phi,u)} \quad & \Big\{\beta_1 \, (\alpha(\bar{x}_2 - \bar{x}_1)\bar{y})^2 \int_{\Gamma_2} (u(x) - \phi(x,\bar{z}))^2 \, dx \\ & + \beta_2 \int_{\Gamma_2} (u(x) - u_v)^2 \, dx + \beta_3 \, k_f \, \bar{z} \, \bar{y} \int_{\Gamma_6} \tfrac{\partial}{\partial n} \phi(0,z)) \, dz \, . \Big\} \end{aligned} \tag{27}$$

4.2. The Discretized Problem, Numerical Results

A uniform grid is defined on the domain Ω, containing m grid points. The piezometric head, $\phi \in I\!R^m$, is defined on all grid vertices. For the boundary Γ_2 there are k grid points, $k \ll m$, on all of which the control u acts. A second order finite difference discretization is applied to the problem (26) which also yields a discrete formulation of (27). The resulting finite dimensional problem is to minimize a quadratic function under linear constraints on the state and control.

As the size of K and its condition number increase, see Table 1, the performance of iterative solvers on the linear system $K\,x = r$ deteriorates considerably. In our numerical experiments, the KKT system is solved with the Krylov subspace methods MINRES [25] and GMRES [27]. On the original system K, MINRES diverges, and GMRES takes an unacceptably high number of steps. The number of required steps is almost as high as the dimension n of the linear system. This is due to the unfavorable eigenvalue distribution of the system, with small eigenvalues decreasing and large eigenvalues increasing as the mesh is refined. See the first lines in Table 2 for the iteration numbers and in Table 3 for the floating point operation numbers. Note that due to the increasing work and storage requirements of GMRES iterations the machine was able only to perform the computations up to the discretization $n_z = 32$. The high computational effort of the unpreconditioned iterative solution, due to the unfavorable eigenvalue distribution of K motivates the use of preconditioners. Discretizations up to $n_z = 128$ can be considered in the preconditioned cases.

TABLE 1. Condition numbers of K and A and number of GMRES iterations on the original K.

n_z	n	m	N	$\kappa(K)$	$\kappa(A)$	#it(K)
4	4	45	94	1356.03	47.974	88
8	8	153	314	11238.04	192.475	294
16	16	561	1138	94713.87	770.496	1028
32	32	2145	4322	832836.00	3082.586	3666

The implementations of MINRES and GMRES that we are using requires only to furnish a routine for matrix–vector multiplication and not the use of the explicitly stored system matrix. Thus, we proceed like it is described in Sections 2 and 3. The stopping criterion for all iterative procedures is a threshold value of 10^{-6} for the residual. All computations are done using *Matlab* on a *Sun Ultra 60*. Iteration counts for GMRES and MINRES, employing the preconditioners P_1, P_2, P_3, and P_4, are given in Table 2 and the corresponding flops in Table 3. The flop counts for the original system are partly in abbreviated form in Table 3. The star $\star$ in Tables 2 and 3 indicates that the machine ran out of memory.

TABLE 2. Iterations of GMRES and MINRES.

n_z	4	8	16	32	64	128
N	94	314	1138	4322	16834	66434
K	88	294	1028	3666	$\star$	$\star$
K_1	18	20	25	30	42	63
K_2	16	19	22	25	34	47
K_3 ($P_H = I$)	5	6	8	8	8	8
K_4 ($P_H = H$)	3	3	3	3	3	3
K_4 ($P_H = I$)	4	4	4	4	4	4

In the computations, exact inverses for the submatrices are used, usually obtained via (complete) LU–decompositions. For P_4, we consider both the ideal case $P_H = H$, and, due to cost issues, also the cheap version $P_H = I$.

As a first observation we note that, as predicted by theory, only three GMRES iterations are necessary for the ideal version of P_4 with $P_A = A$ and $P_H = H$. Since the ideal version of P_4 requires the reduced Hessian H, it is desirable to use more accessible versions of P_H also. Upon use of $P_H = I$, the iteration count is only marginally higher, and still constant over all considered discretizations. The preconditioner P_4 allows to solve the system with a smaller number of iterations

TABLE 3. Megaflops of GMRES and MINRES.

n_z	4	8	16	32	64	128
N	94	314	1138	4322	16834	66434
K	2.73	93.44	$4K$	$191K$	$\star$	$\star$
K_1	0.12	0.73	6.41	64.48	779.12	10,226.02
K_2	0.11	0.71	5.94	58.66	708.18	9,118.92
K_3 ($P_H = I$)	0.05	0.42	4.33	44.86	528.83	6,850.00
K_4 ($P_H = H$)	0.03	0.25	2.90	38.52	548.30	8,091.43
K_4 ($P_H = I$)	0.04	0.27	2.61	30.77	418.49	5,976.46

than the other preconditioners. Similarly, a constant number of iterations seems to be sufficient upon use of P_3. The number of iterations increases with mesh refinement when P_1 and P_2 are used. The application of P_3 allows faster solutions than P_1 and P_2, but it cannot beat the benefits of P_4. Besides iterations, consider also the computational effort. The computational effort is measured in flops, see Table 3. Although only three steps are taken with P_4 employing $P_H = H$, this is not the recommended option. The choice $P_H = I$, requiring four steps, is considerably cheaper. Also cheaper than P_4 with $P_H = H$ is P_3 with eight steps. Nevertheless, P_4, especially in its realistic version with $P_H = I$, compares favorably to the symmetric positive definite preconditioners.

5. Conclusions

Block preconditioners for linear systems of a specific block structure are considered in this work. The linear systems are specifically KKT systems arising in the sequential solution of optimal control problems governed by PDEs. In admitting preconditioners for submatrices with which a solve has to be performed, the preconditioners for the full system are tailored to the application and can be efficiently applied. An indefinite preconditioner and symmetric positive definite preconditioners are compared in this work. Numerical experiments from a problem in ground water modeling substantiate the analysis. While all preconditioners are very effective in reducing overall computational effort, the indefinite preconditioner performs best.

References

[1] E. Arian, A. Battermann, and E. W. Sachs, *Approximation of the Newton step by a defect correction process*, Tech. Rep. 99-12, ICASE, NASA Langley Research Center, Hampton, VA 23681, USA, 1999, submitted.

[2] O. Axelsson, *Iterative Solution Methods*, Cambridge University Press, Cambridge, 1994.

[3] A. Battermann, *Preconditioning of Karush–Kuhn–Tucker Systems Arising in Optimal Control Problems*, Master's thesis, Virginia Polytechnic Institute & State University, Blacksburg, VA 24061, June 1996.

[4] A. Battermann, *Mathematical Optimization Methods for the Remediation of Ground Water Contaminations*, PhD thesis, Univ. of Trier, Germany, 2001.

[5] A. Battermann and M. Heinkenschloss, *Preconditioners for Karush–Kuhn–Tucker Systems Arising in the Optimal Control of Distributed Systems*, in Optimal Control of Partial Differential Equations, Vorau 1996, W. Desch, F. Kappel, and K. Kunisch, eds., Birkhäuser Verlag, 1996, pp. 15–32.

[6] A. Battermann and E. W. Sachs, *An Indefinite Preconditioner for KKT Systems Arising in Optimal Control Problems*, Tech. Rep. XX-XX, Univ. of Trier, Germany, 2000, submitted.

[7] G. Biros and O. Ghattas, *Parallel SQP algorithms for PDE–constrained optimization*, Tech. Rep., Computational Mechanics Laboratory, Carnegie Mellon University, Pittsburgh, PA 15213, USA, 1999, submitted to SC99.

[8] S. L. Campbell, I. C. F. Ipsen, C. T. Kelley, and C. D. Meyer, *GMRES and the Minimal Polynomial*, BIT, **36** (1996), pp. 664–675.

[9] T. D. Choi and C. T. Kelley, *Estimates for the Nash–Sofer preconditioner for the reduced Hessian for some elliptic variational inequalities*, SIAM J. Optim., **9** (1999), pp. 327–341.

[10] I. S. Duff, *The solution of augmented systems*, in Numerical Analysis 1993, D. Griffiths and G. Watson, eds., Longman Sci. Tech. Publ., Harlow, U.K., 1994, pp. 40–55.

[11] M. Embree, *How descriptive are GMRES convergence bounds?*, Tech. Rep. Numerical Analysis Group Research Report NA-99/08, Oxford University Computing Laboratory, Oxford, England, 1999.

[12] B. Fischer, A. Ramage, D. Silvester, and A. Wathen, *Minimal residual methods for augmented systems*, BIT, **38**(3) (1998), pp. 527–543.

[13] R. W. Freund, *Preconditioning of Symmetric, but Highly Indefinite Linear Systems*, Tech. Rep. 97–3–03, Lucent Technologies, Bell Laboratories, Murray Hill, New Jersey 07974–0636, 1997.

[14] R. W. Freund and N. Nachtigal, *QMR: a quasi–minimal residual method for non–Hermitian linear systems*, Numer. Math., **60** (1991), pp. 315–339.

[15] ———, *A New Krylov–Subspace Method for Symmetric Indefinite Linear Systems*, in Proceedings of the 14th IMACS World Congress on Computational and Applied Mathematics, W. Ames, ed., IMACS 1994, 1994, pp. 1253–1256, also available as Tech. Rep. 94–4–07, Lucent Technologies, Bell Laboratories.

[16] P. E. Gill, W. Murray, D. B. Ponceleón, and M. A. Saunders, *Preconditioners for indefinite systems arising in optimization*, SIAM J. Matrix Anal. Appl, **13** (1992), pp. 292–311.

[17] P. E. Gill, W. Murray, M. A. Saunders, and M. H. Wright, *Inertia–controlling methods for general quadratic programming*, SIAM Review, **33** (1991), pp. 1–36.

[18] G. H. Golub and C. F. van Loan, *Matrix Computations*, The Johns Hopkins University Press, 1989.

[19] A. Greenbaum, *Iterative Methods for Solving Linear Systems*, SIAM, Philadelphia, 1997.

[20] M. Heinkenschloss, *Formulation and analysis of a sequential quadratic programming method for the optimal Dirichlet boundary control of Navier–Stokes flow*, in Optimal Control: Theory, Algorithms, and Applications, W. Hager and P. Pardalos, eds., Kluwer Academic Publishers B.V., 1998, pp. 178–203.

[21] C. Keller, N. Gould, and A. Wathen, *Constraint preconditioning for indefinite linear systems*, SIAM J. Matrix Anal. Appl., **21** (2000), pp. 1300–1317.

[22] M. F. Murphy, G. H. Golub, and A. J. Wathen, *A Note on Preconditioning for Indefinite Linear Systems*, SIAM J. Sci. Comp., **21** (2000), pp. 1969–1972.

[23] N. M. Nachtigal, S. C. Reddy, and L. N. Trefethen, *How fast are nonsymmetric matrix iterations?*, SIAM J. Matrix Anal. Appl., **13** (1992), pp. 778–795.

[24] S. Nash and A. Sofer, *Preconditioning reduced matrices*, SIAM J. Mat. Anal. Appl., **17** (1996), pp. 47–68.

[25] C. C. Paige and M. A. Saunders, *Solution of sparse indefinite systems of linear equations*, SIAM J. Numer. Anal., **12** (1975), pp. 617–629.

[26] T. Rusten and R. Winther, *A preconditioned iterative method for saddlepoint problems*, SIAM J. Matrix Anal. Appl., **13** (1992), pp. 887–904.

[27] Y. Saad and M. H. Schultz, *GMRES: A Generalized Minimal Residual Algorithm for solving nonsymmetric linear systems*, SIAM J. Sci. Stat. Comput., **7** (1986), pp. 856–869.

[28] V. Schulz, *Incomplete indefinite decompositions as multigrid smoothers for KKT systems*, Internat. Ser. Numer. Math., **133** (1999), pp. 257–266.

[29] V. Schulz and G. Wittum, *Transforming smoothers for optimization saddlepoint problems*, Tech. Rep. 99-07, Interdisciplinary Center for Scientific Computing, Univ. of Heidelberg, Germany, 1998.

[30] D. Silvester and A. Wathen, *Fast iterative solution of stabilized Stokes systems part I: using simple diagonal preconditioners*, SIAM J. Numer. Anal., **30** (1992), pp. 630–649.

[31] ———, *Fast iterative solution of stabilized Stokes systems part II: using general block preconditioners*, SIAM J. Numer. Anal., **31** (1994), pp. 1352–1414.

[32] L. N. Trefethen, *Pseudospectra of Matrices*, in Numerical Analysis 1991, D. Griffiths and G. Watson, eds., Longman Sci. Tech. Publ., 1991, pp. 234–266.

[33] M. H. Wright, *Some linear algebra issues in large-scale optimization*, in Linear Algebra for Large-Scale and Real-Time Applications, M. Moonen, G. Golub, and B. DeMoor, eds., Kluwer Academic Publishers, 1993, pp. 315–337.

Iterative Multiscale Methods for Process Monitoring*

Thomas Binder, Luise Blank, Wolfgang Dahmen, and Wolfgang Marquardt

Abstract. Process monitoring via state estimation is accomplished by the solution of an appropriately posed optimization problem on a receding horizon. The need of a fast solution of the optimization problem is additionally restricted by the a priorily unknown available computation time. Therefore, we propose a multiscale approach with the following features. It provides a coarse approximation after a very short time and upgrades iteratively approximations from coarser levels through further refinement steps. At each refinement step the error is to be reduced by a fixed fraction according to the discretization error while the corresponding computational cost is to remain proportional to the current number of degrees of freedom. Our multiscale discretization is based on suitable wavelet bases. It will be presented along with numerical results as well as with open questions.

1. Problem formulation

Monitoring chemical processes requires real time estimates of the states, which are typically not accessible by measurement, and of the modeled output functions from a few given, possibly noisy data. In case of nonlinear dynamic modeling state estimation can be formulated as a *nonlinear constrained optimization problem*, where the difference between modeled output and measured data on a receding, moving horizon is to be minimized [14, 16, 17]. The dynamic behaviour of the plant is usually modeled by a system of differential–algebraic equations (DAEs) which together with bounds on selected variables form the constraints. Figure 1 illustrates the situation for a simple one–dimensional setting. The present time is denoted by t_k, and we assume that the noisy discrete data, which are known for the receding time, are processed into denoised signal functions $\boldsymbol{z}$. The control functions are denoted by $\boldsymbol{u}$ and the parameters by $\boldsymbol{p}$. For state estimation both, $\boldsymbol{u}$ and $\boldsymbol{p}$ are given. The goal is to estimate on the receding fixed time interval $[t_{k-m}, t_k]$ the state functions $\boldsymbol{x}$ and the output functions $\boldsymbol{y}$, which correspond to the signals. The solution has to be determined in a computation interval, which is given by the

*This work has been supported by the "'Deutsche Forschungsgemeinschaft"' under grant no. MA1188/6.

dynamic of the process. Then, the time horizon will be moved to $[t_{k-m+1}, t_{k+1}]$ and the estimation process will start again with new additional data. Moreover, additive model correction terms are introduced into the DAEs as functions $\boldsymbol{v}$ and $\boldsymbol{w}$ which have to be estimated as well. The introduction of model correction terms actually may enhance ill–posedness. Therefore, at any rate, regularization should include the correction terms $\boldsymbol{v}$ and $\boldsymbol{w}$ as quadratic terms in the cost functional [2].

FIGURE 1. Receding horizon

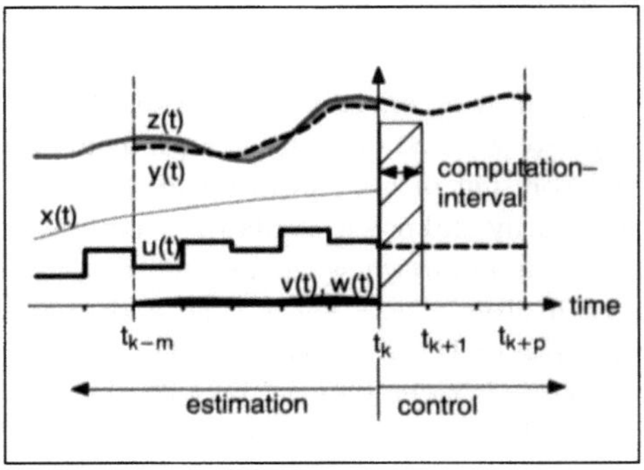

For given $\boldsymbol{z} \in (L_2)^{n_y}$, $\boldsymbol{p}$ and $\boldsymbol{u} \in (L_2)^{n_u}$ and unknown functions $\boldsymbol{x} \in \left(H^1\right)^{n_x}$, $\boldsymbol{y} \in (L_2)^{n_y}$, $\boldsymbol{v} \in (L_2)^{n_v}$ and $\boldsymbol{w} \in (L_2)^{n_w}$ with $n_w \leq n_x$, $n_v \leq n_y$, the resulting optimization problem has the form

$$\min_{\boldsymbol{x},\boldsymbol{y},\boldsymbol{v},\boldsymbol{w}} \int_{t_{k-m}}^{t_k} \left\{(\boldsymbol{y}-\boldsymbol{z})^T \boldsymbol{Q}(\boldsymbol{y}-\boldsymbol{z}) + \boldsymbol{v}^T \boldsymbol{R_v} \boldsymbol{v} + \boldsymbol{w}^T \boldsymbol{R_w} \boldsymbol{w}\right\} \, d\tau \tag{1}$$

with constraints

$$\dot{\boldsymbol{x}} - \boldsymbol{f}(\boldsymbol{x},\boldsymbol{u},\boldsymbol{p}) - \boldsymbol{W}\boldsymbol{w} = 0 \tag{2}$$

$$\boldsymbol{y} - \boldsymbol{g}(\boldsymbol{x},\boldsymbol{u},\boldsymbol{p}) - \boldsymbol{V}\boldsymbol{v} = 0 \tag{3}$$

$$\boldsymbol{l_x}(t) \leq \boldsymbol{x}(t) \leq \boldsymbol{s_x}(t) \tag{4}$$

$$\boldsymbol{l_y}(t) \leq \boldsymbol{y}(t) \leq \boldsymbol{s_y}(t)\,. \tag{5}$$

The dimension of the problem is dominated by the number of state functions n_x. We want to emphasize that state estimation is an ill–posed inverse problem. In the above regularized problem the initial data are still not given but have to be estimated. Additionally, the observability measure of the model, the stiffness of the DE's, the incorporated model error functions and the sensible choice of their regularization terms $\boldsymbol{R_v}$ and $\boldsymbol{R_w}$ have a major influence on the conditioning of the formulated optimization problem. In Section 8 these influences are shortly sketched. For more details we refer the reader to [3, 5].

2. Real time requirement and conceptional framework

In practice, typically several software processes are running simultaneously on one work station. As a consequence the available computation time is a priorily unknown and the process dynamic only gives an upper limit. This suggests a new view on **real time requirement**: we have to *provide an approximate solution at any time* in $[t_k, t_{k+1}]$. Typically, available solvers are based on a conservative estimation of computation costs and a fixed complexity and are therefore not designed to fulfill this restriction. The conceptional framework we want to present now hinges on this real–time requirement.

Firstly, a *unified framework* for signal analysis and the optimization problem (1)–(5) facilitates a very fast data exchange between these two blocks. Secondly, we start with a very coarse approximation and then *successively refine* it by an *adaptive strategy*, where the refinement costs should remain proportional to the current number of freedom, and the acquired informations should be exploited as much as possible for the refinement step. Figure 2 sketches our approach. The data input is processed by denoising, compressing and interpolation schemes to produce signal functions $\boldsymbol{z}$. This step provides also an initial mesh, respectively a first set of basis functions Λ^0, for discretizing the optimization problem. An NLP–solver determines a coarse grid solution of the optimization problem. Then, if time permits, this solution is updated in an adaptive way. Therefore, the coarse grid solution of the optimization problem as well as the denoised signals function as indicator for the new, refined mesh (basis index set $\Lambda^{\ell+1}$). The optimization problem is resolved on this refined mesh by exploiting additional signal information as well as any available information from previous stages. As long as time permits this refinement procedure is repeated, until eventually the time horizon is moved according to the dynamics of the process. The subsequent calculations exploit again the information of the previous time horizon.

FIGURE 2. Concept

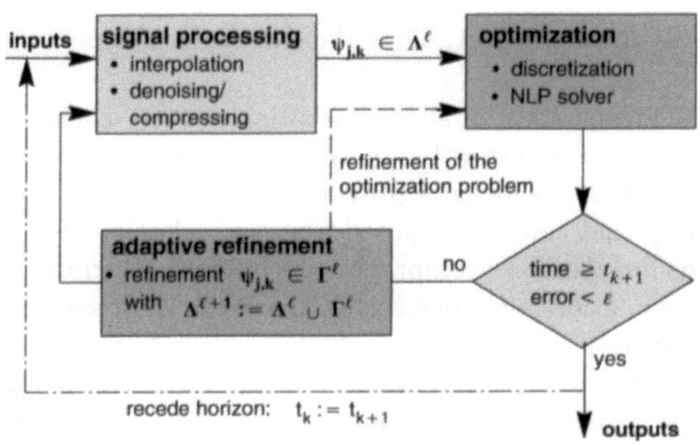

We propose to realize the above concept with the aid of wavelet methods for following reasons. Firstly, in signal processing wavelets are a very efficient tool for denoising and compression [8, 9]. Secondly, multiscale representation avoids the change of basis in the updating procedure. Moreover, wavelet properties can be exploited for example for preconditioning and adaptive refinement. In the next section, we shortly introduce biorthogonal wavelets and some of their basic properties and indicate their significance in the present context. For a more detailed treatment of wavelets we refer the reader to [9, 10, 11, 13].

3. Wavelets

Apart from boundary effects wavelet functions $\psi_{j,k}$ are obtained by dilation and translation of a suitable *mother wavelet* ψ, i.e. $\psi_{j,k} = 2^{j/2}\psi(2^j \cdot -k)$, where the normalizing factor $2^{j/2}$ keeps the L_2–norm of the $\psi_{j,k}$ independent of j, k. Thus when ψ is chosen to have bounded support, as will always be the case in our applications, one has $\operatorname{diam}\operatorname{supp}\psi_{j,k} \sim 2^{-j}$, $j \to \infty$. The key feature is that for suitable ψ the collection $\boldsymbol{\Psi} = \{\psi_{j,k}\}$ forms a *Riesz basis* of L_2, i.e., every $f \in L_2$ has a unique expansion

$$f \quad = \quad \boldsymbol{\Psi}^T \boldsymbol{d}_f = \sum_{j,k} d_{j,k}\psi_{j,k} \tag{6}$$

and

$$c_1 \parallel \boldsymbol{d}_f \parallel_{\ell_2} \leq \parallel f \parallel_{L_2} \leq c_2 \parallel \boldsymbol{d}_f \parallel_{\ell_2} \tag{7}$$

holds for some positive constants c_1, c_2 independent of f. Obviously in the special case when the $\psi_{j,k}$ form an orthonormal basis, (7) holds with $c_1 = c_2 = 1$ and hence equality holds between coefficient and function norm. In general (7) implies the existence of a *dual basis* $\tilde{\boldsymbol{\Psi}}$ in L_2 which is *biorthogonal* to $\boldsymbol{\Psi}$, i.e.

$$\langle\langle \boldsymbol{\Psi}, \tilde{\boldsymbol{\Psi}} \rangle\rangle = \boldsymbol{I} \tag{8}$$

where we use the shorthand notation $\langle\langle \boldsymbol{\Psi}, \tilde{\boldsymbol{\Psi}} \rangle\rangle = \Big(\langle \psi_{j,k}, \tilde{\psi}_{i,l} \rangle \Big)_{(j,k),(i,l)}$ and $\langle \cdot, \cdot \rangle$ is the standard L_2 inner product. Thus the wavelet coefficients $d_{j,k}$ are given by $\langle f, \tilde{\psi}_{j,k} \rangle$.

Clearly, (7) means that small perturbations of the coefficients cause only small changes of f in the norm $|| \cdot ||_{L_2}$ and vice versa. In particular, retaining only the first N largest coefficients provides up to a uniform constant the best approximation to f that can be composed from any selection of N basis functions, which is the basis for nonlinear compression techniques. Accordingly the objective of *adaptive* approximation is to successively track the most significant coefficients of the searched function.

Furthermore, properly scaled wavelet bases turn out to form Riesz bases for other function spaces such as Sobolev spaces for a certain regularity range depending on ψ.

The starting point for the construction of wavelet bases is usually a so called *multiresolution sequences* of nested spaces $S_{j_0} \subset \ldots \subset S_j \subset S_{j+1} \subset \ldots$ whose union is dense in L_2. Here j_0 stands for some coarsest discretization level. The spaces S_j are spanned by scaling functions, $S_j = \text{span}\{\varphi_{j,k} : k \in I_j\}$, which are obtained by a suitable, compactly supported function φ. Due to the compact support of φ we can view 2^{-j} as the uniform mesh size of S_j. Given a dual sequence $\tilde{S}_j$, a successive decomposition of S_j leads to the multiscale splitting $S_J = S_{j_0} \bigoplus_{l=j_0}^{J-1} W_l$ with $W_j = \text{span}\{\psi_{j,k} : k \in I'_j\}$ fulfilling the biorthogonality condition (8).

All above properties can be realized when choosing φ as the cardinal B–Spline of order m. Moreover, one can construct for any $\tilde{m} \in \mathbb{N}, \tilde{m} \geq m, m + \tilde{m}$ even, a compactly supported dual scaling function $\tilde{\varphi}$ such that the dual multiresolution spaces $\tilde{S}_j$ contain all polynomials up to order $\tilde{m}$. As a consequence of biorthogonality, one has the *moment conditions*

$$\langle t^r, \psi \rangle = 0, \qquad r = 0, \ldots, \tilde{m} - 1 \,. \tag{9}$$

Likewise one has $\langle t^r, \tilde{\psi} \rangle = 0$ for $r = 0, \ldots, m - 1$. Therefore, smooth functions have wavelet expansions with rapidly decaying wavelet coefficients so that levelwise truncation of wavelet expansions provide good approximations. However, as soon as singularities occur different nonuniform selections of basis functions are expected to provide more economical approximations. In fact, combining the above statements with the locality of wavelets, large wavelet coefficients reflect a large local change of the function. The adaptive choice of Λ will be based on this fact. Figure 3 illustrates a typical way of space–scale representation of the wavelet coefficients for an adaptive set of coefficients Λ. Each block represents one coefficient $d_{j,k}$ to a basis function $\psi_{j,k}$ where j is the scale (frequency) and k is the locality.

FIGURE 3. Space–scale representation

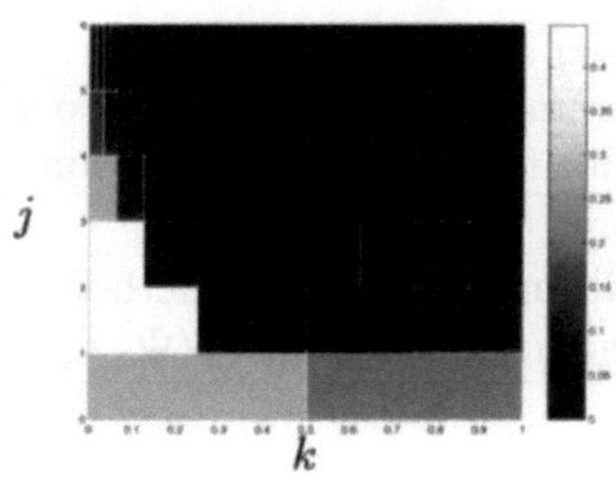

4. Discretization: wavelets as trial– and test–functions

For the discretization of the optimization problem we scale the time horizon to $[0, 1]$ and formulate the equality constraints in a weak sense. We obtain:

$$\min_{\boldsymbol{x},\boldsymbol{y},\boldsymbol{v},\boldsymbol{w}} \int_0^1 \{(\boldsymbol{y} - \boldsymbol{z})^T \boldsymbol{Q}(\boldsymbol{y} - \boldsymbol{z}) \quad + \quad \boldsymbol{v}^T \boldsymbol{R_v} \boldsymbol{v} \; + \; \boldsymbol{w}^T \boldsymbol{R_w} \boldsymbol{w}\} \; dt \tag{10}$$

$$\text{s.t.} \qquad \langle \boldsymbol{F}(\dot{\boldsymbol{x}}, \boldsymbol{x}, \boldsymbol{y}, \boldsymbol{v}, \boldsymbol{w}), \boldsymbol{\nu} \rangle \;=\; 0 \qquad \text{for all } \boldsymbol{\nu} \in L_2^{n_x + n_y} \tag{11}$$

$$\boldsymbol{n}(\boldsymbol{x}, \boldsymbol{y}, \boldsymbol{v}, \boldsymbol{w}) \;\leq\; 0 \tag{12}$$

Discretization is then given by the representation of each function with respect to an appropriately chosen wavelet basis $\boldsymbol{\Psi}$, i.e. we represent $\boldsymbol{x}$ as $\boldsymbol{x} = \boldsymbol{\Psi}^T_{\Lambda_x} \boldsymbol{d_x}$, $\boldsymbol{y}$ as $\boldsymbol{y} = \boldsymbol{\Psi}^T_{\Lambda_y} \boldsymbol{d_y}$ etc.. We obtain an equivalent, *infinite dimensional* but *discretized* optimization problem for the wavelet coefficients $\boldsymbol{\zeta}_\Lambda = (\boldsymbol{d}_{\boldsymbol{x}}^T, \boldsymbol{d}_{\boldsymbol{y}}^T, \boldsymbol{d}_{\boldsymbol{v}}^T, \boldsymbol{d}_{\boldsymbol{w}}^T)^T$:

$$\min_{\boldsymbol{\zeta}_\Lambda} \; \boldsymbol{\zeta}_\Lambda^T \boldsymbol{H}_\Lambda \boldsymbol{\zeta}_\Lambda \; + \; \boldsymbol{b}_\Lambda^T \boldsymbol{\zeta}_\Lambda \tag{13}$$

$$\text{s.t.} \quad \boldsymbol{F}_\Lambda(\boldsymbol{\zeta}_\Lambda) \; = \; \boldsymbol{0} \tag{14}$$

$$\boldsymbol{n}_\Lambda(\boldsymbol{\zeta}_\Lambda) \; \leq \; \boldsymbol{0}\,. \tag{15}$$

In our particular case we use piecewise linear, continuous functions for $\boldsymbol{x}$, $\boldsymbol{y}$ and $\boldsymbol{v}$ and the test–functions for the output equations (3), while we employ piecewise constant wavelets for $\boldsymbol{w}$ and the test–functions for the differential equations (2). This corresponds to the minimal regularity for a conforming discretization.

While the above formulation yields still the exact solution, for numerical treatment we have to choose finite dimensional approximations. Hence, for each involved function we have to choose a finite set of basis functions, described by their index set $\Lambda_{\boldsymbol{x}}$ etc.. As mentioned, we will start with some initial possibly small index sets $\Lambda^0_{\boldsymbol{x}}$ etc. and *adaptively* enlarge these index sets to $\Lambda^{\ell+1}_{\boldsymbol{x}} = \Lambda^{\ell}_{\boldsymbol{x}} + \Gamma_{\boldsymbol{x}}$ in each refinement step.

5. Treatment of the Optimization Problem

To realize the overall concept, which is designed for the particular real time requirement, we face many new problems on each level of the solver. The following sketch gives a short overview of the different levels and our current approach.

Optimization problem
adaptive refinement employing
- sensitivity of the cost functional with respect to $\boldsymbol{w}$ and $\boldsymbol{v}$ $\longrightarrow \Gamma_{\boldsymbol{w}}, \Gamma_{\boldsymbol{v}}$
- local error estimator for the differential equation $\longrightarrow \Gamma_{\boldsymbol{x}}$

NLP
- black–box algorithm using a SQP–method
- efficient functional evaluation for $\langle f(\boldsymbol{\Psi}^T_\Lambda \boldsymbol{d_x}), \psi_{j,k}\rangle$ for given $\boldsymbol{d_x}$

QP
treatment of the inequality constraints
- inner–point methods for the first approximations
- active–set methods with initialization of the active sets for the refinement steps

Linear system of equations
iterative solver (simplified Uzawa)
- employing previous solution as initial value
- using adapted Jacobi preconditioner motivated by wavelet theory

The outermost level correspond to the discretization of the whole optimization problem and the adaptive refinement. The goal is to spend minimal computational effort for realizing a fixed decay rate of the current error in each refinement step. The next subsection will shortly present our algorithm and some numerical results. Then follows the treatment of the nonlinearity. The optimization problem arising at each stage is currently solved by a black–box algorithm using an SQP–method. Due to the nature of the discretization we have to evaluate inner products of the form $\langle f(\boldsymbol{x}_\Lambda), \psi_{j,k}\rangle$ where $\boldsymbol{x}_\Lambda$ is a current wavelet expansion and f is a nonlinear function. This task arises in many other applications of wavelet concepts to variational problems. While straightforward quadrature would spoil the complexity gain by adaptive refinements more sophisticated efficient evaluation schemes have meanwhile been developed, see, e.g., [12]. On the level of the linearized optimization problem the outer loop is concerned with the treatment of the inequality constraints. Here, we considered the question whether inner–point methods or active–set methods can exploit the information of the previous refinement step in a more efficient way. Results will be given in Subsection 5.2. The main focus of the remaining paper will be however on the solution process of the arising linear systems. The goal is that the refinement cost for a fixed error reduction should remain proportional to the current size of the problem. This immediately suggests the idea to use an iterative solver rather than a direct solver.

5.1. Adaptive Refinement

Briefly we sketch now the adaptive refinement step and refer the reader for a detailed discussion to [6]. We use the following *adaptive refinement algorithm*, where Γ^ℓ is the set of newly added indices:

1. solve the problem with index set Λ^ℓ
2. refine $\boldsymbol{z}$ and $\boldsymbol{u}$ by *thresholding* $\longrightarrow \Gamma^\ell_{z_i}, \Gamma^\ell_{u_i}$
3. apply *local error* estimator for $\boldsymbol{x}$, $\boldsymbol{y}$ $\longrightarrow \Gamma^\ell_{x_i}, \Gamma^\ell_{y_i}$
4. use *sensitivity* of Lagrange function for $\boldsymbol{v}$, $\boldsymbol{w}$ $\longrightarrow \Gamma^\ell_{v_i}, \Gamma^\ell_{w_i}$
5. *compress* current approximation by thresholding $\longrightarrow \tilde{\Lambda}^\ell_{x_i}$ etc.
6. take *interactions* into account and use $\tilde{\Lambda}^\ell$, $\Gamma^\ell \longrightarrow \Lambda^{\ell+1}$
   then go to 1.

Step 2 is based on the norm equivalences of the form (7). They imply that using the largest wavelet coefficients gives nearly the best N–term approximation in the sense explained above. Hence, we sort the wavelet coefficients of $\boldsymbol{z}$ and $\boldsymbol{u}$ by their absolute value and increase the number of active coefficients in each step. In step 3 we use a local trapezoidal rule and evaluate the error at the midpoints. Where the largest error appears we increase the set of wavelet basis functions by adding to the current set the neighbouring functions in time and in frequence. In contrast to the local error estimator used for the states and output functions, the problem formulation allows us to employ the sensitivity of the Lagrange function with respect to the model error functions $\boldsymbol{v}$ and $\boldsymbol{w}$, since it facilitates easy–to–compute matrix–vector multiplications. While the first steps increase the index set, step 5 reduces the index set by compressing the current approximation. Basis functions with very small coefficients can be discarded thanks to the norm equivalence (7). Finally the different index sets for the various functions have to be combined to take their close interactions into account.

Figure 5.1 illustrates for an example process the comparison between the scalewise choice, i.e. uniform meshes, and an adaptive choice of index sets. The L_2-error is plotted versus the number of trial functions. We see, for example, that for an error 0.105 we need only roughly 550 basis functions with an adaptive approach while around 1000 are necessary with a uniform mesh (mesh size 2^{-10}). Obviously adaptivity outperforms uniform discretization with regard to work/accuracy balance.

FIGURE 5.1. Scalewise versus adaptive refinement

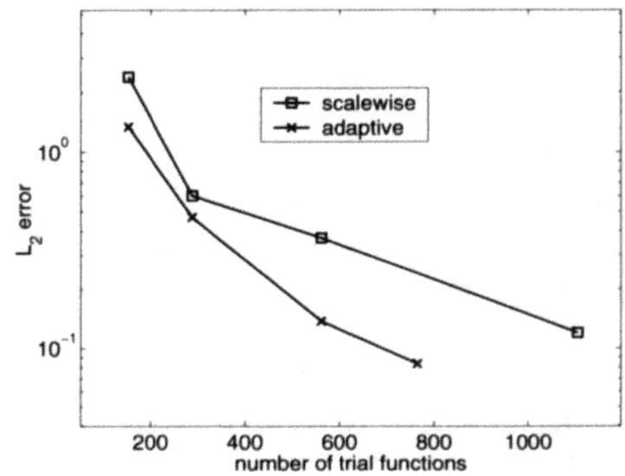

5.2. Inequality constraints

As on all levels of the solver we would like to exploit the information from the previous refinement step for treating the inequality constraints. We compare two common approaches. On one hand, interior–point methods are often applied to very large systems. They usually need a fixed number of iterations independently of the dimension of the system. Hence they suggest themselves in the present context. On the other hand, the active–set methods need to solve smaller linear systems in each iteration step. Nevertheless, the number of iterations typically increases with the size of the optimization problem and the number of inequality constraints. This typically happens if one solves the optimization problem in a single run. In our refinement algorithm, though, this will be the case only in the first loop. For the next refinement we can employ the information from the previous step. In case of the interior–point method we can use the solution as initial guess for the refined optimization problem. Nevertheless, the current solution has most likely an active set and the iterates of the interior–point method tend to stay away from the boundary. So, it could happen, that the current solution tends to diverge during the iteration first away from the actual solution on this mesh, even if they are initially very close.

In case of the active–set method we can initialize the active–set with the set for which the current solution is active on the refined mesh. Since the current solution should be a good approximation of the refined solution, we do not expect significant changes of the active set and, therefore, we do not expect a large increase of iteration numbers.

Numerical examples show that this approach is indeed very effective. In contrast, for interior–point methods the initial value does not effect the iteration number much. The following Figure 5.2 illustrates a typical behaviour of the iteration numbers for the interior–point methods with and without setting the initial value, as well as the iteration numbers for the active–set method with and without initializing the active set. In this study we employ scalewise refinement.

FIGURE 5.2. Treatment of the inequalities

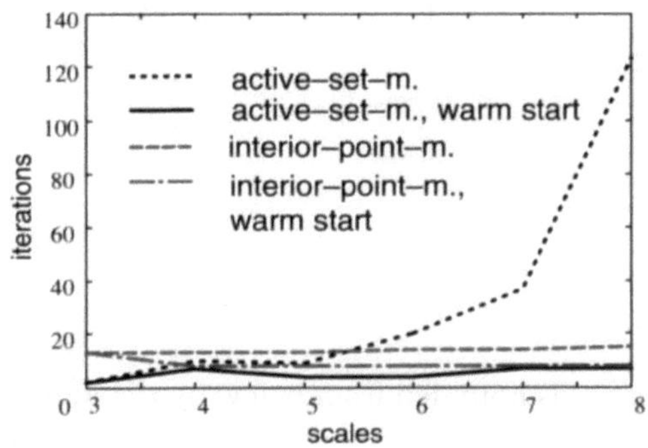

Of course, the dimensions of the systems to solve in each iteration are different for the active–set method, while they are constant for the interior–point method. In the latter one (much) larger systems arise though. Initialization drops the iteration number for the interior–point method only by two, while the numbers for the active–set method turn to be independent of the size of the system and are even less than for the interior–point method. Therefore, our strategy is to use an interior–point method for the first approximate solution of the optimization problem and for all following refinements we use the active–set method with initialization based on the previous solution. Details for our approach can be found in [4].

6. Linear algebra solver

The saddle point problem arising from the discretized optimization problem for active set methods takes the form:

$$\boldsymbol{L}_\Lambda \begin{pmatrix} \boldsymbol{\Upsilon}_\Lambda \\ \boldsymbol{\lambda}_\Lambda \end{pmatrix} := \begin{pmatrix} \boldsymbol{A}_\Lambda & \boldsymbol{B}_\Lambda{}^T \\ \boldsymbol{B}_\Lambda & \mathbf{0} \end{pmatrix} \begin{pmatrix} \boldsymbol{\Upsilon}_\Lambda \\ \boldsymbol{\lambda}_\Lambda \end{pmatrix} = \begin{pmatrix} \boldsymbol{h}_\Lambda \\ \mathbf{g}_\Lambda \end{pmatrix}. \tag{16}$$

For convenience we consider an optimization problem without inequality constraints and nonlinearity, although the ideas can be extended.

$$\min_{\boldsymbol{x},\boldsymbol{y},\boldsymbol{v},\boldsymbol{w}} \int_0^1 \{(\boldsymbol{y}-\boldsymbol{z})^T\boldsymbol{Q}(\boldsymbol{y}-\boldsymbol{z}) \quad + \quad \boldsymbol{v}^T\boldsymbol{R}_v\boldsymbol{v}+\boldsymbol{w}^T\boldsymbol{R}_w\boldsymbol{w}\}\, d\tau \tag{17}$$

$$\text{s.t.} \quad \dot{\boldsymbol{x}} - \boldsymbol{A}\boldsymbol{x} - \boldsymbol{W}\boldsymbol{w} = \boldsymbol{B}\boldsymbol{u} \tag{18}$$

$$\boldsymbol{y} - \boldsymbol{C}\boldsymbol{x} - \boldsymbol{V}\boldsymbol{v} = \boldsymbol{0}\,. \tag{19}$$

Figure 6 shows the typical structure of the matrix $\boldsymbol{L}_\Lambda$. It exhibits the finger block structure and diagonal block matrices arising from wavelet scalar products. Each block corresponds to entries of the regularization and model matrices.

FIGURE 6. $\boldsymbol{L}_\Lambda$

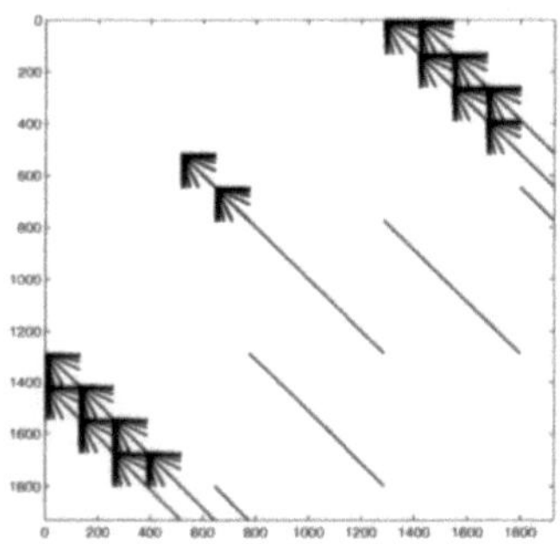

$\boldsymbol{S}_\Lambda$

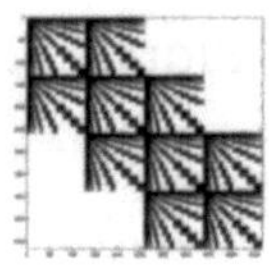

The discretization error is bounded from below by the approximation error of the functions in the adaptively chosen wavelet space. Hence, it is sufficient to solve the above system of equations only up to an error of this order. The goal is now to use an efficient solver up to an *accuracy* which *corresponds* to the *discretization error on each refinement step.* For the efficiency it is important to exploit the previous approximation. Furthermore, successive error reduction is in the spirit of the overall concept. Hence, while most optimization solvers use direct methods for the linear algebra problem, we focus on *iterative solvers.* Due to the indefiniteness of the system one could think of using GMRES, or, exploiting the symmetry, of using SYMMLQ or MINRES [1, 15]. Nevertheless, to exploit the arising structure as much as possible, we reduce the system and obtain a matrix $\tilde{\boldsymbol{L}}_\Lambda$, where we can apply the Uzawa algorithm (see [7]). This algorithm is particularly designed for saddle point problems with positive definite $\boldsymbol{A}_\Lambda$. It basically applies the PCG–method to the Schur complement and, to avoid inversion of $\boldsymbol{A}_\Lambda$, also applies the PCG–method to $\boldsymbol{A}_\Lambda$. In our case the last mentioned, inner application of the PCG–method is not necessary, since it is very efficient to actually set up $\boldsymbol{A}_\Lambda^{-1}$. This means, that we apply essentially the PCG–method to the Schur complement $\boldsymbol{S}_\Lambda$ of the reduced system. For a detailed discussion of this method we refer to [3]. In Figure 6 we can see the reduction of $\boldsymbol{L}_\Lambda$ to the Schur complement $\boldsymbol{S}_\Lambda$. The problem size is reduced from a size corresponding to all functions $\boldsymbol{x}$, $\boldsymbol{y}$, $\boldsymbol{v}$, $\boldsymbol{w}$ and the Lagrange parameter functions to the size of the discretization of $\boldsymbol{x}$.

Furthermore, S_Λ is still very sparse and has finger structure. Nevertheless, it will never be set up explicitly to avoid matrix multiplications.

In connection with the applications of iterative methods there arise, of course, two essential questions: the stopping criteria and the preconditioning. These two questions will be discussed in the next section in the context of nested iterations, which is applied to the sequence of refined optimization problems.

7. Nested Iteration

The refinement concept leads to a sequence of equations: for $j = j_0, j_0 + 1, \ldots$

$$\tilde{L}_{\Lambda_j} \begin{pmatrix} \tilde{\lambda}_{\Lambda_j} \\ \tilde{\Upsilon}_{\Lambda_j} \end{pmatrix} = \begin{pmatrix} \tilde{h}_{\Lambda_j} \\ \tilde{g}_{\Lambda_j} \end{pmatrix} \tag{20}$$

with increasing complexity.

The *stopping criteria* for the iterative solver hinges on three properties: firstly, the maximal reduction rate α of the approximation error in the energy norm; secondly, the convergence order of the PCG–method, which can be estimated in terms of $\kappa = \mathrm{cond}_2(M^{1/2} S_\Lambda M^{1/2})$ where M is the preconditioner; and finally on the norm equivalences induced by the wavelets leading to a constant c. These properties can be used to determine a maximal number of iteration

$$max_{it} \approx log\,(2c/\alpha) \,/\, \left(-log\left((\sqrt{\kappa}-1)/(\sqrt{\kappa}+1)\right)\right) \tag{21}$$

and a maximal relative error

$$\alpha_{rel} \approx \alpha/(c\sqrt{\kappa}) \tag{22}$$

to obtain the best possible convergence order during the refinement process (see [3]). In general the exact error reduction factor α is not computable. For our particular discretization we expect at best a first order convergence in the energy norm. Hence, in terms of degrees of freedom we take $\alpha = \#\Lambda_{j-1}/\#\Lambda_j$, which is $1/2$ for uniform mesh refinement. Moreover, to determine α_{rel} and max_{it} an upper bound for κ has to be estimated. Choosing $\alpha_{rel} = 10^{-3}$ for all j would cover condition numbers κ up to 1700. Nevertheless, to play safe we set $\alpha_{rel} = 10^{-4}$. Moreover, since max_{it} predicts the worst case we mostly use $max_{it} = 20$ as a pragmatic choice, which rigorously covers only condition numbers up to around 100.

The minimization problem has differential equations as constraints. This implies condition numbers increasing with the refinement level. Norm equivalences based on wavelets suggest a *level depending scaling* 2^{-j}, where j denotes the level, to obtain an upper bound for the eigenvalues. However, since the discretization is a Galerkin discretization of a saddle point problem, we require that the LBB–condition is fulfilled, to prove the existence of a lower bound. Additionally, we need the well posedness of the optimization problem with respect to the H^1–norm.

Then, in [3] it is shown, that scale dependent preconditioning leads to eigenvalues bounded independently of the used index set Λ

$$0 < c_1 \leq \lambda(\boldsymbol{M}_{\Lambda_j}^{1/2} \boldsymbol{S}_{\Lambda_j} \boldsymbol{M}_{\Lambda_j}^{1/2}) \leq c_2 \,, \tag{23}$$

and therefore to bounded condition numbers. Instead of 2^{-j} we use an *approximate Jacobi preconditioner*, i.e. an approximation of the diagonal of the $\boldsymbol{S}_{\Lambda_j}$ which can be set up very efficiently. Their entries reflect the necessary level dependent scaling. Furthermore, numerical experiments show, that it is more efficient than the scaling by powers of two, which corresponds to normalizing the wavelet basis in the energy norm and usually gives better results (see [3]).

The algorithm of *nested iteration* reads then as follows:
set $j := j_0$ and solve (20) up to machine precision;

- *update* Λ_j and the representation of $\boldsymbol{z}$ and $\boldsymbol{u}$
- *extend* the discretization *matrices*
- set the initial guess $\tilde{\boldsymbol{\Upsilon}}_{\Lambda_j}^{(0)} \longleftarrow \tilde{\boldsymbol{\Upsilon}}_{\Lambda_{j-1}}^{(k)}$
- apply the *approximate Jacobi* preconditioner $\boldsymbol{M}_{\Lambda_j}$
- iterate the *simplified Uzawa* algorithm until
  - a.) relative residual in ℓ_2–norm is below α_{rel} or
  - b.) maximal number of iteration max_{it} is reached
- go to the first step

We illustrate the effects on condition number (Figure 7) and the results of the nested iteration process (Table 1) with a simple example. We consider the minimization problem $min \int_0^1 \left\{(\boldsymbol{y} - \boldsymbol{z})^2 + \boldsymbol{v}^2 + \boldsymbol{w}^2\right\} \, d\tau$ with the constraints $\dot{\boldsymbol{x}} - \boldsymbol{A}\boldsymbol{x} - \boldsymbol{B}\boldsymbol{u} - \boldsymbol{w} = \boldsymbol{0}$, $\boldsymbol{y} - \boldsymbol{x} = \boldsymbol{0}$ and four state and output functions as well as one control function, and solve it by uniform refinement. In Figure 7 we can clearly see the level dependent increase of $cond(\boldsymbol{S}_\Lambda)$, while preconditioning with approximate Jacobi preconditioner leads to nearly constant behaviour. The corresponding max_{it} determined by the exact condition number is between 15 and 17.

FIGURE 7. Condition numbers

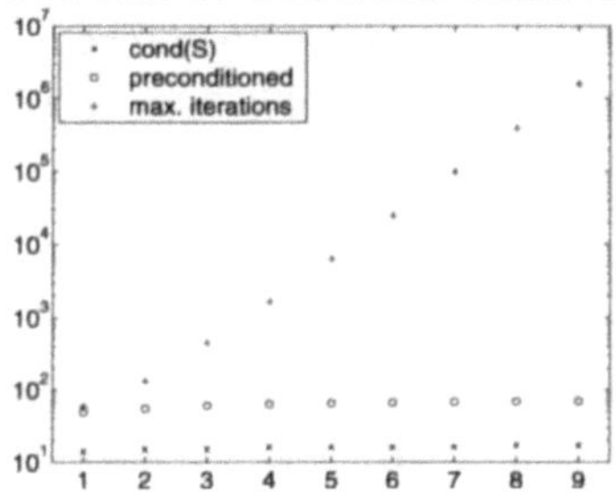

Table 1 lists the used refinement levels, the corresponding size of the system of equations, the number of iterations needed for the nested iteration process stated above, the error e between the wavelet coefficients of the current approximation to the exact solution, the ratio of error reduction e_j/e_{j-1} and the actual residuum of the equation system on each level. The last line gives the results for applying the simplified Uzawa algorithm on level 9 without using any previous information.

TABLE 1. Results corresponding to nested iteration

Level j	dim	k	e_j	e_j/e_{j-1}	$\|res\|_{\ell_2}$
1	12	16	9.4525e−01		6.0166e−15
2	20	17	8.8248e−01	1.07	4.0636e−04
3	36	17	2.4626e−01	3.58	4.9402e−03
4	68	17	6.6437e−02	3.70	3.4785e−03
5	132	17	1.7124e−02	3.87	2.6048e−03
6	260	17	4.3262e−03	3.95	2.3966e−03
7	516	15	1.0811e−03	4.00	2.2025e−03
8	1028	14	2.6419e−04	4.09	2.5533e−03
9	2052	10	6.3861e−05	4.13	2.1355e−03
9	2052	34	5.8380e−05		2.4164e−03

Due to the geometric progression of the problem sizes we see that the computational cost for nested iteration up to scale 9 corresponds to 20 or 21 iterations of the simplified Uzawa algorithm applied to the system on level 9 only. Perhaps more important than the expected higher efficiency of nested iteration is the fact it provides successively improved approximations at a much earlier stage. We can see also that it is not necessary to determine on each scale the exact solution. The best approximation error is closely reached, although the residua are still quite large and in full agreement with the size of the condition numbers in this example. The quotient e_{j-1}/e_j is as good as an exact solver can provide to produce meaningful results associated to discretization accuracy.

Nevertheless, we have also examples with condition numbers up to 10^{20} where equilibration of the states, preconditioning, and choosing more appropriate regularization parameters $\boldsymbol{R_w}$ lead to condition numbers of 10^6. These are still too large to make predictions for the iteration process. The eigenvalue distribution is much more in favour for superconvergence of the method, though, this issue has still to be investigated more thoroughly. In the next section we will briefly outline several sources of the ill conditioning.

8. Inverse Problem – Observability

As mentioned in the beginning, state estimation is an ill–posed inverse problem. Even with our regularized problem formulation (1) – (5), we typically face three different problems:

1. the signal functions $\boldsymbol{z}$ are corrupted by noise and do *not* lie in H^1,
2. the model equations (2), (3) (without $\boldsymbol{v}$ and $\boldsymbol{w}$) have often a (very) low *observability measure* (see [3]),
3. the amount of regularization of $\boldsymbol{v}$ and $\boldsymbol{w}$ may be inadequate.

In particular 2. and 3. play an important role for the constants involved in the estimation of the well posedness with respect to H^1, and hence in the bounds c_1 and c_2 of the eigenvalues (23) of the preconditioned Schur complement. For example, the interrelation of the model matrices $\boldsymbol{A}$ and $\boldsymbol{C}$ for linear constraints is essential in the estimations of the LBB–condition (see [7]), which –ignoring $\boldsymbol{v}$ and $\boldsymbol{w}$– leads to an estimation

$$\left\| \begin{pmatrix} -\dot{\boldsymbol{x}} + \boldsymbol{A}\boldsymbol{x} \\ \boldsymbol{C}\boldsymbol{x} \end{pmatrix} \right\|_{(L_2)^{n_x} \times (L_2)^{n_y}} \geq \beta \|\boldsymbol{x}\|_{(H^1)^{n_x}}$$

where β could be very small. Moreover, we immediately see in case of a one dimensional problem with $\boldsymbol{C} = \delta$ and $\boldsymbol{A} = \alpha$ that large eigenvalues of $\boldsymbol{A}$ are unfavourable, since for $\boldsymbol{x} = e^{\alpha t}$ we have

$$\left\|(-\dot{\boldsymbol{x}} + \boldsymbol{A}\boldsymbol{x}, \boldsymbol{C}\boldsymbol{x})^T\right\|_{L_2 \times L_2} = \frac{\delta}{1+|\alpha|} \|\boldsymbol{x}\|_{H^1} \ .$$

This reflects the fact, that no initial condition is included in the constraints, since for state estimation the initial data have to be estimated.

Numerical considerations confirm, that large in modulus eigenvalues of $\boldsymbol{A}$ and low observability measures lead to very small lower bounds of the eigenvalues of the preconditioned and unpreconditioned Schur complement. This means that the high condition numbers are mainly problem inherent and will occur for any discretization.

Possible improvements of this situation may result from a more sophisticated choice of $\boldsymbol{R_v}$, $\boldsymbol{R_w}$ and/or an additional *scale depending regularization* of $\boldsymbol{x}$ with

$$\int_{\Omega_R} \left\{ (\boldsymbol{x} - \bar{\boldsymbol{x}})^T \boldsymbol{R_x} (\boldsymbol{x} - \bar{\boldsymbol{x}}) \right\} \, d\tau$$

in the cost functional (1), which should fade out for progressing levels. On the other hand, currently we develop and investigate a *preconditioner*, which eliminates the smallest eigenvalues based on the *adaptive refinement algorithm.*

9. Conclusions

We have developed and investigated several components of an iterative multiscale concept, that aims at realizing the real–time requirement encountered in state estimation. The relevant main ingredients for the adaptive treatment of the linear problem formulation with inequality constraints are now essentially available. We had to face in the development of these components a diversity of problems, some of which do only occur due to this particular real–time requirement whereas others are problem inherent to process monitoring. Nevertheless, there are still many open tasks and questions: the above mentioned choice of regularization and the improvement of the preconditioner, the extension to DAE's with appropriate wavelets as test functions, the development and extension of the current method to a nonlinear model, which is in the spirit of our conceptional framework, the improvement and thorough analysis of the adaptive refinement, as well as moving the horizon, while exploiting efficiently the solution on the previous horizon. Moreover, due to problem 1., we have to investigate the length of the horizon and its influence on the estimation quality given noisy data. In this context we discussed in [5] the regularization effect of the discretization. Here, the above adaptive refinement algorithm generates a hierarchy of discretization grids until a best compromise between data and regularization error is achieved.

References

[1] O. Axelsson, *Iterative Solution Methods*, Cambridge University Press, New York, 1994.

[2] T. Binder, L. Blank, W. Dahmen, and W. Marquardt, *Towards Multiscale Dynamic and Data Reconciliation*, Nonlinear Model Based Process Control, NATO ASI, in: R. Berber and C. Kravaris, eds., Kluwer Academic Publishers, 1998, pp. 623–665.

[3] T. Binder, L. Blank, W. Dahmen, and W. Marquardt, *Iterative Algorithms for Multiscale Dynamic Data Reconciliation*, RWTH Aachen, IGPM–Report, **186**, 2000.

[4] T. Binder, L. Blank, W. Dahmen, and W. Marquardt, *Grid Refinement in Multiscale Dynamic Optimization*, Proceedings: ESCAPE-10, Florence, Italy, May 7 – 10, 2000.

[5] T. Binder, L. Blank, W. Dahmen, and W. Marquardt, *Regularization of Dynamic Data Reconciliation Problems by Projection*, Proceedings: ADCHEM 2000, International Symposium on Advanced Control of Chemical Processes, Pisa, Italy, June, 14 – 16, 2000.

[6] T. Binder, L. Blank, W. Dahmen, and W. Marquardt, *An Adaptive Multiscale Method for Real-Time Moving Horizon Optimization*, Proceedings: American Control Conference 2000, Chicago, USA, June, 28 – 30, 2000.

[7] D. Braess, *Finite Elemente*, Springer–Verlag, Berlin, 1992.

[8] A. Chambolle, R. DeVore, N. Y. Lee, and J. Lucier, *Nonlinear Wavelet Image Processing: Varational Problems, Compression, and Noise Removal through Wavelet Shrinkage*, IEEE Image Processing, **7** (1998), pp. 319–335.

[9] C. K. Chui, *An Introduction to Wavelets*, Academic Press, Boston, 1992.

[10] A. Cohen, I. Daubechies, and J. C. Feauveau, *Biorthogonal Bases of Compactly Supported Wavelets*, Comm. Pure and Appl. Math., **45** (1992), pp. 485–560.

[11] W. Dahmen, *Wavelet and Multiscale Methods for Operator Equations*, Acta Numerica, 1997, pp. 55–228.

[12] W. Dahmen, R. Schneider, and Y. Xu, *Nonlinear Functionals of Wavelet Expansions – Adaptive Reconstruction and Fast Evaluation*, Numer. Math., **86**, No. 1 (2000), pp. 49–101.

[13] I. Daubechies, *Ten Lectures on Wavelets*, Society for Industrial and Applied Math., Philadelphia, 1992.

[14] H. Michalska and D. Q. Mayne, *Moving Horizon Observers*, IFAC Symposium NOLCOS'92, Bordeaux, France, 1992.

[15] G. Meurant, *Computer Solution of Large Linear Systems*, Elsevier Science B.V., Amsterdam, The Netherlands, 1999.

[16] K. R. Muske and J. B. Rawlings, *Nonlinear Receding Horizon State Estimation*, in: R. Berber, ed., Methods of Model–Based Control, NATO-ASI Series, Kluwer Press, Dordrecht, The Netherlands, 1995, pp. 489–504.

[17] D. Robertson, J. H. Lee, and J. B. Rawlings, *A Moving Horizon–Based Approach for Least–Squares Estimation*, AIChE J., **42**, No. 8 (1996), pp. 2209–2223.

Three Dimensional Shape and Topology Optimization in Cauchy-Navier's PDE

Thomas Borrvall and Joakim Petersson

Abstract. We consider shape and topology optimization of elastic continua where equilibrium is governed by the elliptic Cauchy-Navier's system of PDEs. The domain of integration, Ω, is fixed, and the design parametrization is done by letting the elasticity tensor depend linearly on a design function ρ. To get binary values of ρ, representing material or not at each point of Ω, we use a regularized penalty function which also guarantees existence of solutions.

The finite element discretized optimization problem is solved by sequential convex programming and equilibrium by an iterative preconditioned conjugate gradient method. Large scale 3D instances of the problem statement are considered and parallel computing and domain decompositions are used. Several numerical results from computations performed on Cray T3E are presented.

1. Introduction

This paper considers optimization in Cauchy-Navier's system of partial differential equations that governs the equilibrium deformation of an elastic continuum when loads are applied. The goal is to modify the shape of the domain to obtain a maximum of stiffness without the available material volume exceeding a given limit value, say V– an important goal for designers of load-carrying mechanical structures. The aim is to, not only modify initial or "known" boundaries to the domain, but also to obtain new previously "unknown" boundaries, i.e. new holes, that promote stiffness. The optimization problem could therefore be termed "variable topology shape optimization" or just "topology optimization". This is a very active field in the engineering structural optimization community, see, e.g., [4]. The standard approach is to keep the actual computational domain Ω fixed, but to introduce a design, or density, function ρ defined on Ω. This function should ideally be selected among all characteristic functions corresponding to structures with volume V or less. Posed as such, the problem generally lacks solutions, and will therefore be prone to mesh dependent numerical results, also possibly polluted by checkerboard patterns [20]. These problems are in this report remedied by a regularized penalty function P.

If the design parametrization is done by letting the elasticity tensor depend linearly on a design function ρ, and if ρ is allowed to give all values between 0 and 1, then the optimization problem possesses solutions and is convex, and it can hence be solved successfully. Whereas it is then possible in two space dimensions to interpret ρ as a thickness (see, e.g., [18, 15]), in three dimensional space, which is the main interest in this paper (cf. also, e.g., [2, 3, 10, 11, 14]), it is necessary to get binary values of ρ, representing material or not at each point of $\Omega \subset \mathbb{R}^3$. One can then allow intermediate values of ρ, but penalize these by some suitable penalization function – an advantage from an algorithmic point of view to avoid integer programming and to continually tune the "nonconvexity" of the optimization problem. Typically such penalty functions will however result in nonexistence of solutions. As described in section 2, a common penalty function g is regularized by composing it with a compact linear operator S, i.e. by choosing $P = g \circ S$, to obtain a well-posed problem that generates results with sufficient degree of discrete values from ρ.

Section 3 describes the numerical methods and procedures. The infinite-dimensional optimization problem is discretized by finite elements: trilinear displacement approximation and piecewise constant design approximation. The resulting discretized optimization problem is solved by sequential convex approximations, and the equilibrium equations by an iterative (Jacobi) preconditioned conjugate gradient method. To get a good resolution of "real life" structures in three dimensional space, large scale instances of the problem statement are considered. Therefore, parallel computing is used: Both the optimization and equilibrium parts are parallelized, and a one-dimensional cartesian static domain decomposition is used.

The final section 4 contains several numerical results from computations performed on Cray T3E.

2. Problem formulations

2.1. Some preliminaries and the Cauchy-Navier's elasticity equations

Let $\Omega \subset \mathbb{R}^d$ $(d \leq 3)$ be bounded, open and connected with Lipschitzian boundary $\partial\Omega$. We assume that, unless otherwise stated, $d = 3$. We define the set of feasible design functions according to

$$\mathcal{H} = \left\{ \rho \in L^\infty(\Omega) \;\middle|\; 0 \leq \underline{\rho} \leq \rho \leq \overline{\rho}, \text{ a.e. in } \Omega, \quad \int_\Omega \rho \, d\mathbf{x} \leq V \right\}.$$

This reflects that the amount of material density must be between $\underline{\rho}$ and $\overline{\rho}$ at (almost) every point, and the structural volumes must not exceed V. The involved parameters are chosen such that $\mathcal{H}$ is nonempty.

The set of kinematically admissible displacements, $\mathbf{V}$, is assumed to be a closed subspace of $\mathbf{H}^1(\Omega)$ such that (4) holds. The space $\mathbf{H}^1(\Omega)$ denotes the standard Sobolev space of functions where (for each of the d components) the first order weak derivatives, as well as the function itself, belong to $L^2(\Omega)$. We use

$\|\cdot\|_{\mathbf{H}^1(\Omega)}$ for any of the standard norms in $\mathbf{H}^1(\Omega)$ and $\|\cdot\|_{L^p(\Omega)}$ for the usual norm in $L^p(\Omega)$, $1 \le p \le \infty$. For more details on the notations concerning these spaces, we refer to [1].

Let the infinitesimal strain be defined by

$$\epsilon_{ij}(\mathbf{u}) = \frac{1}{2}(\frac{\partial u_i}{\partial x_j} + \frac{\partial u_j}{\partial x_i}),$$

for any $\mathbf{u} \in \mathbf{V}$. For fixed non-negative $\rho \in L^\infty(\Omega)$, we assume, for simplicity, isotropic homogeneous body, and denote by $a(\rho,\cdot,\cdot)$ the internal work symmetric bilinear form on $\mathbf{V} \times \mathbf{V}$:

$$a(\rho, \mathbf{u}, \mathbf{v}) = \int_\Omega \rho \left(\lambda \operatorname{div}(\mathbf{u})\operatorname{div}(\mathbf{v}) + 2\mu \sum_{i,j=1}^{d} \epsilon_{ij}(\mathbf{u})\epsilon_{ij}(\mathbf{v}) \right) d\mathbf{x} \tag{1}$$

where the Lamé moduli satisfy $\mu > 0$, $2\mu + 3\lambda > 0$ to guarantee (4). They can be expressed in terms of Young's modulus E and Poisson's ratio ν,

$$\lambda = \frac{E\nu}{(1+\nu)(1-2\nu)}, \quad \mu = \frac{E}{2(1+\nu)}, \tag{2}$$

or, inverted,

$$E = \frac{\mu(3\lambda + 2\nu)}{\lambda + \mu}, \quad \nu = \frac{E}{2(\mu + \lambda)},$$

where $E > 0$ and $-1 < \nu < 1/2$, again to guarantee (4).

We have defined $a(\rho,\cdot,\cdot)$ such that $\rho \mapsto a(\rho, \mathbf{u}, \mathbf{v})$ is linear. When $d = 2$, choosing the material parameters according to plane state of stress, this corresponds to a sheet of variable thickness where ρ is the thickness function. In true topology optimization in 3D, ρ should take only the values 0 and 1, (or, more generally, $\underline{\rho}$ and $\overline{\rho}$). Then, the *effective* Lamé moduli are λ and μ at points $\mathbf{x} \in \Omega$ where $\rho(\mathbf{x}) = 1$, and zero at points where $\rho(\mathbf{x}) = 0$. The former signifies *presence of structure* (with Lamé moduli λ, μ) and the latter signifies *void*, cf. Figure 1.

Concerning the bilinear form (for the constant design $\rho \equiv 1$), we have the following bicontinuity feature:

$$|a(1, \mathbf{u}, \mathbf{v})| \le M\|\mathbf{u}\|_{\mathbf{H}^1(\Omega)}\|\mathbf{v}\|_{\mathbf{H}^1(\Omega)}, \qquad \forall \mathbf{u}, \mathbf{v} \in \mathbf{V}, \tag{3}$$

and $\mathbf{V}$-ellipticity:

$$a(1, \mathbf{u}, \mathbf{u}) \ge \alpha\|\mathbf{u}\|^2_{\mathbf{H}^1(\Omega)}, \qquad \forall \mathbf{u} \in \mathbf{V} \tag{4}$$

for some strictly positive constants M and α. The external load linear form $\ell(\cdot)$ maps $\mathbf{V}$ into $\mathbf{R}$ and is assumed to satisfy the following boundedness feature:

$$\|\ell\| = \sup_{0 \ne \mathbf{v} \in \mathbf{V}} \frac{|\ell(\mathbf{v})|}{\|\mathbf{v}\|_{\mathbf{H}^1(\Omega)}} < +\infty. \tag{5}$$

For instance, it can take the form $\ell(\mathbf{v}) = \int_{\Gamma_F} \mathbf{F}\cdot\mathbf{v}\, ds$ where $\mathbf{F}$ is the given boundary traction on $\Gamma_F \subset \partial\Omega$.

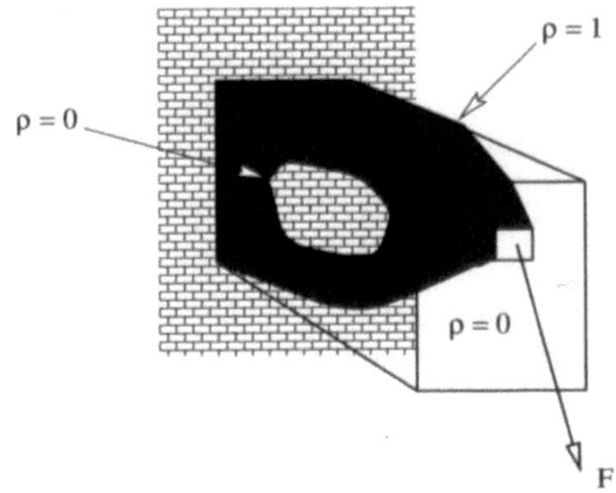

FIGURE 1. An example of boundary conditions and a feasible design in the 3D topology optimization problem.

Equilibria for any non-negative design function $\rho \in L^\infty(\Omega)$ are the solutions of

$$\mathbf{u} \in \mathbf{V} : \quad a(\rho, \mathbf{u}, \mathbf{v}) = \ell(\mathbf{v}), \quad \forall \mathbf{v} \in \mathbf{V}, \tag{6}$$

which is the variational statement of *Cauchy-Navier's elasticity equations*. In solid mechanics (6) is usually referred to as the *principle of virtual work*.

In the design optimization presented in the next subsection the quantity $\ell(\mathbf{u})$ will be minimized. Let E_1 and E_2 be two different strictly positive Young's moduli, and let $\mathbf{u}_1$ and $\mathbf{u}_2$ be corresponding solutions to (6) for any $\rho \in \mathcal{H}$. Then, by (2) and (1), the bilinear form depends linearly on E, so by (6) it follows that $\mathbf{u}_2 = E_1/E_2\mathbf{u}_1$, and therefore $\ell(\mathbf{u}_2) = E_1/E_2\ell(\mathbf{u}_1)$. Hence, the optimization objectives are the same up to a scaling, so it follows that the value of the Young's modulus E is not important. (The Poisson's ratio ν, however, will be chosen to 0.29 which corresponds to carbon steel).

Since ρ can be zero at parts of Ω, the set of possible equilibrium displacements may be both empty and unbounded. To overcome this difficulty the usual approach is to enforce a strictly positive lower design bound. Then, for each such design, by (3)–(5) there exists a unique displacement function in $\mathbf{V}$ which satisfies equilibrium. An important issue is then continuity with respect to this lower design bound.

Let ρ be any non-negative design function in $L^\infty(\Omega)$, and let $\mathbf{U}(\rho)$ be the set of all solutions to (6) for this design ρ. Let $\{\mathbf{u}_\varepsilon\}$ be the sequence of solutions to (6) for the designs $\{\rho_\varepsilon\}$, $\rho_\varepsilon = \rho + \varepsilon$ where ε is any strictly positive real number. Then, assuming $\mathbf{U}(\rho) \neq \emptyset$, $\{\mathbf{u}_\varepsilon\}$ converges strongly in $\mathbf{V}$ to the element of $\mathbf{U}(\rho)$ with minimal energy norm $\|\cdot\|_E = \sqrt{a(1,\cdot,\cdot)}$, [13, Chap. III, Sec. 2]. For a further investigation of continuity with respect to $\underline{\rho}$, including $\underline{\rho} = 0$, we refer to [16]. From now on we assume that $\underline{\rho} > 0$.

2.2. Design optimization

2.2.1. STATEMENTS OF $(\mathbb{P})$ AND $(\mathbb{P})_N$. The *design problem* now consists of minimizing the *compliance* ℓ, subject to *equilibrium* and *design constraints*:

$$(\mathbb{P}) \quad \begin{cases} \underset{(\rho,\mathbf{u})}{\text{Minimize}} & \ell(\mathbf{u}) + k_P P(\rho) \\ \text{subject to:} & a(\rho,\mathbf{u},\mathbf{v}) = \ell(\mathbf{v}),\ \forall \mathbf{v} \in \mathbf{V} \\ & (\rho,\mathbf{u}) \in \mathcal{H} \times \mathbf{V}. \end{cases}$$

We have included an explicit penalty term $k_P P(\rho)$, $k_P \geq 0$, in the objective function. It will be specified in section 2.2.2. By eliminating $\mathbf{u}$ through (6) we have the *nested formulation* of $(\mathbb{P})$:

$$(\mathbb{P})_N \quad \begin{cases} \underset{\rho \in L^\infty(\Omega)}{\text{Minimize}} & \tilde{\ell}(\rho) + k_P P(\rho) \\ \text{subject to:} & \underline{\rho} \leq \rho \leq \overline{\rho}, \quad \text{a.e. in } \Omega \\ & \int_\Omega \rho \leq V. \end{cases}$$

It holds that $\tilde{\ell}(\cdot) = \ell(\mathbf{u}(\cdot))$, where $\mathbf{u}(\rho)$ is the unique solution of (6), is convex and $L^\infty(\Omega)$ weakly* lower semi-continuous, (see, e.g., [6]). Therefore it follows that the optimization problem $(\mathbb{P})_N$ is convex for $k_P = 0$. This problem will be referred to as the ***linear topology optimization problem***, e.g., the variable thickness sheet problem in 2D (choose $d = 2$, $\lambda = E\nu/(1-\nu^2)$ and $\mu = E/(2(1+\nu))$ in (1)), see, e.g., [15, 18, 9], or so-called free material optimization (choose $\lambda = 0$ in (1)), see, e.g., [17, 22, 5]. It is computationally tractable since there exist solutions (due to compactness and lower semi-continuity in the weak* topology), hence one does not expect mesh-dependent finite element based solutions, and global and local optima coincide due to convexity, cf. [20].

2.2.2. THE PENALTY FUNCTIONS. Whereas for $d = 2$ it is possible to interpret ρ as a thickness, when $d = 3$ it is necessary to strive for $\rho(\mathbf{x}) \in \{\underline{\rho}, \overline{\rho}\}$ almost everywhere in Ω, i.e. "black and white" designs. This is mathematically equivalent to the constraint $g(\rho) = 0$ where g is a suitably chosen function, e.g., the one given in (7).

Definition 2.1. (Non-regularized and regularized penalties) *The function* $g : \mathcal{H} \to \mathbb{R}^+$ *defined by*

$$g(\rho) = \int_\Omega (\overline{\rho} - \rho)(\rho - \underline{\rho})\, d\mathbf{x} \tag{7}$$

is called the (non-regularized) penalty function, and

$$P = g \circ S \tag{8}$$

is called the regularized penalty function, where the integral operator $S : L^2(\Omega) \to L^2(\Omega)$ *is given by*

$$S(\rho) = \int_\Omega \phi(\cdot, \mathbf{y})\rho(\mathbf{y})\, d\mathbf{y} \tag{9}$$

where the kernel $\phi : \Omega \times \Omega \to \mathbb{R}$ *fulfils*

$$\phi(\mathbf{x}, \mathbf{y}) \geq 0, \quad \phi(\mathbf{x}, \mathbf{y}) = \phi(\mathbf{y}, \mathbf{x}), \quad \forall(\mathbf{x}, \mathbf{y}) \in \Omega \times \Omega, \tag{10}$$

$$\|\phi(\mathbf{x}, \cdot)\|_{L^2(\Omega)} \leq \overline{\phi} < \infty, \quad \int_\Omega \phi(\mathbf{x}, \mathbf{y})\, d\mathbf{y} = 1, \quad \forall \mathbf{x} \in \Omega. \tag{11}$$

From [6] we have the following:

Theorem 2.2. (Relation between penalties and solvability of $(\mathbb{P})_N$) *The integral operator S given in Definition 2.1 maps $\mathcal{H}$ into $\mathcal{H}$ so the composition $P = g \circ S : \mathcal{H} \to \mathbb{R}^+$ makes sense, and the penalty is overestimated by the regularized penalty, i.e. for any $\rho \in \mathcal{H}$,*

$$0 \leq g(\rho) \leq P(\rho). \tag{12}$$

Finally, with the regularized penalty P given in Definition 2.1, solutions to $(\mathbb{P})_N$ are guaranteed to exist for all $k_P \geq 0$.

In the examples to be presented in section 4, an instance of (7)–(11) based on the linear convolution $S(\rho) = \tilde{\phi} * \rho$, where

$$\tilde{\phi}(\mathbf{x}) = \frac{3}{\pi R^3} \max(0, 1 - \frac{|\mathbf{x}|}{R}), \tag{13}$$

(however modified to maintain (10) and (11) also for points within the distance R from the boundary [6]) was used, cf. [19, 8, 7].

3. Numerical methods

This section describes how the topology optimization problem is solved numerically. Basically, it is solved by means of successive iterations where a large system of linear equations and a small dual optimization problem have to be solved in each iteration. Both parts have been implemented using domain decomposition and parallelization, see section 3.4.

3.1. FE-discretized optimization problem

Let Ω be partitioned into N_e mutually disjoint finite elements Ω_i, such that $\Omega_i \subset \Omega$ and $\cup_{i=1}^{N_e} \overline{\Omega}_i = \overline{\Omega}$ and let h be a characteristic size of the largest element. If L_h denotes the space of elementwise constant functions on Ω, then for each strictly positive $\rho \in L_h$, there is a unique solution to

$$\mathbf{u}_h \in \mathbf{V}_h : \quad a(\rho, \mathbf{u}_h, \mathbf{v}) = \ell(\mathbf{v}), \quad \forall \mathbf{v} \in \mathbf{V}_h, \tag{14}$$

where $\mathbf{V}_h \subset \mathbf{V}$ is a finite dimensional approximation of $\mathbf{V}$ such that $\cup_{h>0}\mathbf{V}_h$ is dense in $\mathbf{V}$. The Ω_i's in section 4 are eight node hexahedrals, and $\mathbf{V}_h$ the space of continuous and elementwise trilinear functions. Now, let $\boldsymbol{\rho}$ denote the N_e-vector of piecewise constant values of a strictly positive function $\rho \in L_h$ and let $\mathbf{u}(\boldsymbol{\rho})$ denote the $3N_n$-vector of nodal values of the solution to (14), where N_n is the number of finite element nodes. With this notation it is well known that the solution of (14) is equivalent to the solution of a system of linear equations,

$$\mathbf{K}(\boldsymbol{\rho})\mathbf{u}(\boldsymbol{\rho}) = \mathbf{f}. \tag{15}$$

Here $\mathbf{K}(\boldsymbol{\rho}) \in \mathbb{R}^{3N_n \times 3N_n}$ is the stiffness matrix depending on the design as

$$\mathbf{K}(\boldsymbol{\rho}) = \sum_{i=1}^{N_e} \rho_i \mathbf{K}_i, \tag{16}$$

where $\mathbf{K}_i \in \mathbb{R}^{3N_n \times 3N_n}$ are element stiffness matrices for the used FE-mesh and $\mathbf{f}$ is the $3N_n$-vector representing the surface and volume tractions (with zeros at the entries corresponding to zero-prescribed displacements).

From $(\mathbb{P})_N$ we can now state the corresponding finite element discretized optimization problem as

$$(\mathbb{P})_N^h \quad \begin{cases} \underset{\boldsymbol{\rho} \in \mathbb{R}^{N_e}}{\text{Minimize}} & \mathbf{f}^T\mathbf{u}(\boldsymbol{\rho}) + k_P P(\boldsymbol{\rho}) \\ \text{subject to:} & \sum_{i=1}^{N_e} c_i \rho_i \leq V \\ & \underline{\rho} \leq \rho_i \leq \overline{\rho} \qquad\qquad i = 1, \ldots, N_e, \end{cases}$$

where $c_i = |\Omega_i|$. Although the argument to the function $P : \mathbb{R}^{N_e} \to \mathbb{R}$ here is a vector, it should be interpreted as the regularized penalty function presented in section 2.2.2 acting on the corresponding function ρ_h in L_h.

From [6] we have the following:

Theorem 3.1. (Convergence of FE-solutions) *Assume that* $\cup_{h>0}\mathbf{V}_h$ *is dense in* $\mathbf{V}$ *and* $\mathbf{V}_h \supset \mathbf{V}_H$ *whenever* $h \leq H$. *Let* $\{\rho_h^*\}$ *be a sequence of solutions to* $(\mathbb{P})_N^h$ *as* $h \to 0$, *(i.e. the corresponding function in* L_h*). Then there exist an element* ρ^* *which is admissible in* $(\mathbb{P})_N$ *and a subsequence, again denoted by* $\{\rho_h^*\}$, *such that*

$$\rho_h^* \to \rho^* \quad \textit{weakly* in } L^\infty(\Omega).$$

Moreover, any such limit ρ^* *solves* $(\mathbb{P})_N$.

It should be noted that the weak* convergence alone cannot guarantee non-prescence of instabilities of checkerboard type in the numerical solutions. However, in [15] it was shown that the convergence is strong on the subset of Ω where the solution to $(\mathbb{P})_N$ attains the upper or lower bound, i.e. wherever the solution is black and white. Now, since we expect the solutions to $(\mathbb{P})_N$ to be mostly black and white, cf. (12), we can expect that such instabilities are *not going to appear.*

In practice the function P has to be approximated through numerical integration, i.e. $P = g \circ S$ where $g : \mathbb{R}^{N_e} \to \mathbb{R}$,

$$g(\boldsymbol{\rho}) = \sum_{i=1}^{N_e} |\Omega_i|(\overline{\rho} - \rho_i)(\rho_i - \underline{\rho}) \tag{17}$$

and $S : \mathbb{R}^{N_e} \to \mathbb{R}^{N_e}$,

$$S(\boldsymbol{\rho}) = \boldsymbol{\Phi}\boldsymbol{\rho}. \tag{18}$$

The matrix $\boldsymbol{\Phi} \in \mathbb{R}^{N_e \times N_e}$ is the convolution matrix corresponding to the integral kernel ϕ in section 2.2.2,

$$(\boldsymbol{\Phi})_{ij} = \frac{\phi(\mathbf{x}_i, \mathbf{x}_j)|\Omega_j|}{\sum_{j=1}^{N_e} \phi(\mathbf{x}_i, \mathbf{x}_j)|\Omega_j|}$$

where $\{\mathbf{x}_i\}_{i=1}^{N_e} \subset \Omega$ is the collection of geometrical centroids of the elements. This choice of numerical approximation is justified because we want the discrete analogue of the second of (11) to hold, i.e. $\sum_{j=1}^{N_e} (\boldsymbol{\Phi})_{ij} = 1$.

3.2. Iterative equilibrium solver

Since we have chosen the nested version of the optimization problem we will need to solve (15) at each iteration. To solve the large instances of (15) we choose an iterative solver. The advantage is not just that asymptotically less computations are required, but also much less memory storage than in direct methods. We have chosen to employ a preconditioned conjugate gradient (PCG) method, a non-stationary iterative method that is suitable for parallelization. The convergence of this method strongly depends on the spectral properties of the coefficient matrix, $\mathbf{K}$. In our situation where we are modeling changes in mechanical structures' topology, ρ is on the lower limit $\underline{\rho}$, as well as on the upper limit $\overline{\rho}$, over large areas in Ω. Since the elements of the stiffness matrix are directly proportional to these density values, cf. (16), there is reason to believe that the condition number can be as large as $\overline{\rho}/\underline{\rho}$, which motivates the introduction of a preconditioner $\mathbf{M}$. Instead of solving (15) we solve the equivalent left preconditioned system

$$\mathbf{M}^{-1}\mathbf{K}\mathbf{u} = \mathbf{M}^{-1}\mathbf{f},$$

where $\mathbf{M}$ is chosen so that $\mathbf{M}^{-1}\mathbf{K}$ is well-conditioned. To avoid substantially extra implementation work, the preconditioner was chosen as a diagonal Jacobi preconditioner. It has worked satisfactorily and is ideal for parallelization.

3.3. Sequential convex programming

The problem $(\mathbb{P})_N^h$ is solved iteratively by sequential explicit approximations. This means that in iteration k, when a design suggestion $\boldsymbol{\rho}^{(k)}$ exists, an approximation of $(\mathbb{P})_N^h$, called $(\mathbb{P})^{(k)}$, is created. The creation of $(\mathbb{P})^{(k)}$ is based on linearization of involved functions. This requires function values and first order derivatives at the current iteration point $\boldsymbol{\rho}^{(k)}$. The idea is to create $(\mathbb{P})^{(k)}$ so that it is *convex* and *separable*[1], hence making it easy to solve, cf. [21, 12]. Once the solution of $(\mathbb{P})^{(k)}$ has been obtained, it defines the new design suggestion $\boldsymbol{\rho}^{(k+1)}$. Functions in $(\mathbb{P})_N^h$ that need to be linearized are $\mathbf{f}^T\mathbf{u}(\cdot)$ and $P(\cdot)$.

[1]The function $f : \mathbb{R}^{N_e} \to \mathbb{R}$ is separable if it can be written as $f(\boldsymbol{\rho}) = \sum_{i=1}^{N_e} f_i(\rho_i)$

3.3.1. Approximation of compliance. Given $\boldsymbol{\rho}$ we have from (16) and implicit differentiation of (15) that

$$\frac{\partial(\mathbf{f}^T\mathbf{u}(\boldsymbol{\rho}))}{\partial \rho_i} = -\mathbf{u}(\boldsymbol{\rho})^T\mathbf{K}_i\mathbf{u}(\boldsymbol{\rho}). \tag{19}$$

Hence, once the system (15) has been solved, the first order derivatives, or sensitivities, of the compliance are obtained without the need for solving another large equation system. The compliance is then linearized with respect to the variables $\frac{1}{\rho_i - l_i}$, i.e.

$$\mathbf{f}^T\mathbf{u}(\boldsymbol{\rho}) \approx r^{(k)} + \sum_{i=1}^{N_e} \frac{q_i^{(k)}}{\rho_i - l_i^{(k)}}. \tag{20}$$

The expansion (20) is due to Svanberg [21] and is commonly referred to as the MMA (Method of Moving Asymptotes). The parameters $l_i^{(k)}$ are moving asymptotes that are dynamically updated and the parameters $r^{(k)}$ and $q_i^{(k)}$ are constants chosen so that function values and first order sensitivities are correct at $\boldsymbol{\rho}^{(k)}$ and so that all terms are convex. In fact, we modify (20) by adding quadratic polynomials $Q_i^{(k)}$ to make the objective function strictly convex to prevent design oscillations during the numerical procedure.

3.3.2. Approximation of regularized penalty function. Let $\tilde{\boldsymbol{\rho}} = S(\boldsymbol{\rho})$. Using (18), the gradient of P is obtained through

$$\frac{\partial P}{\partial \rho_i} = \sum_{j=1}^{N_e} \frac{\partial P}{\partial \tilde{\rho}_j}\frac{\partial \tilde{\rho}_j}{\partial \rho_i} = \sum_{j=1}^{N_e} \frac{\partial P}{\partial \tilde{\rho}_j}(\boldsymbol{\Phi})_{ji}$$

which yields

$$\nabla_{\rho} P = \boldsymbol{\Phi}^T \nabla_{\tilde{\rho}} P \tag{21}$$

where, from (17),

$$(\nabla_{\tilde{\rho}} P)_i = (-2\tilde{\rho}_i + \underline{\rho} + \overline{\rho})|\Omega_i|. \tag{22}$$

The regularized penalty function is then approximated by a linearization with respect to the design variables ρ_i, i.e.

$$P(\boldsymbol{\rho}) \approx s^{(k)} + \sum_{i=1}^{N_e} p_i^{(k)} \rho_i$$

where, again, the involved parameters are chosen so that function values and sensitivities concide with the correct ones at $\boldsymbol{\rho}^{(k)}$.

3.3.3. SOLVING THE SUBPROBLEM. Linearizing as explained in the previous sections one arrives at solving the following subproblem (neglecting constants)

$$(\mathbb{P})^{(k)} \quad \begin{cases} \underset{\rho \in \mathbb{R}^{N_e}}{\text{Minimize}} & \sum_{i=1}^{N_e} \left(\frac{q_i^{(k)}}{\rho_i - l_i^{(k)}} + Q_i^{(k)}(\rho_i) + k_P p_i^{(k)} \rho_i \right) \\ \text{subject to:} & \sum_{i=1}^{N_e} c_i \rho_i \leq V \\ & \underline{\rho}_i^{(k)} \leq \rho_i \leq \overline{\rho}_i^{(k)} \quad i = 1, \ldots, N_e. \end{cases}$$

For details on how to choose the move limits $\underline{\rho}_i^{(k)}$, $\overline{\rho}_i^{(k)}$ and the asymptote value $l_i^{(k)}$, see [21]. The next step is to relax the constraint $\sum_{i=1}^{N_e} c_i \rho_i \leq V$ to obtain the dual problem $(\mathbb{P})_{dual}^{(k)}$ *in one variable only.* The separability of $(\mathbb{P})^{(k)}$ implies that the evaluation of the dual cost function is almost for free, and convexity implies that the dualization procedure gives the global optimum of $(\mathbb{P})^{(k)}$. The problem $(\mathbb{P})_{dual}^{(k)}$ is concave and can be solved by some standard one-dimensional optimization algorithm, e.g., by a bisection method. The time needed to solve $(\mathbb{P})_{dual}^{(k)}$ is negligible compared to the time needed to solve (15).

3.4. Domain decomposition and parallelization

The code was written in C++ and Fortran, based on objective oriented programming, and was fully parallelized. Due to the simplicity of the domain Ω typically used in numerical experiments for topology optimization, the algorithm is based on a one-dimensional cartesian static domain decomposition of the computational domain into subdomains Ω_i. The parallelization is performed so that exactly one processor takes care of the computations and stores data associated to exactly one subdomain. Assume that the number of available processors are p and let Ω be a *parallelepiped* which is discretized in a *uniform* mesh consisting of $N_e = n_x \times n_y \times n_z$ elements and $N_n = (n_x+1) \times (n_y+1) \times (n_z+1)$ nodes. This computational domain is decomposed along the z-axis into p subdomains $\{\Omega_i\}_{i=1}^p$. This decomposition, which reminds of the (equidistant) "slicing of a loaf", is made so that subdomain Ω_i will have a discretization consisting of $N_e^{(i)} = n_x \times n_y \times n_z^{(i)}$ elements and $N_n^{(i)} = (n_x + 1) \times (n_y + 1) \times (n_z^{(i)} + 1)$ nodes where $n_z^{(i)}$ is given by

$$n_z^{(i)} = \begin{cases} \lfloor \frac{n_z}{p} \rfloor + 1 & \text{if } i \leq n_z - p \lfloor \frac{n_z}{p} \rfloor \\ \lfloor \frac{n_z}{p} \rfloor & \text{otherwise} \end{cases} \tag{23}$$

and $\lfloor \frac{n_z}{p} \rfloor$ denotes the largest integer less than or equal to $\frac{n_z}{p}$. In this way one gets an even distribution of computational work and memory storage (ideal when p divides n_z).

4. Numerical examples

4.1. Preliminaries

The computations were performed on Cray T3E in Linköping, Sweden, a distributed memory parallel system where each processor has a clock frequency of 300 MHz and an internal memory of 128 MB. We have for each example chosen $\underline{\rho} = 10^{-6}$, $\overline{\rho} = 1$ and a homogeneous and isotropic material with the Poisson's ratio $\nu = 0.29$. The algorithm is assumed to have converged when the relative difference between two successive objective values is less than 10^{-6} *and* the maximum norm of the difference between two successive designs is less than 0.1.

As an attempt to try to avoid non-global local optima, the following two-step procedure was used:

1. Start at $\boldsymbol{\rho}^{(0)} = \frac{V}{|\Omega|}\mathbf{1}$ and solve for $k_P = 0$, to get $\boldsymbol{\rho}_{k_P=0}$.
2. Start at $\boldsymbol{\rho}^{(0)} = \boldsymbol{\rho}_{k_P=0}$, and solve for

$$k_P = 0.9 \cdot \frac{g(\boldsymbol{\rho}_{k_P=0})}{g_{\max}} \cdot \frac{\mathbf{f}^T\mathbf{u}(\boldsymbol{\rho}_{k_P=0})}{P(\boldsymbol{\rho}_{k_P=0})},$$

to get $\boldsymbol{\rho}^*$.

In step 1, a convex problem is solved, so non-global local optima are avoided. In step 2, a penalization term is solved for. This penalty is hopefully sufficiently small not to introduce many troublesome local optima, and sufficiently large for $\boldsymbol{\rho}^*$ to be enough black and white. More research is certainly needed to exploit efficient and stable *continuation schemes* of this sort. (The term $g_{\max}$ is the maximum penalty value for all ρ in $\mathcal{H}$).

In table 1 we present data for the three numerical experiments that are presented in this section. The penalties are given as the value of g in percentage of $g_{\max}$. The total number of iterations, including steps 1 and 2, were typically around 500, and the solution times range between approximately 3 and 24 hours.

Example name	Crank	"Car" body	Stool
V	$0.5\|\Omega\|$	$0.2\|\Omega\|$	$0.5\|\Omega\|$
R	$0.1L$	$0.15L$	$0.05L$
$g(\boldsymbol{\rho}_{k_P=0})$	19%	52%	32%
N_e (millions)	0.19	1.05	2.1
p	20	18	32

TABLE 1. Table showing data for the three numerical examples.

In the upcoming figures, fixed displacements are represented by grey surfaces and the surface tractions are represented by a white patch with an arrow indicating the direction of the force. The visualization of the optimal structures is done with the software AVS/Express where we have truncated the design at $\rho = 0.99$, i.e. the pictures show the elements that have a density value of at least $\rho = 0.99$. In fact, the appearance of the pictures are very insensitive to the trunctation level since $g(\boldsymbol{\rho}^*) \approx 1$–$2\%$ of $g_{\max}$ for all examples.

4.2. The crank

Consider the L-shaped domain with forces and displacements prescribed as shown to the left in Figure 2. In practice this problem was solved by discretizing a paral-

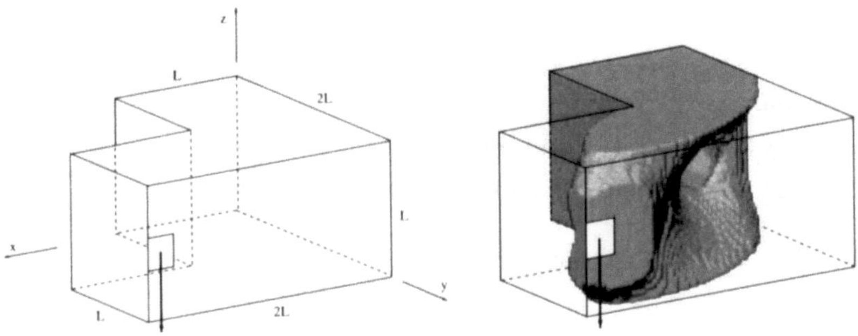

FIGURE 2. Geometry, boundary conditions and optimal design for the crank example.

lelepiped into $40 \times 80 \times 80 = 256000$ elements and letting a quarter of the density values attain the minimum value $\rho = \underline{\rho}$. Moreover, symmetry was used so the computations were only performed on one half of this parallelepiped, i.e. on a mesh consisting of 128000 elements. A number of 20 processors were used and after 11.0 hours the solution shown to the right in Figure 2 was obtained. Two additional views of the crank, based on a rotation about the z-axis, are provided in Figure 3.

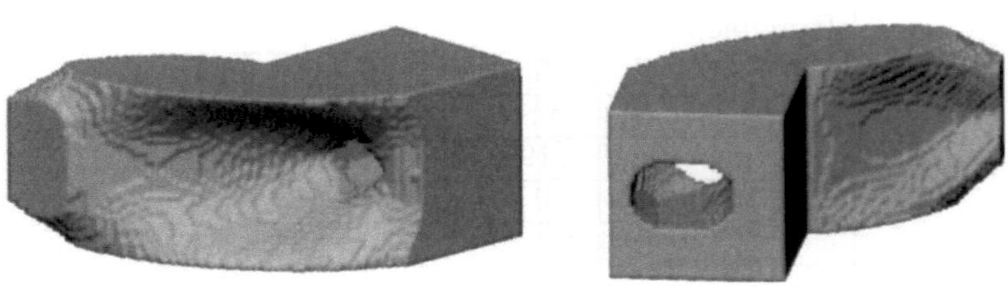

FIGURE 3. The crank from two other views.

The load case of this example can be seen as a combination of bending and torsion. Pure bending typically results in a design of I-beam character. This crank example has also a slanting I-beam section due to the bending part, and a tubular section due to the torsional part.

4.3. The "car" body

Consider the domain in Figure 4 with four square regions of zero-prescribed displacements and a downward force in the center. Symmetry was used so only one eighth of the domain was used in the computations. The obtained numerical solution is depicted in Figure 5.

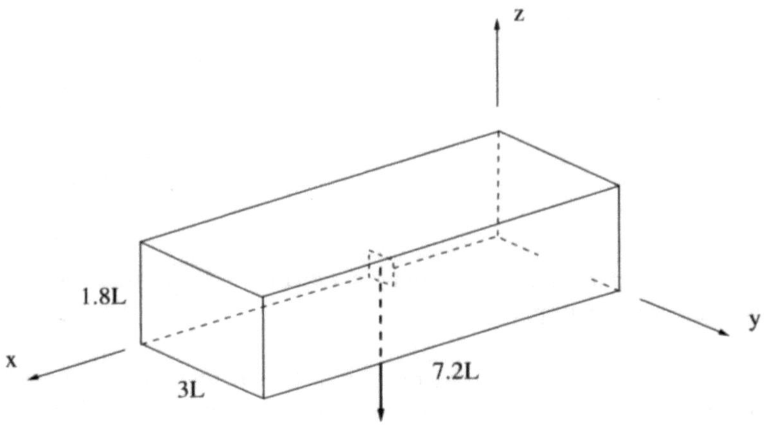

FIGURE 4. Geometry and boundary conditions for the "car" body example.

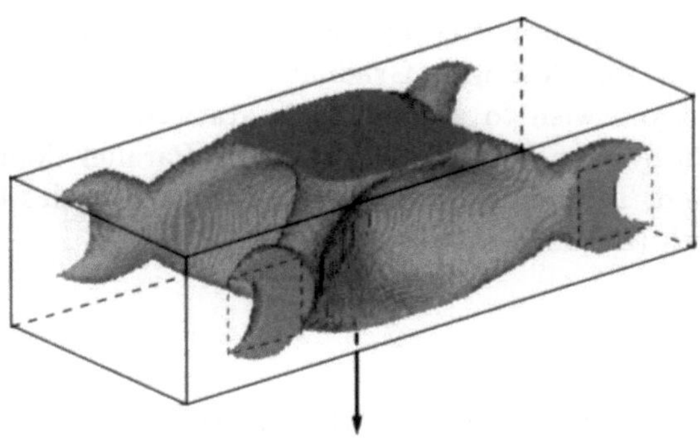

FIGURE 5. Optimal design for the "car" body example.

In this example, unlike in the other two, no topological change has appeared in the final design.

4.4. The stool

Consider the boundary conditions shown to the left in Figure 6 where the grey zones this time indicate that the displacements at these surfaces are fixed only in the z-direction. Only a quarter was used for the computations, and a less strict

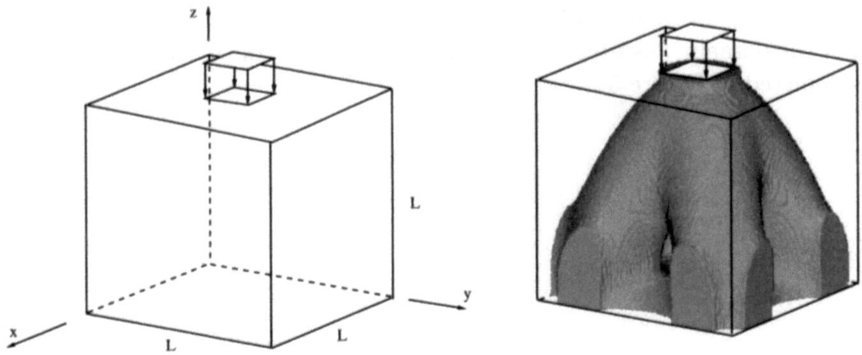

FIGURE 6. Geometry, boundary conditions and optimal design for the stool example.

convergence criterion was used. The optimal solution, shown to the right in Figure 6, consists of four legs that are kept together by a plate on the floor. This plate would of course not be present if the zero-prescribed displacements included also the displacements in the x and y directions.

5. Acknowledgements

This work was financially supported by the National Network in Applied Mathematics (NTM) and the National Graduate School in Scientific Computing (NGSSC). The authors wish to express their gratitude to the National Supercomputer Centre in Sweden (NSC), the Center for Parallel Computers (PDC) at the Royal Institute of Technology in Stockholm, and the High Performance Computing Center North (HPC2N) in Umeå.

References

[1] R. A. Adams, *Sobolev Spaces,* Academic Press, New York, 1975.

[2] G. Allaire, E. Bonnetier, G. Francfort, and F. Jouve, *Shape optimization by the homogenization method,* Numer. Math., **76** (1997), pp. 27–68.

[3] M. Beckers, *Topology optimization using a dual method with discrete variables,* Struct. Optim., **17** (1999), pp. 14–24.

[4] M. P. Bendsøe, *Optimization of Structural Topology, Shape, and Material,* Springer-Verlag, Berlin, 1995.

[5] M. P. Bendsøe, J. M. Guedes, R. B. Haber, P. Pedersen, and J. E. Taylor, *An analytical model to predict optimal material properties in the context of optimal structural design,* J. Appl. Mech., **61** (1994), pp. 930–937.

[6] T. Borrvall and J. Petersson, *Topology optimization using regularized intermediate density control,* Preprint, 2000, to appear in Comput. Methods Appl. Mech. Engrg.

[7] B. Bourdin, *Filters in topology optimization*, Preprint, 1999.

[8] T. E. Bruns and D. A. Tortorelli, *Topology optimization of nonlinear elastic structures and compliant mechanisms,* Preprint, 2000.

[9] J. Céa and K. Malanowski, *An example of a max-min problem in partial differential equations,* SIAM J. Control, **8** (1970), pp. 305–316.

[10] A. Diaz and R. Lipton, *Optimal material layout for 3D elastic structures,* Struct. Optim., **13** (1997), pp. 60–64.

[11] P. Fernandes, J. M. Guedes, and H. Rodrigues, *Topology optimization of three-dimensional linear elastic structures with a constraint on "perimeter",* Comp. Struct., **73** (1999), pp. 583–594.

[12] C. Fleury, *CONLIN: An efficient dual optimizer based on convex approximation concepts,* Struct. Optim., **1** (1989), pp. 81–89.

[13] D. Kinderlehrer and G. Stampacchia, *An Introduction to Variational Inequalities and Their Applications,* Academic Press, New York, 1980.

[14] N. Olhoff, E. Rønholt, and J. Scheel, *Topology optimization of three-dimensional structures using optimum microstructures,* Struct. Optim., **16** (1998), pp. 1–18.

[15] J. Petersson, *A finite element analysis of optimal variable thickness sheets,* SIAM J. Numer. Anal., **36** (1999), pp. 1759–1778.

[16] J. Petersson, *On continuity of the design-to-state mappings for trusses with variable topology,* Preprint, 1999, to appear in Internat. J. Engrg. Sci.

[17] U. Ringertz, *On finding the optimal distribution of material properties,* Struct. Optim., **5** (1993), pp. 265–267.

[18] M. P. Rossow and J. E. Taylor, *A finite element method for the optimal design of variable thickness sheets,* AIAA J., **11** (1973), pp. 1566–1569.

[19] O. Sigmund, *Design of material structures using topology optimization,* Ph.D. thesis, 1994, DCAMM, Technical University of Denmark.

[20] O. Sigmund and J. Petersson, *Numerical instabilities in topology optimization: A survey on procedures dealing with checkerboards, mesh-dependencies and local minima,* Struct. Optim., **16** (1998), pp. 68–75.

[21] K. Svanberg, *The method of moving asymptotes – a new method for structural optimization,* Internat. J. Numer. Methods Engrg., **24** (1987), pp. 359–373.

[22] J. Zowe, M. Kocvara, and M. P. Bendsøe, *Free material optimization via mathematical programming,* Math. Programming, **79** (1997), pp. 445–466.

A Multigrid Method for Optimal Control of Time-Dependent Reaction Diffusion Processes

Alfio Borzì and Karl Kunisch

Abstract. A finite difference multigrid technique that efficiently solves an optimal control problem associated to a time-dependent reaction diffusion model with exponential nonlinearity is presented.

This nonlinear multigrid method solves the nonlinear optimality system of forward/backward parabolic reaction diffusion equations to the level of truncation error in only a few cycles. The contraction factors of the multigrid method result to be independent of the values of the weights in the cost functional and of the number of grid points.

1. Introduction

We present a finite difference nonlinear multigrid method that solves efficiently optimal control of a representative combustion problem [1, 5, 6, 8, 9, 10]. The uncontrolled dynamics exhibit a finite-time blow-up phenomena. Control can be used to avoid blow-up or to guarantee existence of solutions for parameter values for which the uncontrolled equation may not have a solution. The choice of cost functional involves the type of the nonlinearity of the direct problem. The necessary optimality conditions can be expressed as a coupled system of nonlinear partial differential equations. The multigrid algorithm with the Gauss-Seidel-Newton scheme as smoother has contraction factors independent of the number of grid points and of the weights in the cost functional.

1.1. The Solid Fuel Ignition Model

Consider a solid explosive material. We assume that this material is contained in a bounded two-dimensional domain Ω. The ignition process is characterized by the appearance of a localized warm region in which heat is produced due to a chemical reaction process. By considering a solid fuel we study the point-wise balance between chemically generated heat and heat transfer due to conduction; see, e.g., [1]. By considering first-order changes of the temperature T, that is, $T = T_0 + \epsilon u$, the following model for u is obtained [1]:

$$\begin{aligned}\partial_t u = \Delta u + \delta \exp(u) &- f, \quad \delta > 0, \text{ in } Q = \Omega \times (0,T),\\ u(x,0) &= 0, \quad x \in \Omega, \\ u(x,t) &= 0, \quad x \in \partial\Omega, \quad t \in (0,\infty),\end{aligned} \tag{1}$$

which is called the ***solid fuel ignition model.*** Here we introduced a continuous source term function $f = f(x,t)$ which will be used to describe the control mechanism. The stationary version of (1) is called the ***steady state solid fuel ignition model*** and it is given by:

$$\begin{aligned}\Delta u + \delta \exp(u) &= f \text{ in } \Omega,\\ u &= 0 \text{ on } \partial\Omega.\end{aligned} \tag{2}$$

It is recognized that the solutions of (1) and (2) provide many features of the full solid fuel ignition problem; see [1] and references therein.
Let us consider the steady state equation (2), then there exists a critical value δ^* such that: If $\delta > \delta^*$ there is no solution. If $\delta \in]0,\delta^*]$ ($\delta \in]0,\delta^*[$) then there exists at least one and at most two solutions which belong to $W^{2,p}(\Omega)$, $p \geq 1$. If $\delta = \delta^*$ there exists one solution $u^* \in H_0^1(\Omega) \cap W^{2,p}(\Omega)$, $p \geq 1$ (δ^* is referred to as the turning point). For the case of $\Omega = [0,1] \times [0,1]$ and $f = 0$ the turning point has been estimated, based on continuation techniques [6]. It is $\delta^* \simeq 6.80$ and $u(0.5,0.5) \simeq 1.39$.

A necessary condition for the existence of a solution to (2) is obtained by applying the Jensen inequality [2]. It implies that the set of solutions becomes empty for Ω sufficiently large. Sufficient conditions for existence of solutions to (2) are proved in [2]. If the set of solutions of (2) is empty, then the solution u of (1) blows up in a finite time or tends to $+\infty$ as $t \to +\infty$ in the sense that $||u(.,t)||_C \to +\infty$ ($t \to +\infty$), where $||.||_C$ denotes the maximum norm over $\overline{\Omega}$ [5]. If the steady state problem (1) has at least a minimal solution $\underline{u}$ and if $\underline{u} \geq 0$, then the solution of (1) converges uniformly to $\underline{u}$ as $t \to +\infty$ (that is $\underline{u}$ is asymptotically stable). Consequently any solution of the steady state problem other than the minimal solution is unstable [5].

2. The Optimal Control Problem

Control of combustion phenomena will be exerted by adding or subtracting thermal energy to the fuel. For this purpose we consider:

$$\begin{aligned}&\min_{f \in L^2(Q)} J(u(f),f),\\ -\partial_t u + \Delta u + \delta \exp(u) &= f \qquad \text{in } Q,\\ u &= 0 \quad \text{on } \partial\Omega, \quad t \in (0,\infty),\end{aligned} \tag{3}$$

where the cost functional $J(u,f)$ is given by

$$J(u,f)=\frac{1}{2}||u-z||^2_{L^2(Q)}+\frac{\beta}{2}||e^u-e^z||^2_{L^2(Q)}+\frac{\nu}{2}||f||^2_{L^2(Q)}, \tag{4}$$

where ν is the weight of the cost of the control, $\beta>0$ is a scaling factor, and $z\in H^2(Q)$ denotes the desired state. Then existence of a solution $(f^*,u^*)=(f^*,u(f^*))$ to the optimal control problem (3) can be established; see [9]. The necessary optimality conditions can be expressed as the optimality system:

$$\begin{aligned} -\partial_t u+\Delta u+\delta\exp(u)-f &= 0,\\ \partial_t\lambda+\Delta\lambda+\delta\exp(u)\lambda+(u-z)+\beta e^u(e^u-e^z) &= 0,\\ \nu f-\lambda &= 0, \end{aligned} \tag{5}$$

with initial condition $u(x,0)=u_0(x)$ for the state equation (evolving forward in time) and terminal condition $\lambda(x,T)=0$ for the adjoint equation (evolving backward in time).

2.1. The Discretization of the Optimal Control Problem

We use the finite difference discretization of (5). Let Ω_{h_k}, $k=1,2,\dots,M$, denotes a sequence of grids on a square domain. The integer k is called the level number. The coarsest grid corresponds to $k=1$ with mesh size $h_1=1/4$, the finest corresponds to $k=M$ with mesh size $h_M=h_1/2^{(M-1)}$. The set of mesh points is denoted by (x_i,y_j), $x_i=(i-1)h_k$ and $y_j=(j-1)h_k$ $i,j=1,...,N_k$, $N_k=2^{(k+1)}+1$. The value of a numerical function on (x_i,y_j) is denoted by u_{ij}. The time-coordinate is discretized into time levels $t^l=(l-1)\Delta t$, $l=1,2,\dots,L$. We take $Q=\Omega\times[0,1]$. With Δ^h we denote the five-point stencil finite difference discretization of the Laplacian operator.

Necessary and sufficient conditions for existence of bounded solutions to the discretized equations are proved in [2].

To solve the parabolic problem (1) we use a fully implicit Euler backward method applied to the optimality system to obtain:

$$\begin{aligned} -\partial_t^{\Delta t}u^h+\Delta^h u^h+\delta\exp(u^h)-f^h &= 0,\\ \partial_t^{\Delta t}\lambda^h+\Delta^h\lambda^h+\delta\exp(u^h)\lambda^h+(u^h-z^h)+\beta e^{u^h}(e^{u^h}-e^{z^h}) &= 0,\\ \nu f^h-\lambda^h &= 0. \end{aligned} \tag{6}$$

with initial condition $u^h_{ij}(l=1)=u_0(x_i,y_j)$ for the state equation and terminal condition $\lambda^h_{ij}(l=L)=0$ for the adjoint equation. Notice that the Euler backward formula applies to the adjoint equation with time reversed.

3. Numerical Solution of the Optimal Control Problem

In the implementation of multigrid methods the development of the relaxation scheme is of crucial importance. Nowadays, theoretical results and practical experience provide general criteria for an appropriate choice of smoothing schemes for multigrid methods applied to the primal problem, which is (1) in our case. However, in the present case of the nonlinear time-dependent coupled system (5) consisting of primal- and adjoint equation and optimality condition there is no previous experience nor theory available. For this reason, previous papers [2, 3] on this topic report experiments conducted using different relaxation schemes. In [2, 3] we demonstrate that a Gauss-Seidel-Newton scheme should be used to obtain an efficient and robust multigrid algorithm. In the following we describe this iterative scheme in detail.

3.1. The Gauss-Seidel-Newton Relaxation

The Gauss-Seidel-Newton (GSN) iteration is defined by a Newton step applied at each grid point to the set of variables $\phi^h = (u^h, \lambda^h, f^h)$,

$$\phi_{ij}^{h(new)} = \phi_{ij}^{h(old)} - [\mathbf{G}'(\phi_{ij})]^{-1}\mathbf{G}(\phi_{ij}). \tag{7}$$

The equation $\mathbf{G}(u^h, \lambda^h, f^h) = \mathbf{0}$ represents (6) and $\mathbf{G}'$ denotes the Jacobian of $\mathbf{G}$.

To help the reader implement our algorithm, we describe how to derive the iterative scheme using *Mathematica.* Define the following functions representing the discrete counterparts of the primal- and adjoint equations (the variable y represents λ):

$$\begin{aligned} G1[u\_, y\_] &= Su - 4*u + h2*d*E^u - h2*y/n - Fu - h2*u/dt + h2*u0/dt \\ G2[u\_, y\_] &= Sy - 4*y + h2*d*E^u*y + h2*u - h2*z \\ &+ h2*bb*E^u*E^u - h2*bb*E^u*E^z - Fy - h2*y/dt + h2*y0/dt \end{aligned}$$

where the optimality condition has been included in the primal equation.

$$JG = \{\{D[G1[u,y],u], D[G1[u,y],y]\}, \{D[G2[u,y],u], D[G2[u,y],y]\}\},$$

$$G[u\_, y\_] = \{G1[u,y], G2[u,y]\},$$

$$K[u\_, y\_] = Inverse[JG].G[u,y].$$

In the input, $Su - 4*u$ represents $h^2\Delta^h$ ($h2 := h^2$), d and bb represent δ and β respectively, Fu and Fy are the right-hand sides of the optimality system, and E means exponential. Clearly, JG represents $\mathbf{G}'$. In the time discretization u and y are defined at the time level l while $u0$ and $y0$ denote the state variable and the co-state variable at time level $l-1$ and $l+1$ respectively; see the following Plot.

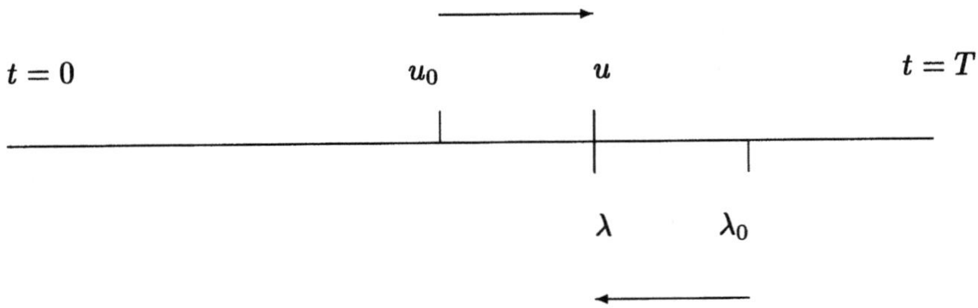

Now $K[u,y]$ (that is, $[\mathbf{G}'(\phi_{ij})]^{-1}\mathbf{G}(\phi_{ij})$) is a two-component column vector. The first component defines the update for u in the GSN iteration. That is, a relaxation sweep for the state variable consists in updating u by a GSN step on all mesh points of a time level (in any order), for all time levels starting from $l=0$ in the forward direction. On the other hand, the adjoint variable evolves backward in time hence λ is being updated using the second component of $K[u,y]$ for the GSN step on all mesh points of a time level, for all time levels starting from $l=L$ in the backward direction. The updating is performed in a Gauss-Seidel fashion.

3.2. A FAS Multigrid Method

In the present work, we apply the multigrid coarse-grid correction strategy in space only. A full coarsening is indeed possible and the present scheme will apply as well. It would be also possible to include more sophisticated multigrid techniques for parabolic problems as described, e.g., in [4, 7]. Based on algebraic (multigrid) considerations [2], we choose I_h^H to be the half-weighted restriction and $\hat{I}_h^H$ as the straight injection. For prolongation I_H^h we use the bilinear interpolation operator. The full approximation scheme (FAS) follows:

FAS method for solving $A^h(\phi^h)=f^h$.

1. Apply ν_1 smoothing steps: $\phi^h = S^{\nu_1}(\phi^h, f^h)$.
2. Transfer the approximate solution: $\phi^H = \hat{I}_h^H\phi^h$.
3. Compute the right hand side of the FAS equation: $f^H = I_h^H f^h + [A^H(\phi^H) - I_h^H A^h(\phi^h)]$.
4. Apply γ times the FAS scheme to $A^H(\hat{\phi}^H) = f^H$. (Use the GSN iteration if $k=1$.)
5. Use coarse level correction: $\phi^h = \phi^h - I_H^h(\phi^H - \hat{\phi}^H)$.
6. Apply ν_2 smoothing steps: $\phi^h = S^{\nu_2}(\phi^h, f^h)$.

In the following we give examples demonstrating efficiency and robustness of the proposed finite difference multigrid method for the optimal control problem. In all experiments we set $\nu_1=2$, $\nu_2=2$, and $\gamma=1$.

The tracking ability of our algorithm will be expressed in terms of the tracking error, $||u^h - z^h||_0$. The discrete L^2-norm of the control function is given by

$||f^h||_{L^2} = ||\lambda^h||_{L^2}/\nu$. Further, the L^2-norm of the residuals of the state and costate equations are reported. The residuals are defined by $r(u)^h = f^h - A^h(u^h)$ and $r(\lambda)^h = f^h - A^h(\lambda^h)$.

4. Numerical Solution of the Time Dependent Optimal Control Problem

In this section we demonstrate the ability of our algorithm to solve the time-dependent optimal control problem associated to combustion processes. Notice that (5) is a system of forward-backward nonlinear parabolic equations for which, in our knowledge, there are no established solution methods available. In the following, some of the results obtained with experiments, that present increasing difficulty, are reported.

First, consider the objective function $z(x,y,t) = \frac{t}{\pi^2}\sin(\pi x)\sin(\pi y)$, $t \in (0,T)$, $T = 1.0$. We use a fixed 5-V-Cycle multigrid with finest level $M = 4$ (with a 4×4 coarsest grid), $\beta = \delta = 6.0$, and $\Delta t = 0.01$.

In Table 1, **Case 1** refers to the situation where $\Omega = [0,1] \times [0,1]$ and the time-dependent solid fuel ignition model admits stable solutions.

In **Case 2** we consider $\Omega = [0,2] \times [0,2]$, that is, when the steady state model has no solution and any solution of the time-dependent model is necessarily unstable. Nevertheless the control is able to stabilize the ignition process and drive it to the desired state: see Table 1.

TABLE 1

		Case 1		
ν	$\|\|u^h - z^h\|\|_0$ (Q)	$\|\|f^h\|\|_{L^2(Q)}$	$\|\|r(u)^h\|\|_{L^2(Q)}$	$\|\|r(\lambda)^h\|\|_{L^2(Q)}$
10^{-3}	4.77(-2)	4.43	1.10(-6)	1.29(-7)
10^{-5}	3.33(-3)	5.24	1.48(-6)	1.27(-8)
10^{-7}	1.59(-4)	5.45	7.42(-14)	4.29(-16)
		Case 2		
ν	$\|\|u^h - z^h\|\|_0$ (Q)	$\|\|f^h\|\|_{L^2(Q)}$	$\|\|r(u)^h\|\|_{L^2(Q)}$	$\|\|r(\lambda)^h\|\|_{L^2(Q)}$
10^{-3}	6.62(-2)	10.77	7.58(-4)	5.70(-5)
10^{-5}	2.98(-3)	11.51	4.49(-9)	3.85(-11)
10^{-7}	5.17(-5)	11.60	3.53(-14)	6.82(-16)

By inspecting Table 1 we see that as values of the weight of the cost of the control ν increases then, as expected, larger values of $||u^h - z^h||_0$ are obtained. An increase of $||f||_{L^2}$ results in a decrease of the value $||u^h - z^h||_0$.

A more challenging series of tests is obtained considering the following objective function: $z(x,y,t) = \frac{1}{\pi^2}\sin(\pi x)\sin(\pi y)\sin(n\pi t)$, in $\Omega = [0,1] \times [0,1]$, $T = 1$. As n increases, the function z will oscillate in time with increasing frequency. Further, we increase the coefficients of the nonlinearity in our model: $\beta = \delta = 10.0$. (A very large range of values for these coefficients is allowed; see [2].) In Table 2, we report results relative to the frequencies $n = 2$ and $n = 4$, respectively. The value

TABLE 2

n	$\|\|u^h - z^h\|\|_0$ (Q)	$\|\|f^h\|\|_{L^2(Q)}$	$\|\|r(u)^h\|\|_{L^2(Q)}$	$\|\|r(\lambda)^h\|\|_{L^2(Q)}$
2	4.12(-3)	9.37	2.02(-6)	2.18(-8)
4	4.12(-3)	9.38	2.02(-6)	2.22(-8)

of the weight of the cost of the control is $\nu = 10^{-5}$. In Figure 1, we represent the values of the state variable u and of the objective function z at $(x, y) = (0.5, 0.5)$ for all time steps. The computed solutions (dashed line) almost coincide with the desired value (circle line).

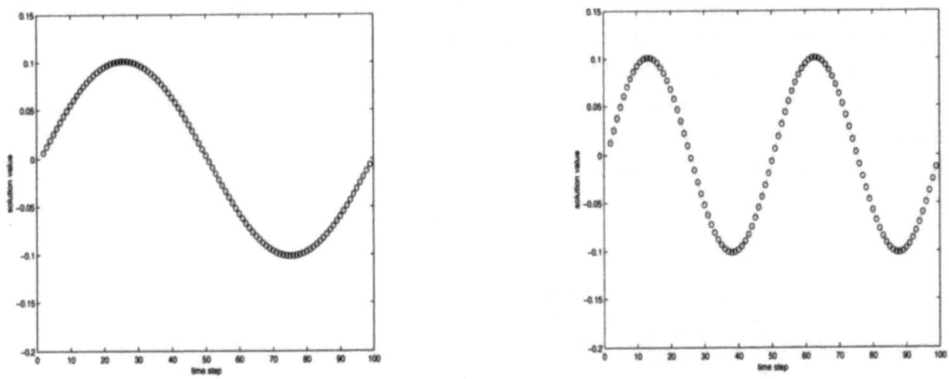

FIGURE 1. Solution values for $n = 2$ (left) and for $n = 4$ (right).

5. Conclusions

An optimal solution- and control algorithm for the time-dependent solid fuel ignition model was presented. It is based on finite difference discretization and multigrid techniques. This nonlinear multigrid method solves the nonlinear optimality system of forward/backward parabolic (exponential-) reaction diffusion equations to the level of truncation error in only a few cycles, independently of the values of the weights in the cost functional.

References

[1] J. Bebernes and D. Eberly, *Mathematical Problems from Combustion Theory*, Springer-Verlag, 1988.

[2] A. Borzì and K. Kunisch, *The numerical solution of the steady state solid fuel ignition model and its optimal control*, SIAM J. Sci. Comp., **22(1)** (2000), pp. 263–284.

[3] A. Borzì, K. Kunisch, and M. Vanmaele, *A Multi-Grid Approach to the Optimal Control of Solid Fuel Ignition Problems*, in: E. Dick, K. Riemslagh, and J. Vierendeels, eds., Lecture Notes in Computer Science and Engineering, European Multigrid Meeting 1999, Springer-Verlag, 2000.

[4] A. Brandt, *Multigrid Techniques: 1984 Guide with Applications to Fluid Dynamics*, GMD-Studien no. 85, St. Augustin, Germany, 1984.

[5] H. Fujita, *On the nonlinear equations* $\Delta u + \exp(u) = 0$ *and* $v_t = \Delta v + \exp(u)$, Bull. Amer. Math., **75** (1968), pp. 132–135.

[6] R. Glowinski, H. B. Keller, and L. Rheinhart, *Continuation-conjugate gradient methods for the least-squares solution of nonlinear boundary value problems*, SIAM J. Sci. Statist. Comput., **6** (1985), pp. 793–832.

[7] W. Hackbusch, *Multi-grid Methods and Applications*, Springer-Verlag, New York, 1985.

[8] K. Ito and K. Kunisch, *Optimal Control of the Solid Fuel Model with* H^1*-Cost*, Preprint, 1998.

[9] A. Kauffmann, *Optimal Control of the Solid Fuel Ignition Model. PhD Thesis*, Technical University of Berlin, Berlin 1998.

[10] C. V. Pao, *Nonlinear Parabolic and Elliptic Equations*, Plenum Press, New York, 1992.

On the Implementation of Multilevel Solvers for Constrained Variational Problems in Dual Formulation

Hinderk Martens Buß

Abstract. Various technical problems are most appropriately described in terms of variational inequalities that involve elliptic differential operators of second or fourth order. Multigrid solvers for variational formulations of this kind are not very well established even though hierarchically refined finite element meshes provide a very natural setting for multilevel techniques. The existing algorithms have been designed with a view to those complementary problems, that arise in the numerical treatment of the primal formulation. In some cases, however, it may be preferable or even unavoidable to consider the dual formulation of a variational inequality.

A multigrid solver is presented for those quadratic programs that result from a conforming discretisation of the dual formulation. Using mixed finite elements of Raviart-Thomas type nested trial spaces are constructed and the multilevel scheme of Mandel is adapted to compute the Lagrange multipliers of the admissibility constraints. Two variants of this solver are derived to cover the case that a hybrid mixed ansatz must be employed. The convergence of all three schemes is discussed and numerical experiments are reported.

1. Introduction

While there exists a vast amount of literature on multilevel techniques in general, comparatively few articles have been written with a view to constrained variational problems. Moreover, all of them consider the special case of box constraints only. A generic example for the type of constrained variational setting, that multilevel algorithms have been proposed for, is known as the obstacle problem: Given two functions $\psi \in W^{1,2}(\Omega)$ and $f \in L^2(\Omega)$ that are defined on some bounded domain $\Omega \subset \mathbb{R}^d$ find the minimiser of the functional

$$J(v) \;:=\; \int_\Omega \left\{ \frac{1}{2} |\nabla v|^2 - f\,v \right\} dx \quad ; \quad v \in W_0^{1,2}(\Omega) \tag{1}$$

under the additional assumption: $v \geq \psi$ almost everywhere on Ω. (The notation will be introduced in the next section.) A conforming finite element scheme which employs linear shape functions to approximate the minimiser was analysed in 1974 by Falk [17]. Today this method can still be regarded as the standard discretisation for the primal formulation. The use of quadratic shape functions in a conforming finite element framework was discussed in [8]. In 1978 a mixed finite element method was proposed for the obstacle problem [9], that relied on the following dual formulation: Minimise the functional

$$J^*(\sigma) \;:=\; \int_\Omega \left\{ \frac{1}{2}\sigma^2 \;+\; \psi\,\nabla\cdot\sigma \right\} dx \quad ; \quad \sigma \in H(\mathrm{div},\Omega) \tag{2}$$

under the condition: $\nabla\cdot\sigma + f \leq 0$ almost everywhere on Ω. For the same reasons, that mixed finite element methods can be preferable to more standard techniques, if unconstrained variational problems must be solved, dual mixed formulations of variational inequalities are sometimes superior to a straightforward treatment of the primal formulation. A conspicuous application of a dual mixed discretisation is presented in [10]: the computation of reliable a posteriori error estimates for finite element approximations to the solutions of certain variational inequalities. Therefore, the development of fast multilevel solvers for the dual problem (2) is definitely not a purely academic pastime.

It seems however, that only one paper [23] has been published to describe an actual implementation of a dual mixed discretisation scheme. The multilevel methods that have been developed to solve the primal formulation (1) either use suitably adapted algorithmic components of a conventional linear multigrid solver [5, 21] or carry out combinatorial procedures employing linear multilevel iterations to update the set of active constraints (see for example [19, 20]). The algorithm proposed in [23] falls into the latter category. However, its multigrid component is merely a device to evaluate a Schur complement resulting from a transformation [2] of the dual problem back into a primal setting: the active set is actually updated by a simple conjugate gradient iteration. As to the efficiency of this approach no conclusive numerical experiments are reported.

Replacing the dual mixed discretisation of an elliptic differential equation by a nonconforming discretisation of its primal formulation is an artifice to be found also elsewhere (for example [6]). In the design of multilevel iterations for unconstrained problems such a substitution is appropriate, as it permits the use of established multigrid technologies. In a constrained setting, however, returning to a primal formulation only leads to a convoluted algorithm without breaking the saddle point structure of the algebraic problem. The introduction of penalty terms is fraught with difficulties even in the unconstrained case [12] and fails likewise in lifting the saddle point character of the problem. Hence, it seems reasonable to treat the dual mixed discretisation in the most immediate manner possible. While a direct approach is easily pursued with the help of Vanka-type smoothers [27] in the case of the Stokes problem, the construction of analogous smoothing iterations for the dual formulation (2) is inherently difficult. Elimination of the dual variables and solving the resulting complementary problem for the Lagrange multipliers of the admissibility constraints is a possible resort: but conventional projected point relaxation methods (see e. g. [15]) no longer qualify as smoothers, since the Schur complement matrix involved in such a procedure is dense.

In this paper two technologies are presented that address the aforementioned difficulties. A defect correction scheme is introduced for the Schur complement problem which incorporates successive projections and the possibility to apply a preconditioner. The performance of this smoother depends on the choice of the preconditioner and on the speed the defect is computed with. After scaling the vector valued shape functions a simple CG iteration is sufficient to evaluate the Schur complement with optimal complexity. An effective preconditioner can be obtained after lumping the mass matrix and explicitly calculating the resulting matrix product. Hybridisation of the dual mixed discretisation is considered as a means to obtain an alternative preconditioner. The second technology relies on hybridisation to define an equation for the Lagrange multipliers associated with the continuity constraints on the normal fluxes. The global linear complementary condition is replaced by very small, independent complementary problems, each of them defined on one of the cells compounding the grid. Thus the computation of the defect becomes as inexpensive as a matrix vector multiplication. A defect correction scheme is devised that is accelerated by a multilevel method operating in the so called full approximation storage (FAS) mode. Due to the sparsity patterns of the operators involved any established point relaxation method can be employed to improve the smoothing.

The paper is organised in the following way: The notation is established in the next section. In addition, the primal problem and its dual counterpart are introduced. The discretisation of the dual formulation is specified in Section 3. Subsequently, three different multilevel methods are presented, that implement the numerical technologies briefly described above. In Section 6 these solvers are compared against several implementations of primal-dual interior point methods.

2. Statement of the problem

Let $\Omega \subset \mathbb{R}^d$ denote a bounded, convex polytope. The space of all p-summable functions, that have p-summable generalised derivatives up to order m, is denoted by $W^{m,p}(\Omega)$. It is a Banach space equipped with the following norm:

$$\| u \|_{m,p}^p \quad := \quad \int_\Omega \sum_{|\alpha| \le m} \left| \frac{\partial^\alpha u}{\partial x^\alpha} \right|^p \, dx \qquad ; \qquad u \in W^{m,p}(\Omega) \ . \tag{3}$$

Taking the closure of the space $C_0^\infty(\Omega)$ in the norm (3) provides the subspace $W_0^{m,p}(\Omega) \subset W^{m,p}(\Omega)$. For $m = 0$ both $W^{m,p}(\Omega)$ and $W_0^{m,p}(\Omega)$ coincide with the space $L^p(\Omega)$ of all p-summable functions. In this case the first subscript of (3) will be dropped. The space $W^{m,\infty}(\Omega)$ is defined by introducing the norm:

$$\| u \|_{m,\infty} \quad := \quad \max_{|\alpha| \le m} \ \operatorname*{ess\,sup}_{x \in \Omega} \left| \frac{\partial^\alpha u}{\partial x^\alpha} (x) \right| \qquad ; \qquad u \in W^{m,\infty}(\Omega) \ .$$

The canonical inner product of the space $L^2(\Omega)$ is denoted by $(\cdot,\cdot)_\Omega$, while the scalar product of two vectors $u, v \in \mathbb{R}^n$ is written as $u \cdot v$. In addition, the well known abbreviations ∇u for the gradient of a scalar function $u \in W^{1,p}(\Omega)$ and $\nabla \cdot \sigma$ for the (weak) divergence of a vector field $\sigma \in L^2(\Omega, \mathbb{R}^d)$ will be used. If the divergence of such a field is square integrable, its normal component with respect to some hypersurface (usually $\partial\Omega$) can be defined as a distribution acting on certain traces (see e. g. theorem 2.5 in [18]). This very distribution will be referred to either by σ_n or by $\sigma_{n,\Omega'}$, in case the boundary of some subset $\Omega' \subset \Omega$ is considered. The space $L^2(\Omega, \mathbb{R}^d)$ is equipped with the following inner product:

$$\langle \sigma , \tau \rangle_\Omega \;:=\; \int_\Omega \sigma \cdot \tau \; dx \qquad ; \qquad \sigma, \tau \in L^2(\Omega, \mathbb{R}^d) \; .$$

Another Hilbert space that will be denominated $H(\text{div}, \Omega)$ is obtained by taking the closure of $C^\infty(\Omega, \mathbb{R}^d)$ with respect to the norm

$$\| \sigma \|^2_{Div} \;:=\; \langle \sigma , \sigma \rangle_\Omega + \left(\nabla \cdot \sigma , \nabla \cdot \sigma \right)_\Omega \quad ; \quad \sigma \in H(\text{div}, \Omega) \; .$$

Problem P: The equilibration position of an elastic "membrane", represented by the domain Ω, may be described by a function $u^* \in W^{1,2}(\Omega)$, that denotes its distortion in normal direction under the action of some load $f \in W^{1,1}(\Omega)$. It is assumed, the membrane is obstructed by an "obstacle" $\psi \in W^{3,\infty}(\Omega)$ and has been clamped along $\partial\Omega$ to adopt the values of some function $u_0 \in W^{3,\infty}(\Omega)$ with the property: $u_0 > \psi$ on $\partial\Omega$. The material law is described by the components $C_{ij} \in W^{2,\infty}(\Omega)$ of a symmetric tensor, that satisfies the ellipticity condition

$$C_0 > 0 \quad : \quad \sum_{i=1}^{n} \sum_{j=1}^{n} C_{ij}(x)\, \xi_i \, \xi_j \;\geq\; C_0 \, \|\xi\|^2 \quad ; \quad \xi \in \mathbb{R}^n \, , \; x \in \Omega \; . \tag{4}$$

The set $K \subset W^{1,2}(\Omega)$ of conceivable equilibria is closed and convex. It reads:

$$K \;:=\; \left\{ u \in W^{1,2}(\Omega) \,\middle|\, v \in W^{1,2}_0(\Omega) \;:\; u \,=\, v + u_0 \,\geq\, \psi \right\} \; .$$

The elastic energy J of the membrane may be described by the functional

$$J(v) \;:=\; \tfrac{1}{2} \left\langle \nabla v, C \cdot \nabla v \right\rangle \,-\, \left(f, v \right) \quad ; \quad v \,\in\, W^{1,2}(\Omega) \; ,$$

so the equilibrium position $u^* \in K$ must obey Hamilton's principle in the form:

$$\left\langle C \cdot \nabla u^*, \nabla v - \nabla u^* \right\rangle \,\geq\, \left(f, v - u^* \right) \quad ; \quad v \in K \; .$$

□

Due to condition (4) the inverse of the matrix $C(x)$ exists for every $x \in \Omega$. Hence, the complementary energy J^* is well defined for any $\sigma \in H(\text{div}, \Omega)$. Denoting the components of the inverse by $C^*_{ij} \in W^{2,\infty}(\Omega)$, this functional reads:

$$J^*(\sigma) \;:=\; \tfrac{1}{2} \left\langle \sigma, C^* \cdot \sigma \right\rangle_\Omega + \left(\psi, \nabla \cdot \sigma \right)_\Omega - \left(\sigma_n , u_0 \right)_{\partial\Omega} \quad ; \quad \sigma \in H(\text{div}, \Omega) \; .$$

The dual formulation of problem **P** demands the minimisation of $J^*(\sigma)$ under the assumption, that certain admissibility constraints be satisfied by the minimiser. The solution σ^* is related to the equilibrium position u^* of the membrane by:

$$\sigma^* \;=\; C \cdot \nabla u^* \quad .$$

Problem D: The regularity requirements stated in the formulation of problem **P** be met and the set of admissible stresses be defined by:

$$\Sigma \;:=\; \left\{ \sigma \in H(\mathrm{div}, \Omega) \,\middle|\, \nabla\cdot\sigma + f \;\leq\; 0 \text{ on } \Omega \right\} \; .$$

The equilibrium distribution σ^* of the stress inside the membrane is the solution of the following constrained variational problem: Find $\sigma^* \in \Sigma$, such that:

$$\left\langle C^{*}\!\cdot \sigma^*, \tau - \sigma^* \right\rangle_\Omega + \left(\psi, \nabla\cdot\tau - \nabla\cdot\sigma^*\right)_\Omega \;\geq\; \left(u_0\,, \tau_n - \sigma_n^*\right)_{\partial\Omega} \quad ; \quad \tau \in \Sigma \; . \qquad \square$$

3. Discretisation of the dual formulation

Let $\mathcal{T} = \{T_i\}_{i=1}^m$ denote a simplicial decomposition of the domain Ω, such that the intersection of two elements T_i and T_k is again a simplex of lower dimension. (This includes the case that $T_i \cap T_k$ contains only a vertex.) In the usual way (see for example §3.1 in the book [13]) a parameter $h_\mathcal{T} > 0$ is introduced

$$h_\mathcal{T} \;:=\; \max_{T\in\mathcal{T}} \inf \left\{ r > 0 \,\middle|\, x \in \mathbb{R}^d \;:\; T \subset B_r(x) \right\}$$

to bound the diameter of these simplices. A second parameter $\varrho_\mathcal{T} > 0$ can be defined to assess, whether simplices tend to degenerate and become unduly flat:

$$\varrho_\mathcal{T} \;:=\; \min_{T\in\mathcal{T}} \sup \left\{ r > 0 \,\middle|\, x \in \mathbb{R}^d \;:\; B_r(x) \subset T \right\} \; .$$

A family $\mathfrak{T} = \{\mathcal{T}_h\}_{h>0}$ of simplicial decompositions will be called *regular*, if the following two conditions are met: For each $h > 0$ there is some $\mathcal{T} \in \mathfrak{T}$, such that $h_\mathcal{T} < h$ holds. The quantity $\varrho_\mathcal{T} \,/\, h_\mathcal{T}$ is uniformly bounded from below.

Once such a regular family of decompositions $\mathfrak{T}$ has been supplied, several trial spaces can be constructed for each $\mathcal{T} \in \mathfrak{T}$. Below the set of polynomials of degree less or equal k will be denoted by $\mathbb{P}_k$. The shape functions are defined by:

$$B^k(\mathcal{T}) \;:=\; \left\{ \tau \in L^2(\Omega, \mathbb{R}^d) \,\middle|\, \tau|_T \in (\mathbb{P}_k)^d \;;\; T \in \mathcal{T} \right\},$$

$$P^k(\mathcal{T}) \;:=\; \left\{ u \in L^2(\Omega) \,\middle|\, u|_T \in \mathbb{P}_k \;;\; T \in \mathcal{T} \right\},$$

$$R^k(\mathcal{T}) \;:=\; \left\{ \tau + \vartheta\, x \,\middle|\, \tau \in B^k(\mathcal{T})\,,\, \vartheta \in P^k(\mathcal{T}) \right\} \; .$$

To specify the degrees of freedom associated with these spaces test functions must be considered which live on the interfaces between adjacent simplices:

$$\mathcal{E} \;:=\; \left\{ E \subset \Omega \,\middle|\, T, T' \in \mathcal{T} \;:\; E = T \cap T' \,\wedge\, T \neq T' \right\},$$

$$\mathcal{B} \;:=\; \left\{ E \subset \partial\Omega \,\middle|\, T \in \mathcal{T} \;:\; E = T \cap \partial\Omega \right\} \; .$$

The spaces $P^k(\mathcal{T})$ and $P^k(\mathcal{E} \cup \mathcal{B})$ are of Lagrange type: Their degrees of freedom consist of function evaluations at certain collocation points $\{\hat{x}^{(k)}_{T,i}\}_{T,i}$ (see [13] §2.2 for details). The corresponding shape functions are abbreviated by $\{u^{(k)}_{T,i}\}_{T,i}$. For any simplex $T \in \mathcal{T}$ and any of its faces $E \subset T$ a set of linear functionals

$$\rho_{T,E,j}(\tau) \;:=\; \left(u^{(k)}_{E,j}\,,\,\tau_{n,T}\right)_E \quad ; \quad \tau \,\in\, H(\mathrm{div},\Omega) \tag{5}$$

can be defined. If $e_l \in \mathbb{R}^n$ represents the l-th basis vector of a Cartesian coordinate system, a second set of functionals, necessary only if $k > 0$, is described by:

$$\rho_{T,i,l}(\tau) \;:=\; |T|^{-1/d}\left\langle u^{(k-1)}_{T,i}\, e_l\,,\,\tau\right\rangle_T \quad ; \quad \tau \,\in\, H(\mathrm{div},\Omega)\ . \tag{6}$$

Both sets together constitute the degrees of freedom associated with the ansatz $R^k(\mathcal{T})$. The functionals (5) are shared by the spaces $B^k(\mathcal{T})$. In the case $k > 1$ the other degrees of freedom differ. Let $\{\sigma_{T,i}\}_{T,i}$ denote a basis of the function space

$$B^k_0(\mathcal{T}) \;:=\; \left\{\, \tau \in B^k(\mathcal{T}) \,\middle|\, \nabla\cdot\tau\big|_T = 0 \;\wedge\; \tau_{n,T} = 0\ ;\ T \in \mathcal{T}\,\right\}\,,$$

such that $\|\sigma_{T,i}\|_\infty = 1$ holds for each basis function. Instead of (6) the missing degrees of freedom for the trial space $B^k(\mathcal{T})$ now read:

$$\hat{\rho}_{T,i}(\tau) \;:=\; \left\langle \nabla u^{(k-1)}_{T,i}\,,\,\tau\right\rangle_T \quad ; \quad \tau \in H(\mathrm{div},\Omega)\ ,$$

$$\hat{\rho}^{\,0}_{T,i}(\tau) \;:=\; |T|^{-1/d}\left\langle \sigma_{T,i}\,,\,\tau\right\rangle_T \quad ; \quad \tau \in H(\mathrm{div},\Omega)\ .$$

Neither $B^k(\mathcal{T})$ nor $R^k(\mathcal{T})$ are subspaces of $H(\mathrm{div},\Omega)$: Yet a vector field σ_k from either of these spaces is contained in $H(\mathrm{div},\Omega)$, if and only if the normal fluxes of σ_k are continuous across the element interfaces inside the domain Ω. Hence, two possibilities exist to discretise problem **D**. As the choice of the ansatz does not affect the general procedure, only the space $R^k(\mathcal{T})$ will be considered. Let $\{\sigma_{T,i,l}, \sigma_{T,E,j}, \dots\} \subset R^k(\mathcal{T})$ denote the dual basis of (5) and (6). The first option is to construct a basis $\{\sigma_{T,i,l}, \sigma_{E,j} \dots\}$ for the conforming trial space

$$R^k_c(\mathcal{T}) \;:=\; R^k(\mathcal{T}) \,\cap\, H(\mathrm{div},\Omega)\ .$$

A weight function $\omega : \mathcal{T} \times (\mathcal{E} \cup \mathcal{B}) \longrightarrow \{-1, 0, 1\}$ can be specified, such that

$$\sigma_{E,j} \;:=\; \sum_{T \in \mathcal{T}} \omega(T,E)\,\sigma_{T,E,j} \quad \in \quad H(\mathrm{div},\Omega) \tag{7}$$

holds. The case $\omega(T,E) = 0$ is implied by $E \not\subset T$. To account for the admissibility constraint $\sigma \in \Sigma$, a cone of test functions is defined:

$$\Lambda_k \;:=\; P^k(\mathcal{T}) \,\cap\, L^2(\Omega, \mathbb{R}^+_0)\ . \tag{8}$$

Furthermore, a Lagrangian $L^k_c : R^k(\mathcal{T}) \times \Lambda_k \longrightarrow \mathbb{R}$ is introduced:

$$L^k_c(\sigma, v) \;:=\; \frac{1}{2}\langle \sigma, C^{*}\!\cdot \sigma\rangle_\Omega + (f, v)_\Omega + \sum_{T \in \mathcal{T}} (\nabla\cdot\sigma, \psi + v)_T - \sum_{E \in \mathcal{B}} (u_0, \sigma_n)_E\ .$$

The second option consists in using additional Lagrange multipliers to enforce that the normal fluxes of $\sigma_k \in R^k(\mathcal{T})$ are continuous across the element interfaces. The augmented Lagrangian $L^k : R^k(\mathcal{T}) \times \Lambda_k \times P^k(\mathcal{E}) \longrightarrow \mathbb{R}$ reads:

$$L^k(\sigma, v, w) \quad := \quad L_c^k(\sigma, v) \quad - \sum_{E \in \mathcal{E}} \sum_{T \in \mathcal{T}} |\omega(T, E)| \left(w \, , \, \sigma_{n,T} \right)_E \; . \tag{9}$$

Denoting the set of linear factors for σ_k by $s \in \mathbb{R}^{m_1}$, that for v by $\mu \in \mathbb{R}^{m_2}$ and that for $w \in P^k(\mathcal{E})$ by $\lambda \in \mathbb{R}^{m_3}$, the first order optimality condition for the saddle point $\{s, \mu, \lambda\}$ of the Lagrangian (9) can be written as:

$$\left| \begin{array}{lcl} M\, s \; + \; D^T \mu \; - \; F^T \lambda & = & s_0 \\ D\, s \; + \qquad\qquad\quad r & = & d_0 \\ F\, s & = & 0 \end{array} \right| \; . \tag{10}$$

If the conforming discretisation scheme is employed, the field σ_k can be represented by the set of linear factors $s_c \in \mathbb{R}^{m_4}$. The first order optimality condition for the saddle point $\{s_c, \mu, \}$ of the Lagrangian $L_c^k(\sigma, v)$ reads:

$$\left| \begin{array}{lcl} M_c\, s_c \; + \; D_c^T \mu & = & s_0^c \\ D_c\, s_c \; + \qquad r & = & d_0 \end{array} \right| \; . \tag{11}$$

In the case $k \in \{0, 1\}$ the requirement $v \in \Lambda_k$ gives rise to a very simple linear complementary condition for the vectors $r \in \mathbb{R}^{m_2}$ and μ:

$$r \, , \, \mu \; \geq \; 0 \quad \wedge \quad r^T \mu \; \stackrel{!}{=} \; 0 \quad .$$

Remark: In the case $k \geq 2$ the requirement $v \in \Lambda_k$ no longer translates into the simple algebraic condition $\mu \geq 0$. Hence, additional inequality constraints appear to account for (8). Moreover, testing with $v \in \Lambda_k$ no longer warrants $\sigma_k \in \Sigma$, if some $f \in P^k(\mathcal{T})$ is chosen: a consistency error is thus introduced.

4. The multilevel algorithm

If the linear factors associated with the vector field σ are eliminated from either (10) or (11) a linear complementary problem for the Lagrange multiplier $v \in \Lambda_k$ of the admissibility constraint eventually emerges:

$$\begin{array}{rl} S\, \mu \; := \; C\, \mu + B\, A^{-1} B^T \mu \; \geq \; f & \\ \wedge \quad \mu \geq 0 \quad \wedge \quad \mu^T (\, S\, \mu - f \,) \; = \; 0 \; . & \end{array} \tag{12}$$

If a hybrid discretisation scheme is used, the matrices A, B and C correspond to:

$$A \; := \; -FM^{-1}F^T \quad , \quad B \; := \; DM^{-1}F^T \quad , \quad C \; := \; DM^{-1}D^T \; .$$

If a conforming discretisation is employed instead, these matrices are defined by:

$$A \; := \; M_c \quad , \quad B \; := \; D_c \quad , \quad C \; := \; 0 \; .$$

Since $R^k(\mathcal{T}) \times P^k(\mathcal{T})$ and $B^k(\mathcal{T}) \times P^{k-1}(\mathcal{T})$ satisfy the LBB-condition (see [14, 7]), the matrix S is positive definite both in the hybrid and in the conforming case. Hence, the problem (12) is known (see e.g. [25]) to have a unique solution. Let $\texttt{pSOR}(S, c, f, \mu, \omega)$ denote the result of one sweep of the projected SOR-scheme as described in [15]. Hereby S is the matrix under consideration, c the obstacle, f the right hand side, μ the initial guess and $\omega \in (0, 2)$ the relaxation parameter. By

$$\mu^{(n+1)} \quad := \quad \texttt{pSOR}(\,S, 0, f, \mu^{(n)}, \omega\,) \quad ; \quad n \in \mathbb{N} \tag{13}$$

a smoothing iteration may be defined. However, the matrix S is dense in both the hybrid and the conforming case, even though A, B and C themselves are very sparse. Therefore, the iteration (13) is too expensive. If the constraint $\mu^{(n)} \geq 0$ did never become active, the smoother (13) would be equivalent to:

$$\mu^{(n+1)} \quad := \quad \mu^{(n)} \; + \; \texttt{pSOR}(\,S, -\mu^{(n)}, f - S\mu^{(n)}, 0, \omega\,) \quad ; \quad n \in \mathbb{N} \; .$$

This latter iteration can be viewed as a defect correction scheme with the matrix S acting as a preconditioner. Consequently, it seems reasonable to replace S by a sparse approximation $\hat{S}$. To ensure that the resulting fixed point iteration is still convergent damping parameters $\Theta > 0$ and $\hat{\Theta} \in (0, 1]$ must be introduced:

$$\mu^{(n+1)} \; := \; \mu^{(n)} \; + \; \hat{\Theta}\; \texttt{pSOR}(\,\hat{S}, -\mu^{(n)}, \Theta\,(f - S\mu^{(n)}), 0, \omega\,) \quad ; \quad n \in \mathbb{N} \; . \tag{14}$$

Indeed, the iteration (14) yields a solver for the complementary problem (12), if the preconditioner $\hat{S}$ is positive definite and $\Theta\hat{\Theta}$ sufficiently small [11]. As the matrix A is essentially a mass matrix both in the conforming and in the hybrid case, its condition number $\kappa(A)$ is uniformly bounded in h, the grid parameter. Therefore, the residual can be evaluated with the help of a simple CG iteration and the numerical complexity of this procedure is still optimal.

In default of a canonical approach to approximate A by a diagonal matrix $\hat{A}$ the preconditioner $\hat{S}$ may be supplied either by lumping A or by dropping all its entries off the diagonal. In the hybrid case the resulting matrix need not necessarily be positive definite, however. Hence, it is convenient to introduce a third damping parameter $\tilde{\Theta} > 0$ and define the preconditioner by:

$$\hat{S} \quad := \quad C \; + \; \tilde{\Theta}\; B\hat{A}^{-1}B^T \quad .$$

This matrix can be computed and stored with optimal complexity $\mathcal{O}(m_2)$, as the number of entries per row can be bounded by a power of k times $d+1$.

It will be assumed, a regular family $\{\mathcal{T}_l\}_{l>0}$ of simplicial decompositions has been generated in such a way, that the trial spaces for the Lagrange multiplier μ are nested: $P^k(\mathcal{T}_0) \subseteq P^k(\mathcal{T}_1) \subseteq \ldots$. For each refinement level l an imbedding operator $\iota_l : P^k(\mathcal{T}_l) \longrightarrow P^k(\mathcal{T}_{l+1})$ is introduced. The matrix I_l associated with the imbedding ι_l consists of the following entries:

$$I^{(l)}_{Ti,Uj} \quad = \quad u^{(k)}_{U,j}(\hat{x}^{(k)}_{T,i}) \qquad ; \qquad U \in \mathcal{T}_l\,,\; T \in \mathcal{T}_{l+1} \quad . \tag{15}$$

To each map ι_l corresponds a restriction operator $\varrho_l : P^k(\mathcal{T}_{l+1}) \longrightarrow P^k(\mathcal{T}_l)$ that is defined in such a way, that the coarse grid correction step of the multilevel algorithm warrants an admissible approximation:

$$(\iota_l \circ \varrho_l)^2 (v) \;=\; (\iota_l \circ \varrho_l)(v) \;\geq\; v \quad ; \quad v \in P^k(\mathcal{T}_{l+1}) \;.$$

No linear operator ϱ_l can be discovered which meets this requirement. However, in the special case $I_l \geq 0$ a suitable map can be found easily. Let $\{\mu_{Ti}^{(l+1)}\}_{Ti}$ denote the set of linear factors of some function $v^{(l+1)} \in P^k(\mathcal{T}_{l+1})$. Those factors of the image $\varrho_l(v^{(l+1)})$ may be computed by the formula:

$$\mu_{Uj}^{(l)} \;:=\; \max_{T,i} \{\, \mu_{Ti}^{(l+1)} \mid I_{Ti,Uj}^{(l)} > 0 \,\} \;. \tag{16}$$

Here again, the case $k \geq 2$ will pose additional difficulties, since the requirement $I_l \geq 0$ is violated. A possible solution is detailed in [11].

Depending on the desired level of accuracy a grid $\mathcal{T}_L$ is fixed, on which the solution of problem **D** is to be approximated. On each of the grids $\mathcal{T}_0, \ldots, \mathcal{T}_L$ several matrices must be assembled that define the complementary problem: these will be denoted by $\{A_l, B_l, C_l\}_{l=0}^{L}$. The amount of smoothing to be applied to the data may differ depending on the level l. A set of counters $\{M_l\}_{l=0}^{L}$ specifies how many times in succession relaxations be carried out. On the finest grid level $\mathcal{T}_L$ the algebraic accuracy is monitored, the solution of (13) has been approximated with. Since this quantity cannot be measured by any norm of the residual, a viable alternative [5] is to control the distance between subsequent iterates by means of a parameter ϵ. If the test for convergence fails, another multigrid cycle must be initiated. An overview of the entire algorithm is presented in Figure 1.

5. Avoiding the global complementary condition

Eliminating λ from (10) has the advantage, that the resulting complementary problem for μ involves simple box constraints and allows for a discretisation, that facilitates the use of nested trial spaces and the straightforward implementation of a multilevel solver. The drawback of this procedure is the necessity to solve a global linear problem in each smoothing step. However, for any fixed $\lambda \in \mathbb{R}^{m_3}$ the Lagrange multiplier for the admissibility constraint can be identified as the solution of the following linear complementary problem:

$$\begin{aligned} DM^{-1}D^T\mu \;&\geq\; DM^{-1}(s_0 + F^T\lambda) - d_0 \;=:\; d(\lambda) \\ \wedge \quad \mu \geq 0 \quad &\wedge \quad \mu^T\big(DM^{-1}D^T\mu - d(\lambda)\big) \;=\; 0 \;. \end{aligned} \tag{17}$$

Since the matrix $DM^{-1}D^T$ is a block-diagonal (17) actually describes numerous small, independent complementary problems, that correspond to the individual elements compounding the simplicial decomposition. Let the map which assigns to $\lambda \in \mathbb{R}^{m_3}$ the corresponding solution of (17) be denoted by $\Lambda : \mathbb{R}^{m_3} \longrightarrow \mathbb{R}^{m_2}$. From (10) a nonlinear equation for the Lagrange multiplier λ is obtained:

$$T(\lambda) \;:=\; FM^{-1}F^T\lambda - FM^{-1}D^T\Lambda(\lambda) \;=\; -FM^{-1}s_0 \;. \tag{18}$$

In order to solve this equation by a FAS variant [4] of the multigrid method, it is necessary to specify a smoothing iteration and suitable transfer mechanisms,

FIGURE 1. Outline of the Multigrid Algorithm (pMG)

Supplied: $\{A_l, B_l, C_l, \hat{S}_l, \Theta_l, \hat{\Theta}_l, \omega_l, M_l,\}_{l=0}^{L}$, $\{I_l\}_{l=0}^{L-1}$, $\epsilon > 0$

Definition: pMG(μ, f, ψ, l) {

1) Set: $\mu_{\text{old}} := \mu$.

2) Loop $m \in \{1 \ldots M_l\}$: {

2a) Apply CG to solve: $A_l v = B_l^T \mu$.

2b) Set: $r := \Theta_l (f - C_l \mu - B_l v)$.

2c) Loop i : {

$$\hat{\eta}_i := \frac{1}{\hat{S}_{ii}} \left(r_i - \sum\nolimits_{j<i} \hat{S}_{ij} \eta_j \right) .$$

$$\eta_i := \max \{ \psi_i - \mu_i, \omega_l \hat{\eta}_i \} .$$

}

2d) Set: $\mu := \mu + \hat{\Theta}_l \eta$.

}

3) If $l = 0$ return.

4) Apply CG to solve: $A_l v = B_l^T \mu$.

5) Set: $\tilde{f} := I_{l-1}^T (f - C_l \mu - B_l v)$.

6) Compute $\tilde{\psi}$ from $\psi - \mu$ e.g. by (16).

7) Set: $\tilde{\mu} := 0$.

8) Call: pMG($\tilde{\mu}$, $\tilde{f}$, $\tilde{\psi}$, $l-1$).

9) Set: $\mu := \mu + I_{l-1} \tilde{\mu}$.

10) Repeat step 2.

11) If $l < L$ or $\| \mu_{\text{old}} - \mu \|_\infty < \epsilon$ return.

12) Goto step 1.

}

that map the defect and the approximate solution from a coarser grid $\mathcal{E}_l$ to the next finer one $\mathcal{E}_{l+1}$ and back. Unfortunately, the trial spaces $P^k(\mathcal{E}_0), P^k(\mathcal{E}_1), \ldots$ are not nested and canonical injection or restriction operators are not available. However, the Lagrange multipliers of the normal fluxes approximate the values of the displacement on the element interfaces [2]. Hence, the injection and restriction operators can be defined by treating the components of the vector λ as associated with a Crouzeix-Raviart ansatz [14] for the displacement variable. Consequently, the level transfer techniques described in [3] may be used. A simpler alternative is suggested in [6]. In either case the choice of proper damping parameters is critical to the success of the resulting multilevel scheme. Numerical experiments indicate that a relaxation of the fine grid contribution to the right hand side on the coarser grid may be the most successful strategy:

$$f_{l-1} \ := \ T_{l-1}(R_{l-1}(\lambda_l^{(j)})) \ + \ \vartheta\, I_{l-1}^T(f_l \ - \ T_l(\lambda_l^{(j)}))\ .$$

Hereby, the effect of smoothing has been ignored for the sake of simplicity. The injections $I_0, \dots I_{L-1}$ are defined by (15), while the restrictions $R_0, \dots R_{L-1}$ have proved most effective, if they are constructed in the spirit of (16). The relaxation parameter is denoted by $\vartheta > 0$. Its impact on the coarse grid problem can be compensated, when the coarse grid correction is injected into the finer grid:

$$\lambda_l^{(j+1)} \ := \ \lambda_l^{(j)} \ + \ \vartheta^{-1}\, I_{l-1}(\lambda_{l-1}^{(j+1)} \ - \ R_{l-1}(\lambda_l^{(j)}))\ .$$

Again, the effect of smoothing has been ignored. If the obstacle ψ were sufficiently far removed from the membrane u^*, the Lagrange multiplier μ would stay strictly positive and the complementary problem (17) would be reduced to an equation. Consequently, (18) would turn into a linear system involving the matrix

$$T \ := \ FM^{-1}F^T \ - \ FM^{-1}D^T(DM^{-1}D^T)^{-1}DM^{-1}F^T\ .$$

This matrix has no more than $d+1$ blocks per row, whose size only depends on the order of the ansatz for the Lagrange multiplier λ. As T can be computed and stored with optimal numerical complexity, a fixed point iteration of the form

$$\lambda^{(j+1)} \ := \ \lambda^{(j)} \ - \ \hat{T}^{-1}FM^{-1}\left(s_0 + F^T\lambda^{(j)} - D^T\Lambda(\lambda^{(j)})\right) \tag{19}$$

may be utilised as a smoother, whereby the preconditioner $\hat{T}$ can be based on a regular splitting of T: among the possible smoothing iterations are the damped Gauß-Seidel or the SOR scheme. Due to the special structure of the auxiliary problem (17) the overall numerical complexity of one smoothing step will not exceed $\mathcal{O}(m_3)$. An analysis of the iteration (19) is presented in [11].

6. Numerical experiments

The performance of the multilevel methods outlined in Section 4 and 5 has been compared to the performance of three interior point codes, that are available via the Internet. Since only few software packages, that are able to solve quadratic programming problems, can be obtained free of royalties, the choice was based on opportunity rather than theoretical considerations. In the following these packages will be referred to under the names *BPMPD*, *Loqo* and *Mosek*. The first code is an implementation of the infeasible primal-dual interior point algorithm, that incorporates predictor-corrector and target following techniques [22]. The second software package also relies on an infeasible primal-dual algorithm and employs a predictor-corrector scheme to approximate the central path [26]. The third code implements a homogeneous interior point method based on a self-dual formulation of the optimality conditions [1]. The newest versions currently available have been used in each case: BPMPD 2.21b, Loqo 5.00 and Mosek 1.3 (Build 8). Below, the conforming discretisation scheme introduced in §4 is abbreviated *Method A*, while its hybrid variant is called *Method B*. The last approach, which has been detailed separately in Section 5, is referred to as *Method C*.

To obtain a coarse grid the unit square in $\mathbb{R}^2$ is decomposed into two triangles. By further subdividing this triangulation a sequence of quadratic programs can be generated, whose size is increasing more or less geometrically. The number of successive refinements is called the level l of the grid. The execution times in the fourth row of Table 2 have been normalised using the *linpack1000d* benchmark [16]. This has been necessary, since the Mosek code can only be run on a personal computer. The other experiments have been carried out on a Sparc Ultra-4 server.

FIGURE 2. Execution times of various solvers

Level	5	6	7	8	9
$\dim R^0(\mathcal{T}_l)$	3136	12416	49408	197120	787456
$\dim P^0(\mathcal{T}_l)$	2048	8192	32768	131072	524288
BPMPD	1.65 s	8.69 s	55.66 s	failed	failed
Mosek	1.52 s	90.23 s	2043 s	38726 s	756136 s
Loqo	3.83 s	21.41 s	138.1 s	1096 s	9232 s
Method A	6.46 s	31.01 s	209.4 s	1438 s	7704 s
Method B	7.08 s	40.78 s	269.6 s	1395 s	6676 s
Method C	1.94 s	11.50 s	60.62 s	310.6 s	1412 s

Failed numerical experiments have been caused by insufficient core memory: The code has requested the dynamic allocation of more than 2GB of memory in one batch.

The data reads: $f = -6$, $\psi = -0.5$ and $u_0(x,y) = x^2 + y^2 - 1$. The tensor C is composed of $C_{11} = 1$, $C_{22} = 2$ and $C_{12} = C_{21} = 0.5$. In all experiments the initial guess $\mu^{(0)} = 0$ respectively $\lambda^{(0)} = 0$ and the termination criterion $\epsilon = 5 \cdot 10^{-9}$ have been used. The various damping parameters have been determined empirically: For the method B the values $\omega = \tilde{\Theta} = 0.8$ and $\Theta = \hat{\Theta} = 1$ have been chosen, while method A has been applied with $\omega = 1.4$, $\Theta = 1$ and $\hat{\Theta} = 0.7$. The preconditioner for method C has been patterned after a SOR smoother for linear problems with $\omega = 1.68$. For level transfer purposes the relaxation parameter $\vartheta = 0.7$ has been used. All multilevel algorithms employ 4 smoothing steps before as well as after grid transfer operations (that is: $M_l \equiv 4$). The interior point solvers have been furnished with a description of the quadratic program encoded in an extended MPS data file [22]. Only default settings have been used for the test runs.

FIGURE 3. Solution σ of problem **D** (x-component)

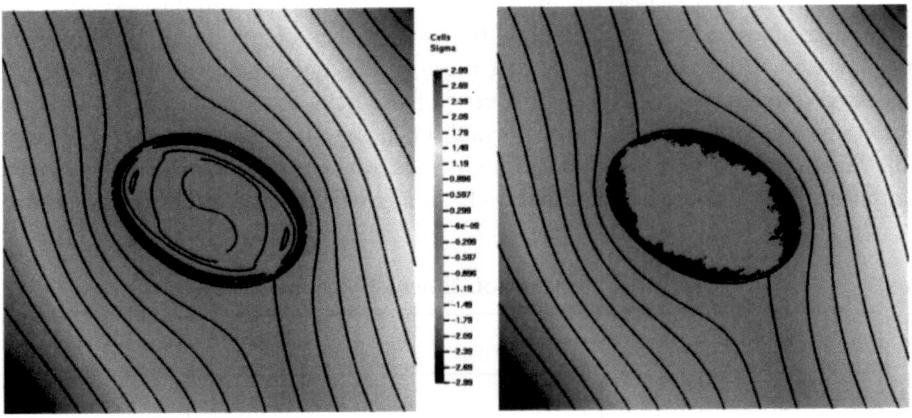

The picture on the left depicts the approximation as it has been obtained from the interior point solver *Loqo* for a grid of level 9. The picture on the right represents the numerical solution found by the algorithm C.

Conclusion

Multilevel methods for constrained variational problems, that rely on projected relaxation schemes for smoothing, can be adapted from conventional multilevel solvers with very little effort. Their performance is competitive if not superior to the performance of sophisticated interior point solvers, especially on finer grids. The requirements of these methods in terms of core memory are moderate, since no sparse matrix factorisations need to be carried out. Using hybrid mixed discretisations the global complementary problem can be replaced by a nonlinear equation for the Lagrange multipliers of the continuity constraints. The resulting multilevel algorithm is more involved, as the trial spaces are no longer nested.

FIGURE 4. Conjectures about the numerical complexity

	Mosek	Loqo	Method A	Method B	Method C
Presumed	m^2	$m^{1.5}$	$m\,(\log m)^2$	$m\,(\log m)^2$	$m\,\log m$
Fitted	$m^{1.96}$	$m^{1.52}$	$m^{1.1}(\log m)^{1.7}$	$m^{1.0}(\log m)^{1.7}$	$m^{1.0}(\log m)^{1.1}$

Its efficiency is exemplary, however. The experiments indicate, that the numerical complexity of all three multilevel methods is comparable to the complexity of a conventional multigrid iteration designed to solve unconstrained elliptic boundary value problems in their primal formulation. The presence of logarithmic factors in Figure 4 may be explained by the fact, that the convergence is monitored in the L^∞-norm (see e. g. [24]).

Acknowledgements

The author would like to acknowledge the financial support of the DFG through the Sonderforschungsbereich 359 in Heidelberg and to thank the company *EKA Consulting ApS, Denmark* for granting him a time limited educational license for their software package *Mosek*.

References

[1] E. D. Andersen and Y. Ye, *A computational study of the homogeneous algorithm for large-scale convex optimization*, Comput. Optim. Appl., **10** (1998), pp. 243–269.

[2] D. N. Arnold and F. Brezzi, *Mixed and nonconforming finite element methods: Implementation, postprocessing and error estimates*, RAIRO, Modelisation Math. Anal. Numer., **19** (1985), pp. 7–32.

[3] D. Braess and R. Verfürth, *Multigrid methods for nonconforming finite element methods*, SIAM J. Numer. Anal., **27(4)** (1990), pp. 979–986.

[4] A. Brandt, *Multi-level adaptive solutions to boundary-value problems*, Math. Comput., **31** (1977), pp. 333–390.

[5] A. Brandt and C. W. Cryer, *Multigrid algorithms for the solution of linear complementarity problems arising from free boundary problems*, SIAM J. Sci. Stat. Comput., **4** (1983), pp. 655–684.

[6] S. C. Brenner, *A multigrid algorithm for the lowest-order Raviart-Thomas mixed triangular finite element method*, SIAM J. Numer. Anal., **29(3)** (1992), pp. 647–678.

[7] F. Brezzi, J. Douglas, and L. D. Marini, *Two families of mixed finite elements for second order elliptic problems*, Numer. Math., **47** (1985), pp. 217–235.

[8] F. Brezzi, W. W. Hager, and P. A. Raviart, *Error estimates for the finite element solution of variational inequalities, Part I, Primal theory*, Numer. Math., **28** (1977), pp. 431–443.

[9] F. Brezzi, W. W. Hager, and P. A. Raviart, *Error estimates for the finite element solution of variational inequalities, Part II, Mixed Methods*, Numer. Math., **31** (1978), pp. 1–16.

[10] H. M. Bußand S. I. Repin, *A posteriori error estimates for boundary value problems with obstacles*, in: ENUMATH99 Proceedings (World Scientific), to be published.

[11] H. M. Buß, *Mixed Discretisations for Variational Inequalities*, Preprint Series of the Sonderforschungsbereich 359, in preparation.

[12] Z. Cai, C. I. Goldstein, and J. E. Pasciak, *Multilevel iteration for mixed finite element systems with penalty*, SIAM J. Sci. Comput., **14(5)** (1993), pp. 1072–1088.

[13] P. G. Ciarlet, *The Finite Element Method for Elliptic Problems*, Studies in Mathematics and its Applications, **4**, Elsevier North-Holland Inc., New York, 1978.

[14] M. Crouzeix and P. A. Raviart, *Conforming and nonconforming finite element methods for solving the stationary Stokes equations I*, Revue Franc. Automat. Inform. Rech. operat., **7(R-3)** (1973), pp. 33–76, published in 1974.

[15] C. W. Cryer, *The solution of a quadratic programming problem using systematic overrelaxation*, SIAM J. Control, **9** (1971), pp. 385–392.

[16] J. Dongarra, *Performance of various computers using standard linear equations software*, Computer Science Technical Report CS-89-85, University of Tennessee, 2000.

[17] R. S. Falk, *Error estimates for the approximation of a class of variational inequalities*, Math. Comp., **28** (1974), pp. 963–971.

[18] V. Girault and P. A. Raviart, *Finite Element Methods for Navier-Stokes Equations*, Springer Series in Computational Mathematics, **5**, Springer, Berlin, 1986.

[19] W. Hackbusch and H. D. Mittelmann, *On multi-grid methods for variational inequalities*, Numer. Math., **42** (1983), pp. 65–76.

[20] R. H. W. Hoppe, *Two-sided approximations for unilateral variational inequalities by multi-grid methods*, Optimization, **18** (1987), pp. 867–881.

[21] J. Mandel, *A multilevel iterative method for symmetric, positive definite complementarity problems*, Appl. Math. Optimization, **11** (1984), pp. 77–95.

[22] C. Mészáros, *The BPMPD interior point solver for convex quadratic problems*, Laboratory of Operations Research and Decision Systems Technical Report WP 98-8, Hungarian Academy of Sciences, 1998.

[23] T. Neunhöffer, *Multigrid methods for mixed finite element discretizations of variational inequalities*, in: P. W. Hemker, ed., Multigrid methods IV, Internat. Ser. Numer. Math., **116** (1994), Birkhäuser, Basel, pp. 257–268.

[24] A. Reusken, *On maximum norm convergence of multigrid methods for elliptic boundary value problems*, SIAM J. Numer. Anal., **31(2)** (1994), pp. 378–392.

[25] H. Samelson, R. M. Thrall, and O. Wesler, *A partition theorem for Euclidean n-space*, Proc. Amer. Math. Soc., **9** (1958), pp. 805–807.

[26] R. J. Vanderbei, *LOQO: An Interior Point Code for quadratic programming*, Statistics and Operations Research Technical Report SOR-94-15, Princeton University, 1998.

[27] S. P. Vanka, *Block-implicit multigrid solution of Navier-Stokes equations in primitive variables*, J. Comput. Phys., **65** (1986), pp. 138–158.

Nonlinear Boundary Control for the Heat Equation Utilizing Proper Orthogonal Decomposition

Franz Diwoky and Stefan Volkwein

Abstract. The proper orthogonal decomposition (POD) is a method to derive reduced-order models for dynamical systems. In this paper POD is utilized to solve a nonlinear boundary control problem for the heat equation. The relative simplicity of the equation allows comparison of the POD based algorithm with numerical results obtained from finite element discretization of the optimality system.

1. Introduction

Recently the application of reduced-order models to optimal control problems for partial differential equations has received an increasing amount of attention. The reduced-order approach is based on projecting the dynamical system onto subspaces consisting of basis elements that contain characteristics of the expected solution. This is in contrast to, e.g., finite element (FE) techniques, where the elements of the subspaces are uncorrelated to the physical properties of the system that they approximate. The reduced basis method, as developed in [12] and utilized efficiently in [13], is one such reduced order method with the basis elements corresponding to the dynamics of expected control regimes. In this paper we shall investigate POD as a procedure to determine the basis elements for the approximating subspaces on which to solve the optimal control problems. POD was successfully used in different fields including signal analysis and pattern recognition (see, e.g., [7]), fluid dynamics and coherent structures (see, e.g., [6, 19]) and more recently in control theory, see for instance [2, 3, 5, 14, 16], and inverse problems [4]. Surprisingly good approximation properties are reported for POD based schemes in several articles, see [23] for example. Error estimates for Galerkin POD based methods for parabolic systems were proved in [15]. The resulting error bounds depend on the number of POD basis functions and on the time discretization.

In this article we consider the optimal control problem

$$\min J(y,u) = \frac{1}{2}\int_\Omega |y(\cdot,T) - z|^2\,dx + \frac{\kappa}{2}\int_\Sigma |u|^2\,dsdt \tag{1a}$$

subject to

$$\begin{aligned} y_t(x,t) &= k\,\Delta_x y(x,t) && \text{a.e. in } Q=\Omega\times(0,T),\\ \partial_n y(x,t) &= b(y(x,t))+u(x,t) && \text{a.e. on } \Sigma=\Gamma\times(0,T),\\ y(x,0) &= y_0(x) && \text{a.e. in } \Omega \end{aligned} \tag{1b}$$

and

$$u \in U_{\mathsf{ad}} = \{v \in L^\infty(\Sigma) : \underline{u} \le v \le \overline{u} \text{ a.e. on } \Sigma\}, \tag{1c}$$

where Ω is a bounded and open subset of $\mathbb{R}^d$ ($d \ge 2$) with boundary $\Gamma = \partial\Omega$ of class $C^{2,\beta}$, $0 < \beta < 1$, i.e., Γ is a $(d-1)$-dimensional manifold of class $C^{2,\beta}$ such that Ω lies locally on one side of Γ. Recall that a function is of class $C^{2,\beta}$ if it is of class C^2 and its partial derivatives of second order are Hölder-continuous of order β. We suppose that $T, \kappa, k > 0$, that $\underline{u} \le \overline{u}$ are fixed real numbers and $y_0, z \in C(\overline{\Omega})$. By ∂_n we denote the (outward) normal derivative on Γ. The function $b = b(y)$ belongs to $C^{2,1}(\mathbb{R})$ and is monotone non-increasing.

In this article we use two different POD strategies to solve (1) numerically. One ansatz is based on the variational formulation of the state equation. The second one includes not only information of the state variable, but also of the adjoint equation into the POD modeling. The POD based solutions are compared with numerical solutions obtained from a FE discretization. It turns out that the second strategy leads to better numerical results. If we combine these POD techniques with a dynamical update of the POD basis, see [1, 2, 11], the reduced-order modeling is very efficient for nonlinear boundary control of the heat equation.

Let us mention the articles [9, 10, 20], where (1) was solved by SQP techniques. The relationship between the SQP method applied to (1) and the generalized Newton method applied to the first-order necessary optimality conditions was studied in [21, 22].

The paper is organized as follows. In the second section, relevant results on problem (1) are summarized. The method of POD is introduced in Section 3. In Section 4 the reduced-order modeling is presented. Numerical examples are given in the fifth section, and in Section 6, we give some conclusions.

2. Theoretical results

In this section we provide some important results for problem (1). First we discuss the state equation (1b). The following result guarantees the existence of a unique solution to (1b) in the Banach space $Y = W(0,T) \cap C(\overline{Q})$ with the norm

$$\|y\|_Y = \|y\|_{W(0,T)} + \|y\|_{C(\overline{Q})} \quad \text{for } y \in Y,$$

where $W(0,T) = \{\phi \in L^2(0,T;H^1(\Omega)) : \phi_t \in L^2(0,T;H^1(\Omega)')\}$. For a proof we refer the reader to [18].

Proposition 2.1. *For each $u \in L^\infty(\Sigma)$ there exists a unique solution $y \in Y$ of the state equation satisfying $y(\cdot\,,0) = y_0$ in $\overline{\Omega}$ and*

$$\langle y_t(\cdot,t),\varphi\rangle_{(H^1)',H^1} + k\int_\Omega \nabla y(\cdot,t)\cdot\nabla\varphi\,dx = k\int_\Gamma \big(b(y(\cdot,t)) + u(\cdot,t)\big)\varphi\,ds$$

for all $\varphi \in H^1(\Omega)$ and $t \in (0,T)$ a.e. Here $\langle\cdot\,,\cdot\rangle_{(H^1)',H^1}$ denotes the dual pair associated with $H^1(\Omega)$ and its dual and ds is the surface measure on Γ.

Using standard techniques we derive that (1) has at least one optimal control, because the control appears linearly, J is convex and continuous, and U_{ad} is weakly-star compact.

Since (1) is a non-convex programming problem, there are probably different local minima. In the following we assume that a reference solution $(y^*,u^*) \in Y \times L^\infty(\Sigma)$ is given. It can be proved [18] that there exists a unique Lagrange multiplier $\lambda^* \in Y$ such that the triple (y^*,u^*,λ^*) satisfies the first-order necessary optimality conditions, which consist of the state equation (1b), the control constraints (1c), the adjoint equation

$$\begin{aligned} -\lambda_t^*(x,t) &= k\,\Delta_x\lambda^*(x,t) && \text{a.e. in } Q,\\ \partial_n\lambda^*(x,t) &= b'(y^*(x,t))\lambda^*(x,t) && \text{a.e. on } \Sigma,\\ \lambda^*(x,T) &= -(y^*(x,T) - z(x)) && \text{a.e. in } \Omega \end{aligned} \tag{2}$$

and the variational inequality

$$\int_\Sigma (\kappa u^* - k\lambda^*)(u - u^*)\,dsdt \geq 0 \quad \text{for all } u \in U_{\mathsf{ad}}. \tag{3}$$

Let us introduce the Lagrange functional $L : Y \times L^\infty(\Sigma) \times Y \to \mathbb{R}$ associated with (1):

$$\begin{aligned} L(y,u,\lambda) &= J(y,u) + \int_0^T \langle y_t(\cdot,t),\lambda(\cdot,t)\rangle_{(H^1)',H^1} + k\int_\Omega \nabla y(\cdot,t)\cdot\nabla\lambda(\cdot,t)\,dxdt\\ &\quad - k\int_\Sigma \big(b(y(\cdot,t)) + u(\cdot,t)\big)\lambda(\cdot,t)\,dsdt. \end{aligned}$$

The function L is of class C^2 with respect to $(y,u) \in Y \times L^\infty(\Sigma)$. In what follows, we shall not distinguish between a functional in the dual and its Riesz representation in the Hilbert space. The first and second Fréchet-derivative with respect to (y,u) are denoted by L' and L''. Then (2)–(3) can be written as

$$L'(y^*,u^*,\lambda^*)(y,u-u^*) \geq 0 \quad \text{for all } (y,u) \in W(0,T) \times U_{\mathsf{ad}} \text{ with } y(\cdot,0) = 0.$$

At (y^*, u^*, λ^*) we introduce the following second-order sufficient optimality condition:

$$\left\{\begin{array}{l} \text{there exists a } \sigma > 0 \text{ such that} \\ \quad L''(y^*, u^*, \lambda^*)\big((y,u),(y,u)\big) \geq \sigma \; \|(y,u)\|^2_{W(0,T)\times L^2(\Sigma)} \\ \text{for all } (y,u) \in W(0,T) \times L^2(\Sigma) \text{ satisfying} \\ \begin{array}{rcll} y_t(x,t) & = & k\Delta_x y(x,t) & \text{a.e. in } Q, \\ \partial_n y(x,t) & = & b'(y^*(x,t))y(x,t) + u(x,t) & \text{a.e. on } \Sigma, \\ y(x,0) & = & 0 & \text{a.e. in } \Omega. \end{array} \end{array}\right. \tag{SSC}$$

If (y^*, u^*, λ^*) satisfies the first-order necessary optimality system and (SSC), then (y^*, u^*) is a locally unique solution to (1). Let us mention that one can use much weaker versions of (SSC) by introducing the set of strong active inequalities, see [8] for instance.

3. The method of POD

In this section, we introduce the POD method. For that purpose, let us consider the semi-linear Cauchy problem

$$\begin{array}{rcll} \dot{w}(t) & = & Aw(t) + f(t, w(t)) & \text{for } t \in (0,T), \\ w(0) & = & w_0, & \end{array} \tag{4}$$

where A is the infinitesimal generator of a C_0-semigroup $\{G(t)\}_{t\geq 0}$ in the Hilbert space X, $f : [0,T] \times X \to X$ is continuous, and $w_0 \in X$. Under appropriate conditions, see [17] for instance, there exists a unique strong solution $w \in C([0,T]; X) \cap C^1((0,T); X)$ of the initial value problem (4) given by the integral equation

$$w(t) = G(t)w_0 + \int_0^T G(t-s)f(s, u(s))\, ds.$$

Suppose that we know the snapshots $w_j = w(t_j)$ at given time instances $0 \leq t_1 < t_2 < \ldots < t_n \leq T$. For example, the w_j's can be given by photographical pictures, by measurements or by the solution of a very accurate numerical integration scheme. We set

$$\mathcal{W} = \operatorname{span}\{w_1, \ldots, w_n\},$$

and refer to $\mathcal{W}$ as the ensemble consisting of the snapshots $\{w_j\}_{j=1}^n$, at least one of which is assumed to be non-zero. Let $\{\psi_i\}_{i=1}^l$ denote an orthonormal basis of $\mathcal{W}$ with $l = \dim \mathcal{W}$. Then each member of the ensemble can be expressed as

$$w_j = \sum_{i=1}^{l} (w_j, \psi_i)_X \psi_i \quad \text{for } j = 1, \ldots, n. \tag{5}$$

The method of POD consists in choosing the orthonormal basis such that for every $\ell \in \{1, \ldots, l\}$ the mean square error between the elements w_j, $1 \leq j \leq n$, and the corresponding ℓ-th partial sum of (5) is minimized on average:

$$\min_{\{\psi_i\}_{i=1}^{\ell}} \frac{1}{n} \sum_{j=1}^{n} \left\| w_j - \sum_{i=1}^{\ell} (w_j, \psi_i)_X \psi_i \right\|_X^2 \tag{6}$$
$$\text{subject to } (\psi_i, \psi_j)_X = \delta_{ij} \text{ for } 1 \le i \le \ell, 1 \le j \le i.$$

A solution $\{\psi_i\}_{i=1}^{\ell}$ to (6) is called a POD basis of rank ℓ. We introduce the correlation matrix $K = ((K_{ij})) \in \mathbb{R}^{n\times n}$ corresponding to the snapshots $\{w_j\}_{j=1}^{n}$ by

$$K_{ij} = \frac{1}{n} (w_j, w_i)_X.$$

The matrix K is positive semi-definite and has rank l. The solution of (6) can be found in [19], for instance.

Proposition 3.1. *Let $\lambda_1 \ge \ldots \ge \lambda_l > 0$ denote the positive eigenvalues of K and $v_1, \ldots, v_l \in \mathbb{R}^n$ the associated eigenvectors. Then a POD basis of rank $\ell \le l$ is given by*

$$\psi_i = \frac{1}{\sqrt{\lambda_i}} \sum_{j=1}^{n} (v_i)_j w_j,$$

where $(v_i)_j$ is the j-th component of the eigenvector v_i. Moreover, we have the error formula

$$\frac{1}{n} \sum_{j=1}^{n} \left\| w_j - \sum_{i=1}^{\ell} (w_j, \psi_i)_X \psi_i \right\|_X^2 = \sum_{i=\ell+1}^{l} \lambda_i.$$

4. Reduced-order modeling with POD

In our numerical algorithm, we will ignore the infinite-dimensional inequality constraint, i.e., we suppose that $\underline{u} < u^* < \overline{u}$. Then, there exists a neighborhood $D = D(u^*)$ of u^* in $L^\infty(\Sigma)$ with $D \subset U_{\mathsf{ad}}$. Due to (SSC) the neighborhood D can be chosen in such a way that u^* is the unique solution to (1) in D. Thus, (y^*, u^*) is also a local unique solution to

$$\min J(y,u) = \frac{1}{2} \int_\Omega |y(\cdot, T) - z|^2 \, dx + \frac{\kappa}{2} \int_\Sigma |u|^2 \, dsdt \tag{7a}$$

subject to

$$\begin{aligned} y_t(x,t) &= k\Delta_x y(x,t) && \text{a.e. in } Q, \\ \partial_n y(x,t) &= b(y(x,t)) + u(x,t) && \text{a.e. on } \Sigma, \\ y(x,0) &= y_0(x) && \text{a.e. in } \Omega. \end{aligned} \tag{7b}$$

In our computations, we directly apply the SQP method to (7). In the first-order necessary optimality conditions, the variational inequality (3) has to be replaced by the equation $\kappa u^* = k\lambda^*$ on Σ. Since (7) is an infinite-dimensional optimal control problem, it requires discretization before it can be realized numerically. The reduced-order approach to optimal control problems such as (7) is based on approximating the nonlinear dynamics by a Galerkin technique utilizing basis functions that contain characteristics of the expected flow. A choice has to be made concerning the dynamics from which the POD basis is computed. Of course, the test functions should contain as much information about the state y^* and the adjoint variable λ^* as possible. In our examples, we choose a guess u_{exp} for the optimal control. Then, we construct two different POD bases, where we take $X = L^2(\Omega)$ in the context of the previous section. Moreover, we illustrate that the numerical results can be improved significantly if the POD basis is updated dynamically.

VARIANT 1: The first POD basis $\{\psi_i\}_{i=1}^{\ell}$ is computed from an ensemble consisting of ten snapshots $y_{exp}(t_j)$, where y_{exp} solves (7b) with $u = u_{exp}$ and $0 = t_1 < t_2 < \ldots < t_{10} = T$ are equidistant time instances in $[0, T]$. Furthermore, the desired state z is added to this ensemble.

VARIANT 2: In the second variant, we use two sets, $\{\psi_i^I\}_{i=1}^{\ell}$ and $\{\psi_i^{II}\}_{i=1}^{\ell}$, of POD formed vectors. $\{\psi_i^I\}_{i=1}^{\ell}$ is calculated from ten snapshots of y_{exp} and the desired state z. Snapshots for $\{\psi_i^{II}\}_{i=1}^{\ell}$ are obtained from the adjoint equation (2) with $y^* = y_{exp}$.

After computing the POD basis, we make a Galerkin approximation of (7). We approximate state and control by finite sums of time dependent modal coefficients multiplied by the POD basis elements:

$$y_{pod}(x,t) = \sum_{i=1}^{\ell} \alpha_i(t)\psi_i(x), \quad u_{pod}(x,t) = \sum_{i=1}^{\ell} \beta_i(t)\ (\tau_\Gamma \psi_i)(x),$$

where $\tau_\Gamma : H^1(\Omega) \to L^2(\Gamma)$ denotes the continuous trace operator.

Let us introduce the matrices

$$\begin{aligned} M &\in \mathbb{R}^{\ell\times\ell} \quad \text{with} \quad M_{ij} = \int_\Omega \psi_j \psi_i \, dx, \\ A &\in \mathbb{R}^{\ell\times\ell} \quad \text{with} \quad A_{ij} = -k \int_\Omega \nabla\psi_j \cdot \nabla\psi_i \, dx, \\ B &\in \mathbb{R}^{\ell\times\ell} \quad \text{with} \quad B_{ij} = k \int_\Gamma (\tau_\Gamma \psi_j)\ (\tau_\Gamma \psi_i) \, ds, \end{aligned}$$

$H = M/2$ and $R = \kappa\ B/2$. We compute the projection of the nonlinearity b onto the POD basis:

$$\big(F(\alpha(t))\big)_i = k \int_\Gamma (\tau_\Gamma \psi_i)\ b(y_{pod}(\cdot,t)) \, ds \quad \text{for } i = 1, \ldots, \ell.$$

Now, the Galerkin approximation of the optimal control problem (7) is given by

$$\min J_{pod}(y_{pod}, u_{pod}) = (\alpha(T) - \hat{z})^\mathsf{T} H (\alpha(T) - \hat{z}) + \int_0^T \beta(t)^\mathsf{T} R \beta(t) \, dt \tag{8a}$$

subject to the ℓ-dimensional system of ordinary differential equations

$$\begin{aligned} M\dot{\alpha}(t) &= A\alpha(t) + F(\alpha(t)) + B\beta(t) \quad \text{for } t \in (0,T), \\ \alpha(0) &= M^{-1}\hat{y}_0, \end{aligned} \tag{8b}$$

where $\hat{y}_0$ and $\hat{z}$ stand for the POD projections of the initial condition y_0 and the desired state z onto the space spanned by the first ℓ POD bases functions.

One of the strength of POD approximations is that in applications ℓ is often very small, for instance $\ell \leq 10$. Then, (8) is a full, but low-dimensional problem.

For both variants, we investigate an adaptive control strategy. Hence, the control arising from the reduced order approach to (7) serves as expected control for the computation of a new POD basis. We use the following scheme, see [1, 11]:

Algorithm 4.1 (POD based adaptive control strategy).

1. Let an expected control u_{exp} be given. Solve (7b) with $u = u_{exp}$ for a set of snapshots. Set $i = 1$.
2. If $i > 1$, then add $j \leq \ell$ POD basis elements from the last iteration to the set of snapshots. Compute ℓ POD basis elements according to Variant 1 or 2, respectively.
3. Solve the reduced optimization problem (8) for y^i_{pod} and u^i_{pod}.
4. We define the state y^i_{check} as the solution to (7b) with $u = u^i_{pod}$.
5. If the relative norm

$$\Upsilon = \frac{\|u^i_{pod} - u^{i-1}_{pod}\|_{L^2(\Sigma)}}{\|u^i_{pod}\|_{L^2(\Sigma)}}$$

is greater than TOL, then build up a new ensemble of snapshots using y^i_{check}. Set $i = i + 1$ and goto 2.

5. Numerical results

The dynamical systems needed for the set of snapshots are evaluated in a finite element space. All computations are executed on a DIGITAL Alpha 21264 computer. We use the Matlab Partial Differential Equation Toolbox to generate a mesh of triangular, piecewise linear finite elements.

The reduced nonlinear control problem (8) is solved with the SQP technique. The resulting linearized equations for state and control are numerically integrated with the Crank–Nicolson method using n time instances with constant step sizes. We use eight POD modes for the reduced problem, whereas for the FE description we take 185 elements.

Example 5.1. In this example we think of a cooling material in the time interval $[0, 1]$. The initial condition is given by $y_0(x_1, x_2) = 1 - (x_1 - 0.5)^2 - (x_2 - 0.5)^2$. Whilst regimes close to the surface are already cooler, its core is still hot. We want to track the temperature distribution to the desired state $z(x_1, x_2) = 1 + x_1$ with a regularization parameter $\kappa = 0.001$. Heat conductivity is $k = 0.1$. The nonlinearity is given by $b(y) = 0.2 \cdot y^4$, which is Stefan–Boltzmann's law describing heat radiation. We discretize the time interval by $n = 60$ points and choose $u_{exp} = x_1$ to build up the test functions.

	Variant 1			Variant 2		
ℓ	$\Lambda(\ell)$	$\Phi_{L^2(\Omega)}$ (%)	Ψ (%)	$\Lambda(\ell)$	$\Phi_{L^2(\Omega)}$ (%)	Ψ (%)
1	$4.89\cdot10^{-3}$	13.86	54.13	$5.02\cdot10^{-1}$	13.86	54.13
2	$1.14\cdot10^{-3}$	11.22	46.52	$4.98\cdot10^{-3}$	12.30	31.35
3	$4.21\cdot10^{-5}$	4.117	15.68	$3.10\cdot10^{-3}$	3.946	12.55
4	$1.55\cdot10^{-6}$	3.848	17.11	$6.15\cdot10^{-4}$	3.759	12.10
5	$5.34\cdot10^{-8}$	3.189	14.97	$6.56\cdot10^{-5}$	3.184	13.09
6	$3.36\cdot10^{-9}$	3.189	15.01	$2.20\cdot10^{-5}$	2.708	9.666
7	$1.92\cdot10^{-10}$	3.175	15.07	$1.68\cdot10^{-6}$	2.428	9.390
8	$5.93\cdot10^{-12}$	2.195	11.74	$7.89\cdot10^{-7}$	2.102	8.890

TABLE 1. Information defect $\Sigma(\ell)$ and time averaged relative L^2 errors Φ and Ψ in percent.

We compare both POD variants to a FE solution, which is calculated with the same algorithm as the POD solution. We apply Algorithm 4.1 to both variants and investigate improvements of the performance.

To examine how precise a POD solution can become, if we feed it with as much information about the system as possible, we run a third variant. This Variant 3 uses the FE solution y_{fe} of the optimal control problem as snapshots for a POD. It follows the idea of Variant 2 with y_{fe} instead of the expected state y_{exp}.

Now, Variant 1 uses ℓ modes which are built up from snapshots describing mainly the state variable. Variant 2 uses $\ell/2$ modes from snapshots of the state and $\ell/2$ from snapshots describing characteristics of the adjoint variable.

Table 1 gives the information defect

$$\Lambda(\ell) = \frac{\sum_{i=\ell+1}^{l} \lambda_i}{\sum_{i=1}^{l} \lambda_i}$$

between ℓ POD modes and the set of snapshots. Note that the set of snapshots for Variant 2 is twice as large as that for Variant 1. Hence, the information defect is larger for the same amount of modes.

The other columns of the table give the time averaged L^2-error of state and control, i.e.,

$$\Phi_X = \frac{1}{n}\sum_{i=1}^{n} \frac{\|y_{pod}(\cdot,t_i) - y_{fe}(\cdot,t_i)\|_X}{\|y_{fe}(\cdot,t_i)\|_X}, \quad \Psi = \frac{1}{n}\sum_{i=1}^{n} \frac{\|u_{pod}(\cdot,t_i) - u_{fe}(\cdot,t_i)\|_{L^2(\Gamma)}}{\|u_{fe}(\cdot,t_i)\|_{L^2(\Gamma)}}$$

for both variants. The averaged L^2-error for the state variable nearly coincides for both variants. The error for the optimal control is 25% smaller for Variant 2. This is furthermore revealed by the temporal evolution of the control's L^2-norm, see Figures 1 and 2. These figures also show the relative difference between the control check y_{check}, computed from (7b) with the control $u = u_{pod}$, and the FE solution of the optimal control problem. Variant 2 clearly outperforms Variant 1.

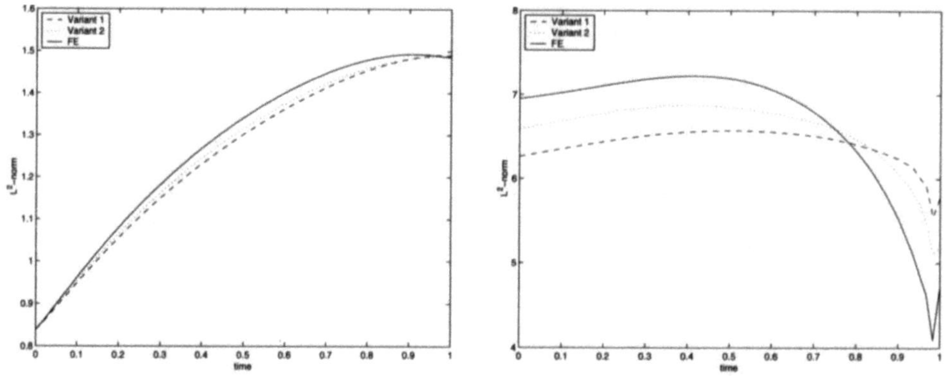

FIGURE 1. Here, Variant 1 and Variant 2 are compared to the FE solution. The left graph shows the L^2-norm evolution of the state solutions. The right graph shows the control norms.

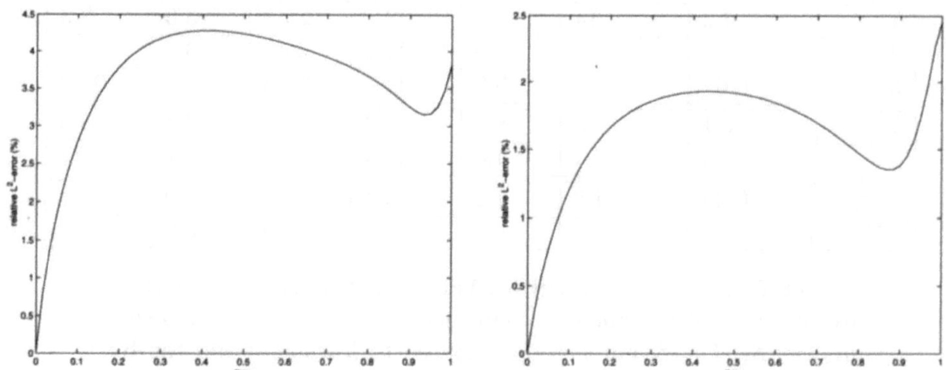

FIGURE 2. L^2-norm deviations of the control check y_{check} for Variant 1 (left) and Variant 2 (right) from the FE solution.

From Figure 3 we actually see that the check $y_{check}(\,.\,,T)$ of Variant 2 is closer to the desired state z than that of Variant 1.

Now we update the POD-bases dynamically for both variants and investigate their performance. We enlarge the set of snapshots with four POD modes from the last iteration. As a stopping criterium for Algorithm 4.1 we take $TOL = 10^{-1}$. In Table 2 we compare both variants by several parameters including the cost functional J and the relative deviation of the final state from the desired state, i.e.,

$$\Theta := \frac{\|y(T,\,.\,) - z\|_{L^2(\Omega)}}{\|z\|_{L^2(\Omega)}}.$$

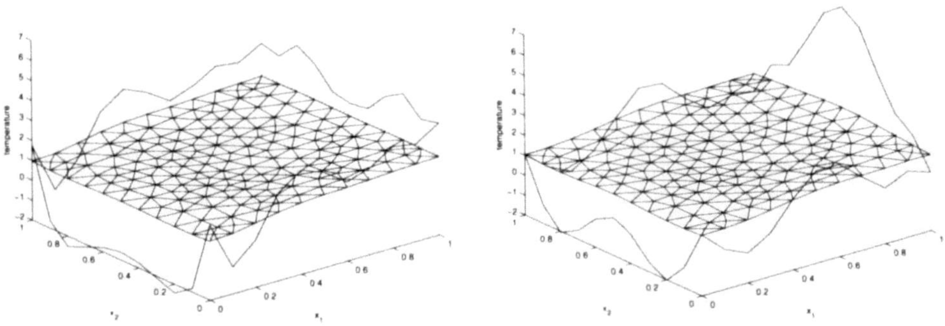

FIGURE 3. $y_{check}(T,\,.)$ and $u_{pod}(T,\,.)$ for Variant 1 (left) and Variant 2 (right).

i	Φ_{L^2} (%)	Φ_{H^1} (%)	Ψ (%)	$\Phi^\star$ (%)	$J \cdot 10^2$	Θ (%)	$\Upsilon \cdot 10^1$
1	2.2	14	12	2.1	2.31	3.3	8.3
2	0.96	7.4	6.9	0.82	2.43	3.7	1.1
3	0.62	6.6	5.7	0.67	2.48	3.8	0.62
1	2.1	16	8.9	1.8	2.36	3.4	8.3
2	0.74	7.4	4.4	0.83	2.48	3.7	0.82
Var. 3	0.44	4.5	1.2	0.63	2.49	3.9	
FE					2.49	3.9	

TABLE 2. Adaptive strategy for Variant 1 (first ensemble of three iterations) and Variant 2 (second ensemble of two iterations) with $u_{exp} = x_1$, $\ell = 8$ and $j = 4$ compared to Variant 3 and the FE solution. Here, $\Phi^\star$ denotes the relative L^2 norm deviation of y_{check} from the FE solution, hence equals Φ_{L^2} with y_{check} instead of y_{pod}.

First, we see that results can be improved by a factor two . Second, the cost functional and the deviation from the desired state is well reached by the adaptive strategy. Here, Variant 1 needs three iterations to reach the tolerance whereas for Variant 2 two iterations are sufficient.

Variant 3, which uses the information of the FE solution to the optimal control problem, lies very close to that. A comparison to the adaptive strategy solutions with Variant 1 and 2 reveals that these POD variants closely approach the benchmark of Variant 3.

Although the nonlinearity and its deviations have to be evaluated in the higher dimensional FE space for each time step and every iteration, the POD

		Variant 1	Variant 2	FE approx.
create	Mflops	9.4	12	
snapshots	s	1.7	2.4	
make a POD	Mflops	0.20	0.35	
	s	$4.9 \cdot 10^{-3}$	$9.8 \cdot 10^{-3}$	
solve the	Mflops	29	29	
red. problem	s	8.0	7.9	
total	Mflops	$3.8 \cdot 10^{1}$	$4.1 \cdot 10^{1}$	$4.0 \cdot 10^{5}$
	s	9.7	$1.0 \cdot 10^{1}$	$3.0 \cdot 10^{4}$

TABLE 3. shows computation effort and time for both POD variants compared to the FE solver.

methods are 3000 times faster than the FE approach and need 10^4 times less additions and multiplications, see Table 3. The computational effort for the adaptive control strategy multiplies by the number of iterations.

Example 5.2. In this example, the initial condition equals $y_0(x_1, x_2) = 10x_1x_2$, the time interval is set to $[0, 1]$. We choose the heat conductivity $k = 0.1$, the nonlinearity $b(y) = \arctan(y)$, the regularization parameter $\kappa = 0.01$ and the desired state $z(x_1, x_2) = 2 + 2 \cdot |2x_1 - x_2|$. We take $n = 55$ time instances.

Here, a concave nonlinearity with upper bound is chosen. We think of heat transition of a hot material into a fluid. For lower surface temperature regimes, linear heat transition into the fluid dominates. This is described by arctan from zero up to around $\pi/2$. If temperature gets above the boiling-point, the fluid steams and a thin gas layer insulates the surface of the material from the fluid. This gas layer decelerates the increase of heat flow when temperature raises. The function arctan models this temperature dependence above a value of around $\pi/2$. The high peak of the initial condition at one edge causes the surface temperature to exhaust almost the whole range of the nonlinearity. The sharp edge of the desired state at $2x_1 = x_2$ challenges the optimal control problem.

Figures 4 to 6 show the impact of Algorithm 4.1 on Variant 1 and 2. The graphs show the errors of the state and the control for all iterations that are needed to reach the stopping criterium $TOL < 10^{-1}$. We compare the performance by the same parameters as in Example 5.1. From Table 4 we see that Algorithm 4.1 improves performance in the state by a factor of three for Variant 1 and a factor of four for Variant 2. The L^2 error in the control is halved for both variants. As in the last example, Variant 2 closely approaches the benchmark of Variant 3.

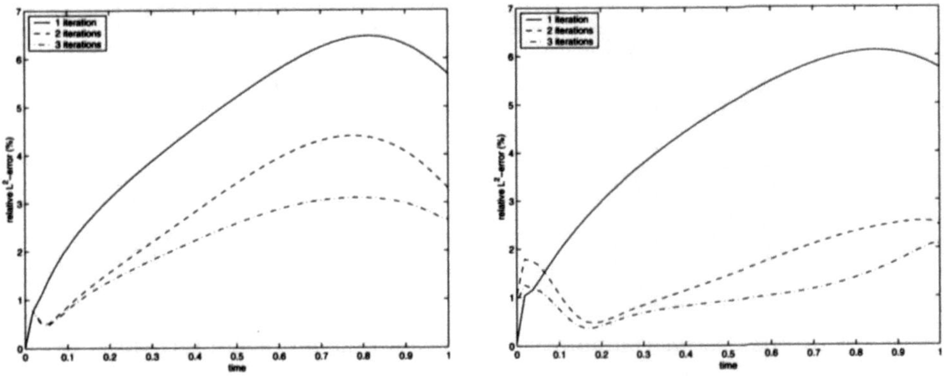

FIGURE 4. shows the relative error Φ_{L^2} of the state y_{pod} compared to the FE solution. It is computed with Variant 1 (left) and Variant 2 (right). The adaptive strategy is used with $j = 5$.

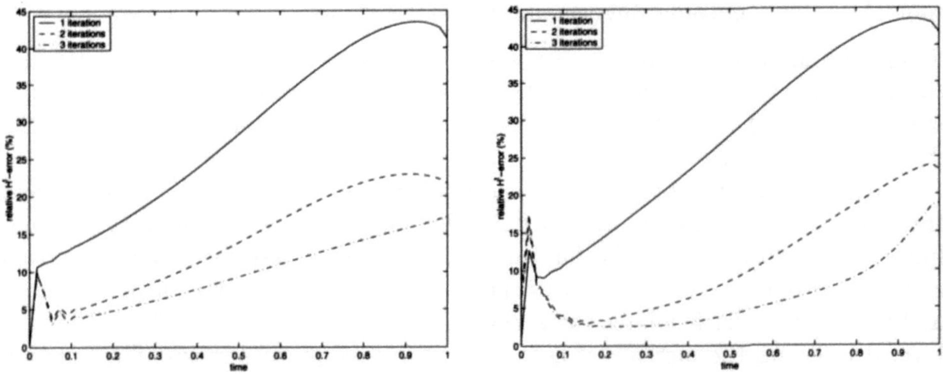

FIGURE 5. gives the H^1 error Φ_{H^1} of y_{pod} for Variant 1 (left) and Variant 2 (right).

The computation effort and time is presented in Table 5.

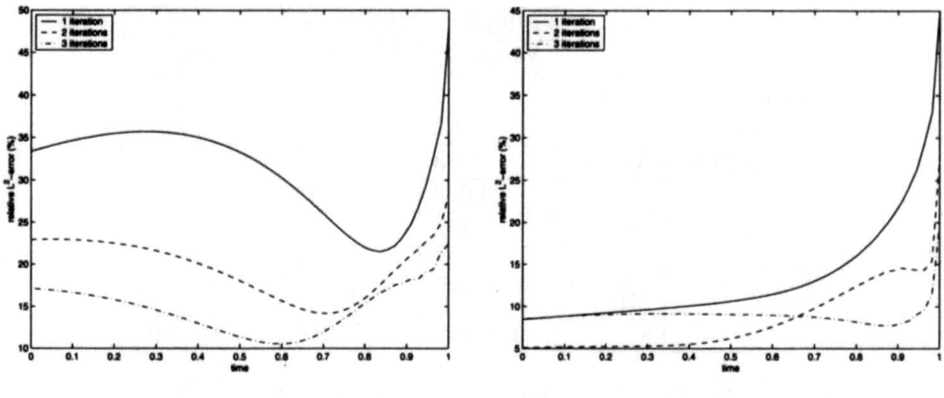

FIGURE 6. The graphs show the relative L^2 error Ψ of the control u_{pod} compared to the FE solution.

i	Φ_{L^2} (%)	Φ_{H^1} (%)	Ψ (%)	$\Phi^\star$ (%)	$J \cdot 10^1$	Θ (%)	$\Upsilon \cdot 10^1$
1	4.9	29	31	5.4	3.38	13.4	10
2	2.3	11	19	1.2	3.66	15.2	2.6
3	1.9	9.2	13	1.1	3.64	15.2	0.96
1	4.4	27	12	4.9	3.58	14.0	10
2	1.0	6.6	8.1	1.4	3.60	15.2	1.3
3	0.93	6.6	6.8	1.1	3.61	15.3	0.83
Var. 3	0.52	4.0	5.7	0.78	3.58	15.0	
FE					3.58	15.0	

TABLE 4. Performance is compared by time-averaged errors. The state is examined in the L^2 norm (Φ_{L^2}) and in the H^1 norm (Φ_{H^1}), the control in the $L^2(\Gamma)$ norm (Ψ) and the control check y_{check} in the L^2 norm ($\Phi^\star$). Furthermore, the cost functional J, the deviation from the desired state Θ and the tolerance TOL are given. The first three rows show the adaptive algorithm with $j = 5$ applied to Variant 1 with $\ell = 8$ and $u_{exp} = 0$. The next three rows show Variant 2 with the same parameters.

6. Conclusions

In this article, the method of POD is utilized to solve nonlinear boundary control problems for the heat equation. If we include information of the state as well as of the adjoint variable into the POD basis, we get an efficient reduced-order method for the optimal control problem. As it is illustrated in the two numerical examples, the results can be improved significantly by updating the POD basis dynamically.

		Adaptive strategy for		
		Variant 1	Variant 2	FE approx.
create snapshots	Mflops	25	18	
	s	4.4	3.3	
make a POD	Mflops	0.83	0.44	
	s	0.023	0.014	
solve the red. problem	Mflops	111	84	
	s	29	22	
total	Mflops	$1.4 \cdot 10^2$	$1.0 \cdot 10^2$	$1.9 \cdot 10^5$
	s	$3.3 \cdot 10^1$	$2.5 \cdot 10^1$	$6.6 \cdot 10^3$

TABLE 5. shows computation effort and time for both POD variants compared to the FE solver.

Finally, we can observe a good agreement between the FE- and the POD-based solutions and a dramatic reduction of the computation time.

References

[1] K. Afanasiev and M. Hinze, *Adaptive control of a wake flow using proper orthogonal decomposition*, in: Shape Optimization & Optimal Design, Lecture Notes in Pure and Applied Mathematics, Marcel Dekker, 2001.

[2] E. Arian, M. Fahl, and E. W. Sachs, *Trust-region proper orthogonal decomposition for flow control*, ICASE Report No. 2000-25, 2000.

[3] J. A. Atwell and B. B. King, *Reduced order controllers for spatially distributed systems via proper orthogonal decomposition*, Technical report, ICAM Report 99-07-01, July 1999.

[4] H. T. Banks, M. L. Joyner, B. Winchesky, and W. P. Winfree, *Nondestructive evaluation using a reduced-order computational methodology*, Inverse Problems, **16** (2000), pp. 1–17.

[5] H. T. Banks, R. C. H. del Rosario, and R. C. Smith, *Reduced order model feedback control design: Numerical implementation in a thin shell model*, Technical report, CRSC-TR98-27, North Carolina State University, 1998.

[6] G. Berkooz, P. Holmes, and J. L. Lumley, *Turbulence, Coherent Structures, Dynamical Systems and Symmetry*, Cambridge Monographs on Mechanics, Cambridge University Press, 1996.

[7] K. Fukunaga, *Introduction to Statistical Recognition*, Academic Press, 1990.

[8] H. Goldberg and F. Tröltzsch, *Second order sufficient optimality conditions for a class of nonlinear parabolic boundary control problems*, SIAM J. Control Optim., **31** (1993), pp. 1007–1025.

[9] H. Goldberg and F. Tröltzsch, *On a Lagrange-Newton method for a nonlinear parabolic boundary control problem*, Preprint 96-8, Technische Universität Chemnitz-Zwickau, 1997.

[10] H. Goldberg and F. Tröltzsch, *On a SQP-multigrid technique for nonlinear parabolic boundary control problems*, Preprint 97-11, Technische Universität Chemnitz-Zwickau, 1997.

[11] M. Hinze and K. Kunisch, *Three control methods for time-dependent fluid flow*, submitted, 1999.

[12] K. Ito and S. S. Ravindran, *A reduced basis method for control problems governed by PDEs*, Control and Estimation of Distributed Parameter Systems, Internat. Ser. Numer. Math., **126** (1998).

[13] K. Ito and J. D. Schroeter, *Reduced order feedback synthesis for viscous incompressible flows*, Technical Report CRSC-TR98-41, North Carolina State University, 1998.

[14] K. Kunisch and S. Volkwein, *Control of Burgers' equation by a reduced order approach using proper orthogonal decomposition*, JOTA, **102** (2) (1999), pp. 345–371.

[15] K. Kunisch and S. Volkwein, *Galerkin proper orthogonal decomposition methods for parabolic problems*, to appear in Numer. Math.

[16] H. V. Ly and H. T. Tran, *Modelling and control of physical processes using proper orthogonal decomposition*, Mathematical and Computer Modeling, **33** (2001), pp. 223–236.

[17] A. Pazy, *Semigroups of Linear Operators and Applications to Partial Differential Equations*, Springer-Verlag, 1983.

[18] J. P. Raymond and H. Zidani, *Hamiltonian Pontryagin's principles for control problems governed by semilinear parabolic equations*, Appl. Math. Optim., **39**, No. 2 (1999), pp. 143–177.

[19] L. Sirovich, *Turbulence and the dynamics of coherent structures, parts I-III.* Quart. Appl. Math., **XLV** (1987), pp. 561–590.

[20] F. Tröltzsch, *An SQP method for the optimal control of a nonlinear heat equation*, Control and Cybernetics, **23** (1994), pp. 267–288.

[21] F. Tröltzsch, *Lipschitz stability of solutions to linear-quadratic parabolic control problems with respect to perturbations*, Dynamics of Continuous, Discrete and Impulsive Systems, 1999.

[22] F. Tröltzsch, *On the Lagrange-Newton-SQP method for the optimal control of semilinear parabolic equations*, SIAM J. Control Optim., **3** (1999), pp. 294–312.

[23] S. Volkwein, *Optimal control of a phase-field model using the proper orthogonal decomposition*, Z. Angew. Math. Mech. (ZAMM), **81**, No. 2 (2001), pp. 83–97.

A Posteriori Error Estimation for the Sourcewise Represented Solution with Application to the Problem of Heat Conductivity

Konstantin Dorofeev and Anatoly Yagola

Abstract. In this paper a method for solving linear ill-posed problems with additional a priori information that the exact solution is sourcewise represented is offered. Application of these methods for inverse problems for the heat conduction equation is considered.

1. Introduction

The theory of ill-posed problems was founded by Tikhonov [14, 15] and developed by many authors (e.g., [1, 2, 3, 5, 7, 8, 9, 11, 16, 17, 19, 20]). When designing stable methods for solving ill-posed problems, there is the understanding that ill-posed problems are generally underdetermined. To solve such problem is to find a stable approximation to exact solution. It can be made by so called regularizing algorithms, if they exist. But the existence of this stable methods is not equivalent with the well-posedness of the problem. We should use all available information when constructing a regularizing algorithms. Sometimes this information is sufficient to reformulate the initial ill-posed problem into a well-posed one. In this report it will be shown that using the information about a sourcewise representation of exact solution with compact operator allows to construct an algorithm with special properties. An a posteriori error estimate can then be found. We propose the application of this method for inverse problems of heat conductivity.

2. The sourcewise representation of solutions and the method of extending compacts

Consider the operator equation

$$Az = u \tag{2.1}$$

where $A : Z \to U$ is a linear bounded injective operator from the normed space Z into the normed space U.

For an exact right-hand side $\overline{u} \in U$ of (2.1) we define the unique exact solution $\overline{z} \in Z$. Instead of $\overline{u}$ we are given $u_\delta \in U$ and $\delta > 0$, such that $\|u_\delta - \overline{u}\| \leqslant \delta$.

The problem of solving equation (2.1) can be ill-posed, e.g. if A is a compact operator.

In general, it is impossible to solve an ill-posed problem without the knowledge of the error level [2, 17], to get the error estimate for an approximate solution [2, 16] and even to get a rate of convergence for approximate solutions on the whole space Z [2, 21]. Therefore, we should use all available a priori information about the exact solution to construct algorithms with special properties. It will be shown below that an additional a priori information about a sourcewise representation of an unknown solution with a compact operator allows to construct regularizing algorithms with properties which are not generally available.

2.1. The case when operators are known exactly

Additionally, assume the a priori information: the exact solution $\overline{z}$ is sourcewise represented with a compact operator B acting from reflexive Banach space V into Z:

$$\overline{z} = B\overline{v}. \tag{2.2}$$

For reasons of simplicity, we suppose that the operator B is injective, otherwise it is possible to consider a factor space.

Let us consider the method of extending compacts [22, 23] for solving problem (2.1) with a priori information (2.2). Set $n = 1$ and define the set Z_n:

$$Z_n = \{z \in Z : z = Bv, v \in V, \|v\| \leqslant n\}. \tag{2.3}$$

Then we minimize the discrepancy $F(z) = \|Az - u_\delta\|$ on the set Z_n. If

$$\min_{z \in Z_n} \|Az - u_\delta\| \leqslant \delta,$$

then the solution is found. We denote it as $n(\delta) = n$. Otherwise, we change n to $(n+1)$ and reiterate the process. If $n(\delta)$ is found, then we define the approximate solution $z_{n(\delta)}$ of (2.1) as an arbitrary solution of the inequality

$$\|Az - u_\delta\| \leqslant \delta \qquad z \in Z_{n(\delta)}.$$

The next theorem takes place:

Theorem 2.1. [22, 23] *The process described above converges: $n(\delta) < +\infty$. There exists $\delta_0 > 0$ (generally speaking, depending on $\overline{z}$) such that $n(\delta) = n(\delta_0)$ for any $0 < \delta \leqslant \delta_0$. Approximate solutions $z_{n(\delta)}$ strongly converge to $\overline{z}$ as $\delta \to 0$.*

Proof. Since $V_n = \{v \in V : \|v\| \leqslant n\}$ is a bounded and closed set in V and operator B is compact, Z_n is a compact in Z for any n. Due to the Weierstrass theorem the continuous functional $F(z) = \|Az - u_\delta\|$ reaches its exact lower bound on Z_n (possibly in a non-unique point $z_n \in Z_n$).

The finiteness of $n(\delta)$ and the existence of δ_0 such that $n(\delta) = n(\delta_0)$ for any $0 < \delta \leqslant \delta_0$ follows from the fact that $\overline{z} = B\overline{v} \in Z_N$, where $N = \|\overline{v}\|$, if $\|\overline{v}\|$ is a positive integer, or $N = [\|\overline{v}\|] + 1$ otherwise. It is easy to see that $n(\delta)$ cannot be less N for arbitrary small positive δ. So there exists δ_0 such that for any $0 < \delta \leqslant \delta_0$ approximate solutions $z_{n(\delta)}$ are in the compact set $Z_{n(\delta_0)}$ and the method of extending compacts coincides with the quasisolutions method [7] for all sufficiently small positive δ. The convergence of approximate solutions to the exact solution follows from the general theory of ill-posed problems [7, 16]. Thus, the method is a regularizing algorithm, according to Tikhonov's definition [16]. □

Remark 2.2. *The results of the theorem can be obtained from the results of* [4].

Theorem 2.3. [22, 23] *For the method described above there exists an a posteriori error estimate. It means that a functional $\Delta(\delta, u_\delta)$ exists such that $\Delta(\delta, u_\delta) \to 0$ as $\delta \to 0$, and $\|z_{n(\delta)} - \bar{z}\| \leqslant \Delta(\delta, u_\delta)$ at least for all sufficiently small positive δ.*

Proof. We define a functional $\Delta(\delta, u_\delta)$ as follows:

$$\Delta(\delta, u_\delta) = \max\left\{\|z - z_{n(\delta)}\| : z \in Z_{n(\delta)}, \|Az - u_\delta\| \leqslant \delta\right\}.$$

The operator A is bounded and $Z_{n(\delta)}$ is a compact set. Thus, the set $\{z \in Z_{n(\delta)}, \|Az - u_\delta\| \leqslant \delta\}$ is a compact set too. Then $\Delta(\delta, u_\delta) < +\infty$. It is evident that for all positive $\delta \leqslant \delta_0$ the inequality $\|z_{n(\delta)} - \bar{z}\| \leqslant \Delta(\delta, u_\delta)$ is valid because $\bar{z} \in Z_{n(\delta)}$. It is easy to see that $\Delta(\delta, u_\delta) \to 0$ as $\delta \to 0$. □

Remark 2.4. *The a posteriori error estimate can be calculated while solving the operator equation (2.1) by the method of extending compacts. We would like to emphasize that the a posteriori error estimate is in reality an upper error estimate for the approximate solution only for $0 < \delta \leqslant \delta_0$ where δ_0 depends on the unknown exact solution $\bar{z}$.*

2.2. Special properties of the method of extending compacts

The sourcewise representation of the kind

$$\bar{z} = (A^*A)^{p/2}\,\bar{v} \qquad \bar{v} \in Z, p = \text{const} > 0$$

where A is the compact linear injective operator acting from Hilbert space Z into U is considered by different authors [2, 5, 6, 13, 18, 19] in order to compare the rate of convergence of their algorithms.

Lemma 2.5. *The operator $(A^*A)^{p/2}$ is compact and injective from Z to Z for any $p > 0$.*

Proof. The operator A^*A is compact and selfadjoint. The compactness of the operator $(A^*A)^{p/2}$ follows immediately from known properties of the eigenvalues of linear compact selfadjoint operators, see [12]. The injectiveness is obvious. □

Consider now the suggested method of extending compacts in the following case: Z, U are Hilbert spaces, $V = Z, A$ is a linear compact injective operator acting from Z to U, $B = (A^*A)^{p/2}, p = const > 0$ (p is given).

Theorem 2.6. *The method of extending compacts in the case that Z, U are Hilbert spaces, $V = Z$ that A is a linear compact injective operator acting from Z to U and that $B = (A^*A)^{p/2}$, is a regularizing algorithm which is of optimal order of accuracy and allows an a posteriori error estimate.*

Proof. It is evident that both Theorems 2.1 and 2.3 are valid in this case. Thus, the method of extending compacts is a regularizing algorithm which allows an a posteriori error estimate for an approximate solution of the problem. Here we need only to prove the optimality in the order of accuracy for such approach. For all $\delta \in (0, \delta_0]$ (δ_0 is defined in Theorem 2.1) the method of extending compacts coincides with the method of quasisolutions on the convex balanced compact set $BV_{n(\delta_0)}$. Thus, the method is of optimal order of accuracy [2]. As it follows from [18], the accuracy of the method is at least $O(\delta^{p/(p+1)}$ for all $p > 0$. □

We need to note here that using the a priori information about the exact solution allows to construct the algorithm with properties which are not available in the general case.

2.3. The case when operators are known approximately

Let us consider now the case, when operators A and B so as the right-hand side u are given with errors. The following cases are possible [22]:

1. A is known approximately and B is known exactly,
2. A is known exactly and B is known approximately,
3. A and B are known approximately.

At first, we suppose that the operator B is known exactly and the operator A is known approximately. Instead of the operator A and the right-hand side $\overline{u}$ we are given approximate data $\{A_h, u_\delta\}$ and the estimates of their errors (or "noise levels") $\{h, \delta\}$ for the problem (2.1). Here $A_h : Z \to U$ is a linear bounded operator such that $\|A - A_h\| \leq h$. The element $u_\delta \in U$ is an approximate right-hand side in (2.1) such that $\|\overline{u} - u_\delta\| \leq \delta$. Error levels h and δ are supposed to be known. Given the data $\{A_h, u_\delta\, h, \delta\}$ we wish to construct an element $z_\eta \in Z, \eta \overset{def}{=} (h, \delta)$, such that $\|z_\eta - \overline{z}\| \to 0$ as $\eta \to 0$. We define the set Z_n as in (2.3) and find the minimal positive integer $n = n(\delta, h)$ such that the following inequality has nonempty set of solutions:

$$\|A_h z - u_\delta\| \leq \delta + h\|B\|n, z \in Z_n. \tag{2.4}$$

If we found $n(\delta, h)$, then the arbitrary fixed solution of the inequality (2.4) can be considered as an approximate solution of (2.1). We denote this solution as $z_{n(\delta,h)}$.

Theorem 2.7. *The problem formulated above has a solution $n(\delta, h) < +\infty$. Approximate solutions $z_{n(\delta,h)}$ strongly converge to $\overline{z}$ as $\delta, h \to 0$. There exists an a posteriori error estimate $\Delta(\delta, h, u_\delta, A_h) \to 0$ as $\delta, h \to 0$, and $\|z_{n(\delta,h)} - \overline{z}\| \leq \Delta(\delta, h, u_\delta, A_h)$ at least for all sufficiently small positive δ, h.*

Proof. It is easy to see that if we define N as in Theorem 2.1 then $\overline{z}$ is a solution of the inequality (2.4), when $n = N$. After this remark it is clear that the proof of the theorem is very similar to the proofs of Theorems 2.1 and 2.3. As an a posteriori error estimate that is valid for sufficiently small positive δ, h we can introduce a functional

$$\begin{aligned}&\Delta(\delta, h, u_\delta, A_h)\\&= \max\left\{\|z - z_{n(\delta,h)}\| : z \in Z_{n(\delta,h)}, \|A_h z - u_\delta\| \leq \delta + h\|B\|n(\delta, h)\right\}.\end{aligned}$$

□

The case when the operator A is known exactly and the operator B is specified with an error, is more complicated. We introduce an additional assumption: let V be a reflexive Banach space. Below, instead of the operator B and the right-hand side $\overline{u}$ we are given approximate data $\{B_h, u_\delta\}$ and the estimates of their errors (or "noise levels") $\{h, \delta\}$ for the problem (2.1). Here, $B_h : V \to Z$ is a linear compact injective operator such that $\|B - B_h\| \leq h$. The element $u_\delta \in U$ is an approximate right-hand side in (2.1) such that $\|\overline{u} - u_\delta\| \leq \delta$. Error levels h and δ are supposed to be known. Given the data $\{B_h, u_\delta, h, \delta\}$ we wish to construct an element $z_\eta \in Z, \eta \overset{def}{=} (h, \delta)$, such that $\|z_\eta - \overline{z}\| \to 0$ as $\eta \to 0$.

For any positive integer n and $h \geq 0$ we define a compact set

$$Z_{n,h} = \{z \in Z : z = B_h v, v \in V, \|v\| \leq n\}.$$

Let us note now that the set $Z_{n,h}$ depends on an operator B_h. That means that then $h \to 0$ compact sets $Z_{n,h}$ are different in general for different h an B_h.

Let us consider the problem: to find the minimal positive integer $n = n(\delta, h)$ such that the following inequality has a nonempty set of solution:

$$\|Az - u_\delta\| \leq \delta + h\|A\|n, z \in Z_{n,h}. \tag{2.5}$$

If we found $n(\delta, h)$ then we can consider an arbitrary fixed solution of the inequality (2.5) as an approximate solution of (2.5). We denote this solution as $z_{n(\delta,h)}$.

Theorem 2.8. *The problem formulated above has a solution $n = n(\delta, h) < +\infty$. Approximate solutions $z_{n(\delta,h)}$ strongly converge to $\overline{z}$ as $\delta, h \to 0$. There exists an a posteriori error estimate $\Delta(\delta, h, u_\delta, B_h)$ such that $\Delta(\delta, h, u_\delta, B_h) \to 0$ as $\delta, h \to 0$, and $\|z_{n(\delta,h)} - \overline{z}\| \leq \Delta(\delta, h, u_\delta, B_h)$ at least for all sufficiently small positive δ, h.*

Proof. It easy to see that if we define N as in Theorem 2.1 then $\overline{z}$ is a solution of the inequality (2.5) when $n = N$:

$$\|AB_h\overline{v} - u_\delta\| = \|AB_h\overline{v} - AB\overline{v} + AB\overline{v} - u_\delta\| \leq \delta + h\|A\|N.$$

We note that $B_h\overline{v} \in Z_{N,h}$ for any h and δ, so the solution of the problem (2.5) $n(\delta, h) \leq N$. Let us fix now any positive $\overline{h}$ and consider $h \in (0, \overline{h}]$ only. Then all compact sets $Z_{n(\delta,h),h}$ are in the ball of the radius $(\|B\| + \overline{h})N$ because

$$\|B_h v\| \leq \|B_h\|\|v\| = \|B_h - B + B\|\|v\| \leq (\|B\| + \overline{h})N.$$

Let us prove that the proposed method is a regularizing algorithm due to Tikhonov's definition. As usually (see, e.g. [16]), we suppose that approximate solutions $z_{n(\delta,h)}$ does not converge to $\overline{z}$ as $\delta, h \to 0$. Then there exist $\epsilon > 0$ and sequences $\delta_k, h_k \to 0$ such that

$$\|z_{n(\delta_k,h_k)} - \overline{z}\| \geq \epsilon. \tag{2.6}$$

We supposed that V is a reflexive Banach space. Thus, $V_N = \{v \in V : \|v\| \leq N\}$ is weakly compact in V (see, e.g. [16]). All elements $v_{n(\delta_k,h_k)} = B_{h_k}^{-1} z_{n(\delta_k,h_k)}$ are in the weak compact V_N. So, we can choose from the sequence $v_{n(\delta_k,h_k)}$ a subsequence that weakly converges to an element $v^* \in V_N$. We can without loss of generality suppose that the sequence $v_{n(\delta_k,h_k)}$ weakly converges to v^*. Then

$$\begin{aligned}&\|ABv_{n(\delta_k,h_k)} - AB\overline{v}\|\\ &\le \|ABv_{n(\delta_k,h_k)} - AB_{h_k}v_{n(\delta_k,h_k)} + AB_{h_k}v_{n(\delta_k,h_k)} - u_{\delta_k} + u_{\delta_k} - AB\overline{v}\|\\ &\le h_k\|A\|N + \delta_k + h_k\|A\|N + \delta_k \to 0 \quad \text{as} \quad k \to +\infty.\end{aligned}$$

This contradicts (2.6). Thus, the proposed method is a Tikhonov regularizing algorithm.

If there exists a sequence $\delta_k, h_k \to 0$ such that $n(\delta_k, h_k) < N$ for all k (it means that $n(\delta_k, h_k) \le N - 1$ for all k), then it is easy to prove the existence of a solution $\overline{z} = B\overline{v}$ with $\|\overline{v}\| \le N - 1$ that contradicts the assumptions of the theorem. Thus, the following inequality is valid for sufficiently small δ and h:

$$\begin{aligned}&\|z_{n(\delta,h)} - \overline{z}\| = \|z_{n(\delta,h)} - B_h\overline{v} + B_h\overline{v} - B\overline{v}\|\\ &\le \max\left\{\|z - z_{n(\delta,h)}\| : z \in Z_{n(\delta,h),h}, \|Az - u_\delta\| \le \delta + h\|A\|n(\delta,h)\right\} + hn(\delta,h)\\ &\stackrel{def}{=} \Delta(\delta, h, u_\delta, B_h).\end{aligned}$$

The functional $\Delta(\delta, h, u_\delta, B_h)$ is an a posteriori error estimate. □

Finally, let us consider the case when both operators A and B are specified with errors. We suppose that V is a reflexive Banach space. Below, instead of the operators A, B and the right-hand side $\overline{u}$ we are given approximate data $\{A_{h_1}, B_{h_2}, u_\delta\}$ and the estimates of their errors (or "noise levels") h_1, h_2, δ for the problem (2.1). Here $A_{h_1} : Z \to U$ is a linear bounded operator such that $\|A - A_{h_1}\| \le h_1, B_{h_2} : V \to Z$ is a linear compact injective operator such that $\|B - Bh_2\| \le h_2$. The element $u_\delta \in U$ is an approximate right-hand side in (2.1) such that $\|\overline{u} - u_\delta\| \le \delta$. Error levels h_1, h_2 and δ are supposed to be known. Given the data $\{A_{h_1}, B_{h_2}, u_\delta, h_1, h_2, \delta\}$ we wish to construct an element $z_\eta \in Z, \eta \stackrel{def}{=} (h_1, h_2, \delta)$, such that $\|z_\eta - \overline{z}\| \to 0$ as $\eta \to 0$.

For any integer positive n and $h_2 \ge 0$ we define a compact set

$$Z_{n,h_2} = \{z \in Z : z = B_{h_2}v, v \in V, \|v\| \le n\}.$$

Consider the problem: to find the minimal positive integer $n = n(\delta, h_1, h_2)$ such that the following inequality has a nonempty set of solutions:

$$\|A_{h_1}z - u_\delta\| \le \delta + (h_1\|B_{h_2}\| + h_2\|A_{h_1}\| + h_1h_2)n, z \in Z_{n,h_2}. \tag{2.7}$$

If we found $n(\delta, h_1, h_2)$ then we can consider an arbitrary fixed solution of inequality (2.7) as an approximate solution of (2.1). We denote this solution as $z_{n(\delta,h_1,h_2)}$.

Theorem 2.9. *The problem formulated above has a solution $n(\delta, h_1, h_2) < +\infty$. Approximate solutions $z_{n(\delta,h_1,h_2)}$ strongly converge to $\overline{z}$ as $\delta, h_1.h_2 \to 0$. There exists an a posteriori error estimate $\Delta(\delta, h_1, h_2, u_\delta, A_{h_1}, B_{h_2})$ such that $\Delta(\delta, h_1, h_2, u_\delta, A_{h_1}, B_{h_2}) \to 0$ as $\delta, h_1, h_2 \to 0$, and at least for all sufficiently small positive δ, h_1, h_2 $\|z_{n(\delta,h_1,h_2)} - \overline{z}\| \le \Delta(\delta, h_1, h_2, u_\delta, A_{h_1}, B_{h_2})$.*

Proof. It is easy to see that, if we define N as in Theorem 2.1 then $\overline{z}$ is a solution of the inequality (2.7) if $n = N$. After this remark, the proof of the Theorem is just the same as the proof of Theorem 2.8. We have to define a functional $\Delta(\delta, h_1, h_2, u_\delta, A_{h_1}, B_{h_2})$ only:

$$\Delta(\delta, h_1, h_2, u_\delta, A_{h_1}, B_{h_2}) \stackrel{def}{=} \max\{\|z - z_{n(\delta,h_1,h_2)}\| : z \in Z_{n(\delta,h_1,h_2),h_2}, \\ \|A_{h_1} z - u_\delta\| \leq \delta + (h_1\|B\| + h_2\|A\| + h_1, h_2)n(\delta, h_1, h_2)\} + h_2 n(\delta, h_1, h_2).$$

□

Remark 2.10. *It is not obligatory for compact set to "extend" around zero. They can "extend" around* Bv_0 *where* v_0 *is any a priori given element.*

Remark 2.11. *In definition of sets* Z_n*, instead of the sequence of positive integers* n*, we can take any sequence of positive numbers* R_n *such that* $0 < R_1 < R_2 < \ldots < R_n \ldots$ *and* $R_n \to +\infty$ *as* $n \to +\infty$.

3. Application to the inverse problems for the heat conduction equation

We now apply the method of extending compacts for solving the inverse problem for the heat conduction equation:

$$\begin{cases} u_t = a^2 u_{xx} & 0 < x < l, 0 < t < \infty \\ u(x,0) = \varphi(x) & 0 \leqslant x \leqslant l \\ u(0,t) = 0 & 0 \leqslant t < \infty \\ u(l,t) = 0 & 0 \leqslant t < \infty. \end{cases} \tag{3.1}$$

It is known that the solution of this problem is:

$$u(x,t) = \int_0^l G(x,\xi,t)\varphi(\xi)d\xi \tag{3.2}$$

where $G(x,\xi,t)$ is the Green function:

$$G(x,\xi,t) = \sum_{n=1}^{\infty} \frac{2}{l} \sin\left(\frac{\pi n}{l}\xi\right) \sin\left(\frac{\pi n}{l}x\right) \exp\left(-\left(\frac{\pi n}{l}\right)^2 a^2 t\right).$$

From (3.2), it follows that for any moment of time $t_\varepsilon > 0$ we have:

$$\varphi_\varepsilon(x) \equiv u(x,t_\varepsilon) = \int_0^l G(x,\xi,t_\varepsilon)\varphi(\xi)d\xi. \tag{3.3}$$

Consider now the next problem:

$$\begin{cases} u_t = a^2 u_{xx} & 0 < x < l, t_\varepsilon \leqslant t < \infty \\ u(x,t_\varepsilon) = \varphi_\varepsilon(x) & 0 \leqslant x \leqslant l \\ u(0,t) = 0 & t_\varepsilon \leqslant t < \infty \\ u(l,t) = 0 & t_\varepsilon \leqslant t < \infty. \end{cases} \tag{3.4}$$

TABLE 1. The illustration of the saturation effect.

δ	1	0.1	0.01	10^{-3}	10^{-4}	10^{-10}	10^{-11}	10–13	10^{-15}	10^{-16}
$n(\delta)$	1	3	4	4	5	5	6	6	6	6

For any moment of time $T : T \geqslant t_\varepsilon$ we have:

$$\psi(x) \equiv u(x,T) = \int_0^l G(x,\xi,T-t_\varepsilon)\varphi_\varepsilon(\xi)\mathrm{d}\xi. \tag{3.5}$$

Now we suppose $\varphi(x), \varphi_\varepsilon(x), \psi(x) \in L^2[0,l]$ and formulate the inverse problem. We have the approximate function $\psi_\delta \in L^2[0,l]$ and the estimate of its error $\delta : \|\psi_\delta - \psi_\varepsilon\| \leqslant \delta$. We should find a stable approximation for $\varphi_\varepsilon(x)$. It can be made easily with the help of the method of extending compacts. Moreover, the a posteriori error estimate exists. Indeed, the unknown solution $\varphi_\varepsilon(x)$ is sourcewise represented with completely continuous operator due to (3.3).

Now we adduce the results (see figure (1)) for the problem with the next parameters: $t_\varepsilon = 0.01$, $T = 0.1$, $\delta = 10^{-11}$, $a = 0.6$, $l = 1$. As a function $\varphi(x)$

$$\varphi(x) = \begin{cases} 0 & 0 \leqslant x \leqslant 0.4 \quad \text{or} \quad 0.6 \leqslant x \leqslant 1 \\ 13.4162 & \text{otherwise} \end{cases}$$

was taken. Then we find functions $\varphi_\varepsilon(x)$ and $\psi(x)$ by formulae (3.3) and (3.5).

FIGURE 1. The exact (- -) and the calculated (—) solutions.

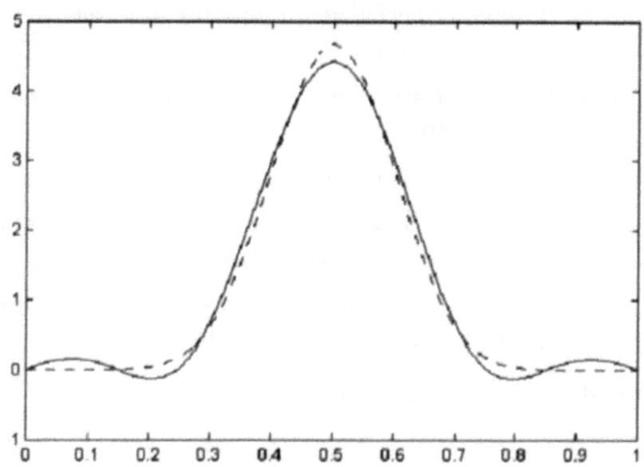

Calculations for several values of errors were made (see table (1)). These calculations confirm the validity of Theorem 2.3, stating the existence of the a posteriori error estimate. From this table, it is easy to see that according to Theorem 2.3 $n(\delta)$ suffers saturation, i.e., generally speaking, $\delta_0 > 0$ exists, depending on $\bar{z}$ such that $n(\delta) = n(\delta_0)$ for any $0 < \delta \leqslant \delta_0$. Therefore, it is possible to find an a posteriori error estimate for this problem.

Acknowledgement. The work was supported by the programme RFFI, grant 99-01-00447.

References

[1] O. M. Alifanov, E. A. Artyukhin, and S. V. Rumyantsev, *Extreme Methods for Solving Ill-Posed Problems With Applications to Inverse Heat Transfer Problems,* Begell House Inc., New York, 1995.

[2] A. B. Bakushinskii and A. V. Goncharskii, *Ill-Posed Problems: Theory and Applications,* Kluwer, Dordrecht, 1994.

[3] A. M. Denisov, *Elements of the Theory of Inverse Problems,* VSP, Utrecht, 1999.

[4] I. N. Dombrovskaya and V. K. Ivanov, *Some Questions to the Theory of Linear Equations in Abstract Spaces,* Sibirskii Mat. Zhurnal, **16** (1965), pp. 499–508 (in Russian).

[5] H. W. Engl, M. Hanke, and A. Neubauer, *Regularization of Inverse Problems,* Kluwer, Dordrecht, 1996.

[6] C. W. Groetsch, *The Theory of Tikhonov Regularization for Fredholm Equations of the First Kind,* MA: Pitman, Boston, 1984.

[7] V. K. Ivanov, V. V. Vasin, and V. P. Tanana, *The Theory of Linear Ill-Posed Problems and Its Applications*, Nauka, Moscow, 1978 (in Russian).

[8] M. M. Lavrentiev, *Some Improperly Posed Problems of Mathematical Physics,* Springer, Berlin, Heidelberg, New York, 1967.

[9] M. Lavrentiev, V. Romanov, and S. Shishatskii, *Ill-Posed Problems of Mathematical Physics and Analysis,* RI: AMS, Providence, 1986.

[10] A. S. Leonov and A. G. Yagola, *Special Regularizing Methods for Ill-Posed Problems with Sourcewise Represented Solutions,* Inverse Problems, **14** (1998), pp. 1539–1550.

[11] V. A. Morozov, *Methods for solving Incorrectly Posed Problems,* Springer, Berlin, Heidelberg, New York, Tokyo, 1984.

[12] F. Riesz and B. Szoekefalvi-Nagy, *Functional Analysis,* Ungar, New York, 1978.

[13] V. P. Tanana, M. A. Rekant, and S. I. Yanchenko, *The Optimization of Methods for Solving Operator Equations,* Ural University Press, Sverdlovsk, 1987 (in Russian).

[14] A. N. Tikhonov, *Solution of incorrectly formulated problems and the regularization method,* Sov. Math. Dokl., **5** (1963), pp. 1035–1038.

[15] A. N. Tikhonov, *Regularization of incorrectly posed problems,* Sov. Math. Dokl., **4** (1963), pp. 1624–1627.

[16] A. N. Tikhonov, A. V. Goncharsky, V. V. Stepanov, and A. G. Yagola, *Numerical Methods for the Solution of Ill-Posed Problems,* Kluwer, Dordrecht, 1995.

[17] A. N. Tikhonov, A. S. Leonov, and A. G. Yagola, *Non-Linear Ill-Posed Problems,* Vol. **1**, 2, Chapman and Hall, London, 1998.

[18] G. M. Vainikko, *Methods for Solving Linear Ill-Posed Problems in Hilbert Spaces,* Tartu University Press, Tartu, 1982 (in Russian).

[19] G. M. Vainikko and A. Yu. Veretennikov, *Iterative Procedures in Ill-Posed Problems,* Nauka, Moscow, 1986 (in Russian).

[20] V. V. Vasin and A. L. Ageev, *Ill-Posed Problems with A-priori Information,* VSP, Utrecht, 1995.

[21] V. A. Vinokurov, *Ill-Posed Problems in the Natural Sciences,* Mir, Moscow, 1987, pp. 53–69 (in Russian).

[22] A. G. Yagola and K. Yu. Dorofeev, *Sourcewise Representation and A Posteriori Error Estimates for Ill-Posed Problems,* in: Operator Theory and Its Applications, Fields Institute Communications, RI: AMS, Providence, **25**, 2000, pp. 543–550.

[23] A. G. Yagola and K. Yu. Dorofeev, *The Method of Extending Compacts for Solving Ill-posed Problems with Sourcewise Represented Solution,* Moscow Univ. Phys. Bull., **2** (1999), pp. 64–66 (in Russian).

Efficient Algorithms for Contact Shape Optimization

Zdeněk Dostál, Vít Vondrák, and John Rasmussen

Abstract. We review our work on the development of efficient algorithms for sensitivity analysis of contact problems with application to the solution of multi-body contact shape optimization problems solved by the gradient methods. The methods presented exploit a special structure of quadratic programming problems arising in dual formulation of the state problem to efficient implementation of semi-analytic sensitivity analysis. The results of numerical experiments are reported that illustrate the efficiency of the algorithms presented.

1. Introduction

In this paper, we review our work related to the development of effective algorithms for the solution of variational inequalities with particular stress on application to the sensitivity analysis and to the solution of differentiable contact shape optimization problems such as minimization of the compliance [1].

We start our exposition by describing the discretized conditions of equilibrium of the system of elastic bodies in contact as an indefinite quadratic programming (QP) problem in nodal displacements with a block diagonal stiffness matrix and general equality and inequality constraints. Then, we show that the difficulties arising from general inequality constraints and possible semi-definiteness of the problem in displacements can be essentially reduced by the application of the duality theory. The matrix of the dual quadratic form turns out to be positive definite with a spectrum that is more favorably distributed for the application of the conjugate gradient based methods than its primal counterpart. The algorithm and the corresponding theoretical results are then reviewed in Section 3.

In Section 4, we show that this algorithm complies well with the semi-analytic method [1, 8] for evaluation of the gradients of the cost function that are necessary for implementation of the feasible direction method. In particular, it turns out that the gradient may be evaluated with only one decomposition of the stiffness matrix, regardless of the number of design variables.

The algorithm has been implemented by the system ODESSY [7] developed by the Institute of Mechanical Engineering in Aalborg and tested on several model problems. Results of numerical experiments presented indicate that there are problems for which the algorithm presented is effective.

2. Discretized conditions of equilibrium

Consider a system of s elastic bodies, each one occupying, in a reference configuration, a domain $\Omega^p, 1 \leq p \leq s$ in $\mathbb{R}^d$, $d = 2, 3$ with sufficiently smooth boundary Γ^p. Suppose that each boundary Γ^p consists of three disjoint parts Γ^p_U, Γ^p_F and $\Gamma^p_C, \Gamma^p = \Gamma^p_U \cup \Gamma^p_F \cup \Gamma^p_C$, and that the displacements $\mathbf{U}^p : \Gamma^p_U \to \mathbb{R}^d$ and forces $\mathbf{F}^p : \Gamma^p_F \to \mathbb{R}^d$ are given. The part Γ^p_C denotes the part of Γ^p that may get into unilateral contact with some other subregion.

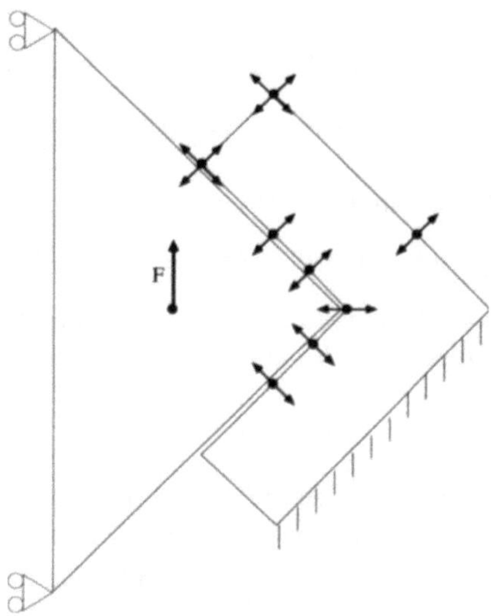

FIGURE 1. State problem with shape parametrization

The linearization of the contact conditions and the finite element discretization of $\Omega = \Omega^1 \cup \ldots \cup \Omega^s$ with suitable numbering of nodes result in the following QP problem

$$\min \frac{1}{2} u^T K u - f^T u \quad \text{subject to} \quad Nu \leq c \tag{1}$$

with a symmetric block-diagonal matrix $K = \text{diag}(K_1, \ldots, K_s)$ of order n, $f \in \mathbb{R}^n$, an $m \times n$ full rank matrix N, $m \leq n$, and $c \in \mathbb{R}^m$. The diagonal blocks K_p that correspond to the subdomains $\Omega^p, p = 1, \ldots, s$ are positive definite or semidefinite sparse matrices. Moreover, we assume that the nodes are numbered in such a way that $K_1, \ldots, K_s$ are banded matrices that can be effectively decomposed by the Cholesky factorization. The vector f describes the nodal forces arising from the volume forces or some other tractions. The matrix N and the vector c describe the linearized incremental non-interpenetration conditions. More details may be found in [6].

Even though (1) is a standard convex QP problem, its numerical solution may be expensive. The reasons are that K is typically ill-conditioned or singular, and that the feasible set is so complex that projections onto it can hardly be effectively computed, so that it would be very difficult to achieve fast identification of the contact interface and fast solution of auxiliary linear problems. These complications may be essentially reduced by applying the duality theory of convex programming (e.g. [3, 6]).

Let us first assume that the matrix K has a nontrivial null space that may also be used to define the natural coarse grid [12]. The Lagrangian associated with problem (1) is

$$L(u, \lambda) = \frac{1}{2}u^T K u - f^T u + \lambda^T (Nu - c), \tag{2}$$

so that we can rewrite the problem (1) as the saddle point problem

$$\text{Find} \quad (\hat{u}, \hat{\lambda}) \quad \text{such that} \quad L(\hat{u}, \hat{\lambda}) = \sup_{\lambda \geq 0} \inf_{u} L(u, \lambda). \tag{3}$$

If we eliminate u from (3), we shall get the minimization problem

$$\min \Theta(\lambda) \quad \text{s.t.} \quad \lambda \geq 0 \quad \text{and} \quad R^T(f - N^T\lambda) = 0, \tag{4}$$

where R denotes a matrix whose columns span the null space of K, $K^\dagger$ denotes any matrix that satisfies $KK^\dagger K = K$, and

$$\Theta(\lambda) = \frac{1}{2}\lambda^T N K^\dagger N^T \lambda - \lambda^T (N K^\dagger f - c). \tag{5}$$

Farhat and Roux proposed to define $K^\dagger$ as the left generalized inverse that satisfies $K_p^\dagger = K_p^{-1}$ whenever K_p is non-singular and $K^\dagger = \text{diag}(K_1^\dagger, \ldots, K_s^\dagger)$. The most important fact is that the product of such $K^\dagger$ with a vector may be carried out effectively by suitable combination of Cholesky and spectral decompositions [11] applied to each K_p. Once the solution $\hat{\lambda}$ of (4) is obtained, the vector u that solves (3) can be evaluated by means of explicit formulas that may be found in [3, 6]. The Hessian of Θ is closely related to that of the basic FETI method by Farhat and Roux, so that its spectrum is relatively favorably distributed for application of the conjugate gradient method.

Even though problem (4) is much more suitable for computations than (1) and has been used for efficient solution of contact problems [6], further improvement may be achieved by adapting simple observations and results of Farhat, Mandel and Roux [12]. It is even possible to achieve that the spectral condition number of the Hessian of the Lagrangian involved in computation is bounded independently of both the penalty parameter and the number of bodies [9].

If the stiffness matrix K is regular, than the same procedure leads to the dual problem

$$\min \Theta(\lambda) \quad \text{s.t.} \quad \lambda \geq 0. \tag{6}$$

3. Algorithm

The problem (6) comprises only bound constraints, so that efficient algorithms using projections and adaptive precision control [4] may be used. To apply this algorithm also for problem (4), we shall apply a variant of the augmented Lagrangian type algorithm proposed by Conn, Gould and Toint [2] for identification of stationary points for more general problems. The algorithm treats each type of constraints separately as the bound constrained QP problems. To solve our problem, the algorithm approximates the Lagrange multipliers for the equality constraints in (4) in the outer loop while QP problems with simple bounds are solved in the inner loop. However, the algorithm that we describe here is modified in order to exploit the specific structure of our problem.

To simplify our notation, let us denote $F = NK^{\dagger}N^T$, $G = R^T N^T$, and $d = R^T f$, and let us introduce the augmented Lagrangian with the penalization parameter ρ and the multiplier μ for the equality constraints for problem (4) by

$$L(\lambda, \mu, \rho) = \frac{1}{2}\lambda^T F\lambda - \lambda^T f + \mu^T(G\lambda - d) + \frac{1}{2}\rho||G\lambda - d||^2.$$

If we denote by $g = g(\lambda, \mu, \rho)$ the gradient of L with respect to λ, then the *projected gradient* $g^P = g^P(\lambda, \mu, \rho)$ of L at λ is given component-wise by

$$g_i^P = g_i \text{ for } \lambda_i > 0 \text{ or } i \notin I \text{ and } g_i^P = g_i^- \text{ for } \lambda_i = 0 \text{ and } i \in I$$

with $g_i^- = \min(g_i, 0)$, where I is the set of indices of constrained entries of λ.

All the parameters that must be defined prior to the application of the algorithm are listed in Step 0.

Algorithm 4.1. (Simple bound and equality constraints)

Step 0. *Initialization of parameters*
Set $0 < \alpha < 1$, $1 < \beta$, $\rho_0 > 0$, $\eta_0 > 0$, $M > 0$, μ^0, λ^0.
Step 1. Find λ^k so that $||g^P(\lambda^k, \mu^k, \rho_k)|| \leq M||G\lambda^k||$.
Step 2. If $||g^P(\lambda^k, \mu^k, \rho_k)||$ and $||G\lambda^k||$ are sufficiently small, then stop.
Step 3. If $||G\lambda^k|| \leq \eta_k$
Step 3a. then $\mu^{k+1} = \mu^k + \rho_k G\lambda^k$, $\rho_{k+1} = \rho_k$, $\eta_{k+1} = \alpha\eta_k$
Step 3b. else $\rho_{k+1} = \beta\rho_k$, $\eta_{k+1} = \eta_k$
end if.
Step 4. Increase k and return to Step 1.

An implementation of Step 1 is carried out by the minimization of the augmented Lagrangian L subject to $\lambda \geq 0$. An efficient algorithm for the solution of convex QP problems with simple bounds can be found in [4]. The algorithm reported in [4] uses projections and a precision control that depends on a prescribed positive parameter Γ. It has been proved that the algorithm converges to the solution for any positive Γ and that the solution is obtained in a finite number of steps provided that Γ is sufficiently large or the problem is not dual degenerate. This particular feature indicates that the algorithm can avoid the oscillations often attributed to the active set based algorithms and that it can efficiently treat the dual degenerate problems. The algorithm may be applied directly to problem (6).

The proposed algorithm has been proved [5] to converge for any set of parameters that satisfy the prescribed relations. Moreover, an estimate of the rate of convergence of the approximations of the Lagrange multipliers has been proved that does not have any term that accounts for inexact solution of the bound constrained problems that are solved in Step 1, and it was proved that the penalty parameter is uniformly bounded. These results give theoretical support to Algorithm 4.1.

4. Discretized contact shape optimization problem

Let us now assume that the shape of the first region Ω^1 depends on a vector of design variables α, so that the energy functional will have the form

$$j(\mathbf{u}, \alpha) = \frac{1}{2}\mathbf{u}^T\mathbf{K}(\alpha)\mathbf{u} - \mathbf{f}^T(\alpha)\mathbf{u}, \tag{7}$$

where the stiffness matrix $\mathbf{K}(\alpha)$ and possibly the vector of nodal forces $\mathbf{f}(\alpha)$ depend on α. The matrix $\mathbf{N}$ and the vector $\mathbf{c}$ that describe the linearized incremental conditions of non-interpenetration also depend on α, so that the solution $\mathbf{u}(\alpha)$ of the contact problem with the region $\Omega^1 = \Omega^1(\alpha)$ satisfies

$$\mathbf{u}(\alpha) = \arg\min\{J(\mathbf{u}, \alpha) : u \in C(\alpha)\}, \tag{8}$$

where

$$C(\alpha) = \{\mathbf{u} : \mathbf{N}(\alpha)\mathbf{u} \le \mathbf{c}(\alpha)\}.$$

We shall consider the contact shape optimization problem to find

$$\min\{J(\alpha) : \alpha \in D_{adm}\}, \tag{9}$$

where $J(\alpha)$ is the cost functional that defines the cost function for design of body $\Omega^1(\alpha)$. The set of admissible design variables D_{adm} defines all feasible designs. For the rest of the paper, we shall consider the cost functional $J(\alpha) \equiv -j(\mathbf{u}, \alpha)$ that defines the minimal compliance problem. The set of admissible design parameters will be defined by

$$D_{adm} = \{-l \le \alpha \le r : \mathrm{vol}(\Omega(\alpha)) \le \mathrm{vol}(\Omega(0))\}, \tag{10}$$

where l, r are given vectors with non-negative entries that define bounds on the design variables, and vol(.) is a mapping that assigns to each domain its volume. It has been proved that the minimal compliance problem has at least one solution and that the functional $J(\mathbf{u}, \alpha)$ considered as a function of α is differentiable under reasonable assumptions [1].

If we want to exploit differentiability of problem (9), we must evaluate effectively the partial derivatives of u with respect to the design variables $\alpha_1, \dots, \alpha_k$. It seems that the most straightforward method to get approximations of these partial derivatives is the application of the forward finite difference method that reads

$$\frac{\partial u(\alpha)}{\partial \alpha_i} \approx \frac{1}{h}\left(u(\alpha_1, ..., \alpha_i + h, ..., \alpha_k) - u(\alpha_1, ..., \alpha_i, ..., \alpha_k)\right), \tag{11}$$

where $h > 0$ is a suitable parameter. To use the formula (11), it is necessary not only to choose properly the perturbation parameter h, but also to solve at each design step $k + 1$ minimization problems in (4) with varying stiffness matrices K_1 which must be decomposed $k + 1$ times.

Let us show that both drawbacks may be relieved by the application of the semi-analytic sensitivity analysis [1]. Let us denote $I = \{1, ..., m\}$ the set of indices of the Lagrange multipliers λ, $I_s = \{i \in I : N_{ij}(\alpha)u_j(\alpha) = d_i(\alpha) \wedge \lambda_i(\alpha) > 0\}$ the set of indices that correspond to couples of nodes in strong contact, and $I_w = \{i \in I : N_{ij}(\alpha)u_j(\alpha) = d_i(\alpha) \wedge \lambda_i(\alpha) = 0\}$ the set of indices that correspond to couples of nodes in weak contact. We have used the standard summing convention. Analysis of the Karush-Kuhn-Tucker conditions [1] enables to evaluate the directional derivative $u'(\alpha, \beta)$ in the direction β by solving the quadratic programming problem

$$\min_{\substack{N_w(\alpha)z \le d_w(\alpha,\beta) \\ N_s(\alpha)z = d_s(\alpha,\beta)}} \frac{1}{2} z^T K(\alpha) z - z^T(f'(\alpha, \beta) - K'(\alpha, \beta)u(\alpha) + N'^T(\alpha, \beta)\lambda(\alpha)), \tag{12}$$

where $K'(\alpha, \beta)$, $f'(\alpha, \beta)$ and $N'(\alpha, \beta)$ denote computable directional derivatives of the stiffness matrix, traction vector and the constraint matrix, respectively. Matrices $N_w(\alpha)$, and $N_s(\alpha)$ are formed by the rows of the matrix $N(\alpha)$ with the indices that belong to I_w and I_s, respectively. Similarly, the vectors $d_w(\alpha, \beta)$ and $d_s(\alpha, \beta)$ are formed by the entries of $d'(\alpha, \beta) - N'(\alpha, \beta)u(\alpha)$ with indices in I_w and I_s, respectively. Solving (12) for $\beta = e_i$, $i = 1, ..., m$, where e_i are the standard unit vectors, we evaluate the gradient of the state problem. Denoting $\bar{f}(\alpha, \beta) = f'(\alpha, \beta) - K'(\alpha, \beta)u(\alpha) + N'^T(\alpha, \beta)\lambda(\alpha)$, we can see that problem (12) has the same structure as problem (1).

Similarly as in Section 2, we can rewrite (12) in dual form

$$\min_{\substack{\mu_w \ge o \\ \bar{E}(\alpha)\mu = \bar{e}(\alpha,\beta)}} \frac{1}{2}\mu^T \bar{N}(\alpha) K^\dagger(\alpha) \bar{N}(\alpha)^T \mu - \mu^T\left(\bar{N} K^\dagger(\alpha) \bar{f}(\alpha, \beta) - \bar{d}(\alpha, \beta)\right), \tag{13}$$

where

$$\bar{N}(\alpha) = \begin{bmatrix} N_w(\alpha) \\ N_s(\alpha) \end{bmatrix}, \quad \bar{d}(\alpha, \beta) = \begin{bmatrix} d_w(\alpha, \beta) \\ d_s(\alpha, \beta) \end{bmatrix} \quad \text{and} \quad \mu = \begin{bmatrix} \mu_w \\ \mu_s \end{bmatrix}. \tag{14}$$

Now solvability requires $\bar{E}(\alpha)\mu = \bar{e}(\alpha, \beta)$, where $\bar{E}(\alpha) = R(\alpha)^T \bar{N}(\alpha)^T$ and $\bar{e}(\alpha, \beta) = R(\alpha)^T \bar{f}(\alpha, \beta)$. The gradient then may be evaluated by modification of the formulae of [3, 6].

The key point is that the semi-analytic sensitivity analysis based on the dual formulation (13) requires only one assembly and decomposition of the stiffness matrix. Moreover, the semi-analytic sensitivity analysis does not require any potentially troublesome perturbation parameter. More information may be found in [1, 14, 8]. In the next section, we use the directional derivatives in the shape optimization by simple sequential linearization. More sophisticated procedures may be found in Kirsch [13] or Fancello and Feijóo [10].

5. Numerical experiments

We have tested our algorithm on the solution of a simple model problem to find the shape of the lower part of the right body of the system of elastic bodies in Figure 1 so that the compliance of the system is minimal while the volume of the modified upper body does not exceed the volume of the body in the initial design. The system has been discretized by the finite element method using 1635 degrees of freedom with 24 couples of nodes that may get in contact.We compared the semi-analytic method with the finite difference method. The cost function was defined by negative value of the total potential energy of the system of bodies, i.e. $-E(u(\alpha)) = -\frac{1}{2}u^T(\alpha)K(\alpha)u(\alpha) + u^T(\alpha)f(\alpha)$. This optimization problem is called "compliance problem". The admissible shape of bodies was restricted by the box constraints on the design variables (depicted in Figure 1) and by the bound on the volume specified above. The initial and optimized designs are displayed on Figures 2a and 2b.

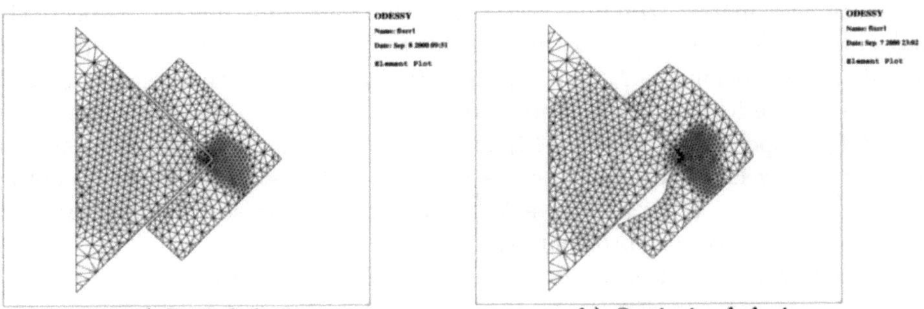

a) Initial design b) Optimized design

FIGURE 2. Comparison of shape designs

The results obtained by the application of the semi-analytic approach (SA) and the finite differences (OFD) can be found in Table 1. We recorded separately data concerning the solution of the state problems that we refer to as analysis, data concerning the solution that involves the solution of the state problems and the sensitivity analysis, and the total solution time that includes everything from the mesh generation to the solution. The number of designs that was necessary to find the approximation to the optimal design by the semi-analytic method differs very little from the number of designs that was necessary to find a very

	SA	OFD, h=8.9e-6	Speedup
# designs	30	28	
Starting cost function	5.9627	5.9627	
Final cost function	0.0952	0.0956	
Total time	498	1393	2.8
Total solution time	96.58	622.75	6.45
Total analysis time	60.97	55.88	0.92
Total sensitivity analysis time	35.61	566.87	15.92
Avg. time per DV	0.12	2.02	16.83
Total cg steps	2568	20029	7.8
Total analysis cg	1936	1804	0.93
Total sensitivity analysis cg	632	18225	28.84
Avg. cg steps per DV	2.11	65.09	30.85

TABLE 1. Performance of sensitivity analysis methods

similar approximation to the optimal design by the finite diference methods. This indicates that both methods perform similarly in terms of accuracy. However, the semi-analytic method turns out to be much more efficient in the sensitivity analysis both in time and numbers of the conjugate gradient steps. In particular, the average number of the conjugate gradient steps per one design variable was approximately 30 times larger in case of the finite difference method than in case of semi-analytical method! The explanation is that the number of the inequality constraints in the auxiliary QP problems (13) solved in the semi-analytic sensitivity analysis is very small as each inequality corresponds to a couple of nodes in a weak contact, so that it is very easy to identify the active set of the solution.

Comments and conclusions

The proposed algorithms for duality-based solution of coercive and semicoercive quadratic programming problems has been applied to the minimization of the compliance of a system of elastic bodies. Theoretical results [9] guarantee the convergence and the robustness of the method. The method has been applied to minimization of compliance of a system of elastic bodies and the efficiency of the method has been confirmed by results of a numerical experiment. The salient feature of the algorithms presented is essentially the reduction in the cost of decomposition in preparing domain decomposition based solutions for the state variational inequality that is enabled by the special structure of the problem considered. The algorithm may be even more efficient in a parallel environment and is closely related to the FETI method.

Acknowledgements

This research has been supported by the grant GA ČR No. 101/98/0535.

References

[1] J. Haslinger and P. Neittaanmäki, *Finite element approximation for optimal shape, material and topology design,* John Wiley and Sons, London, 1996.

[2] A. R. Conn, N. I. M. Gould, and Ph. L. Toint, *A globally convergent augmented Lagrangian algorithm for optimization with general constraints and simple bounds,* SIAM J. Numer. Anal., **28** (1991), p. 545–572.

[3] Z. Dostál, *Duality based domain decomposition with proportioning for the solution of free boundary problems,* J. Comput. Appl. Math., **63** (1995), pp. 203–208.

[4] Z. Dostál, *Box constrained quadratic programming with proportioning and projections,* SIAM J. Optim., **7** (1997), pp. 871–887.

[5] Z. Dostál, A. Friedlander, and S. A. Santos, *Augmented Lagrangians with adaptive precision control for quadratic programming with simple bounds and equality constraints,* to appear in SIAM J. Optim.

[6] Z. Dostál, A. Friedlander, and S. A. Santos, *Solution of coercive and semicoercive contact problems by FETI domain decomposition,* Contemporary Math., **218** (1998), pp. 82–93.

[7] ODESSY System for Optimization, IME Aalborg University, http://www.ime.auc.dk/afd3/odessy/manuals/index.htm

[8] V. Vondrák, Z. Dostál, and J. Rasmussen, *FETI domain decomposition algorithms for sensitivity analysis in contact shape optimization,* Proceedings of 11th International Conference Domain Decomposition in Science and Engineering, Choi-Hong Lai, P. E. Bjorstad, M. Cross, and O. B. Widlund, eds., Domain Decomposition Press, Bergen, 1999, pp. 561–567.

[9] Z. Dostál, F. A. M. Gomes, and S. A. Santos, *Solution of contact problems by FETI domain decomposition with natural coarse-space projections,* to appear in Comput. Methods Appl. Mech. Engrg.

[10] E. A. Fancello and R. Feijóo, *Shape optimization in frictionless contact problems,* Internat. J. Numer. Methods Engrg., **37** (1997), pp. 2311–2335.

[11] C. Farhat and M. Gérardin, *On the general solution by a direct method of a large scale singular system of linear equations: application to the analysis of floating structures,* Internat. J. Numer. Methods Engrg., **41** (1997), pp. 675–696.

[12] C. Farhat, J. Mandel, and F.-X. Roux, *Optimal convergence properties of the FETI domain decomposition method,* Comput. Methods Appl. Mech. Engrg., **115** (1994), pp. 365–385.

[13] U. Kirsch, *Efficient sensitivity analysis for structural optimization,* Comput. Methods Appl. Mech. Engrg., **117** (1994), pp. 143–156.

[14] V. Vondrák, Ph.D. Thesis, VSB-Technical University Ostrava, in preparation.

Fast Algorithms for Positional Optimization of Dynamic Systems

Rafail Gabasov and Faina M. Kirillova

Abstract. The canonical optimal control problem for a linear time-varying dynamic system in the class of discrete controls is under consideration. Using principles of the adaptive method of linear programming, algorithms of open-loop and close-loop optimization are described. Results are illustrated by a fourth order problem.

1. Introduction

The modern statements of optimization problems of dynamic systems have been studied since the 40s. They include both parametric optimization and optimal control. In the paper only optimal control problems will be considered.

As is known there are two principles of control for dynamic systems: 1) open-loop control, 2) closed-loop control. Accordingly, in optimal control theory optimal open-loop control and optimal feedback control are distinguished. In many situations engineers prefer feedback control rather than open-loop control and it was the main reason why the first results on optimal control obtained by engineers at the end of the 40s concerned optimal feedback control systems. Success of engineers in solving the first optimal control problems attracted mathematicians' attention to a new type of extremal problems. Their variational interpretation gave the main direction to the development of optimal control theory due to which almost all efforts of scientists were focused on proving necessary optimality conditions for open-loop controls. The Pontryagin maximum principle represents the most important result of these efforts [1].

At present, after almost 50 years of development of optimal control theory the intensity of investigations in the theory is essentially reduced. One may believe that the main problems in this direction were solved and we can give the positive answer to Dantzig's criterion of a theory's value [2]:

The final test of a theory is its capacity
to solve the problems which originated it.

In reality, the situation is not so optimistic. In fact, hundreds of theorems on necessary optimality conditions have been proved, all conceivable properties of control systems have been studied and linear optimal control problems were investigated in great detail. A lot of papers are concerned with numerical solving of various optimal control problems, important applications of algorithms are made [3 – 9]. But in spite of numerous successful solutions of many applied problems of optimal control, it is difficult to find an effective numerical method in which even for linear optimal control problems their specific features are taken into account. In particular, this concerns optimal positional solutions (optimal feedback controls). A number of important problems of constructive theory of optimal control have not been solved yet. Among them one have to mention 1) reliable fast algorithms for optimal open-loop controls; 2) effective realization of optimal feedbacks.

The aim of this paper is to describe some results obtained in Minsk (Belarus) on the two mentioned problems.

The program of our investigation on constructive theory of optimal control started in the early 70s and was based on the simple logical analysis. A modern extremal problem includes nonlinear systems, partial and ordinary differential equations, conflict situation, vector cost functions, incomplete information, uncertainties. It is clear that effective algorithms for the solution of this problem must be effective if we use them to solve 1) optimization problems with one cost function; 2) optimization problems for ordinary differential equations; 3) optimization problems for linear ordinary differential equations. Continuing the process of selecting more simple problems one inevitably comes to 4) optimization problems for linear discrete processes. If a discrete (multistage) process consists of one stage, we obtain a static object of optimization. Thus, the logic in consideration inevitably leads to 5) linear programming.

Analysis of the classical simplex-method (the most interesting method at the 70s) showed that the simplex-method possessed a number of lacks that were real obstacles at using in optimal control theory. At the end of the 70s a new method of linear programming was created in Minsk, called the adaptive method. The ideas of the adaptive method were used to construct algorithms for the solution of nonlinear programming and optimal open-loop control problems [10]. In the 50–70s the optimal feedback control problem seemed to be an extremely difficult problem and elaborating algorithms we kept in our mind only optimal open-loop solutions. At the end of the 80s after analysis of the classical statements of optimal synthesis problem we concluded that the problem in this form was not solvable and suggested a new concept of solution of the optimal feedback control problem which differed from the traditional one and used an idea of the solution in real-time mode with the help of modern computer tools [11]. A new approach has been used for solving applied problems which had no extremal form (stabilization, amortization, damping and so on).

Considering the optimal synthesis problem we came to the problem of constructing fast algorithms for optimization of dynamic systems. At first, fast algorithms for linear dynamic systems were justified. On this base, algorithms for the fast optimization of nonlinear dynamic systems were created [10]. The paper only deals with linear dynamic systems.

2. Statement of the problems. Optimal open-loop solutions

Consider a linear (canonical) optimal control problem

$$c'x(t^*) \to \max, \quad \dot{x} = A(t)x + b(t)u, \quad x(t_*) = x_0, \tag{1}$$

$$Hx(t^*) = g, \quad |u(t)| \leq 1, \quad t \in T = [t_*, t^*].$$

Here, $x = x(t) \in R^n$ is a state vector of (1), $u = u(t) \in R$ is a control variable, $g \in R^m$, $\operatorname{rank} H = m < n$; $A(t)$, $b(t)$, $t \in T$, are piecewise-continuous matrix and vector functions.

Problem (1) is the simplest problem of optimal control theory but it is nontrivial and includes the main elements of typical optimal control problems.

At first, consider the problem of constructing open-loop solutions in the class of discrete controls.

Definition 2.1. *A function $u(t)$, $t \in T$, is said to be discrete (with a quantization period $h = (t^* - t_*)/N$) if it has the form*

$$u(t) = u(t_* + kh), \quad t \in [t_* + kh, t_* + (k+1)h[, \quad k = \overline{0, N-1}.$$

The use of discrete controls as admissible functions is quite natural because it is impossible to solve the problem in question without digital computers.

Definition 2.2. *A discrete control $u(t)$, $t \in T$, is called admissible if it together with a trajectory $x(t)$, $t \in T$, of system (1) satisfies the geometrical constraint*

$$|u(t)| \leq 1, \quad t \in T,$$

and the terminal condition

$$x(t^*) \in X^* = \{x \in R^n : Hx = g\}.$$

Definition 2.3. *An admissible control $u^0(t)$, $t \in T$, is said to be an optimal open-loop control of problem (1) if the optimal trajectory $x^0(t)$, $t \in T$, satisfies the equality*

$$c'x^0(t^*) = \max c'x(t^*).$$

The open-loop solutions allow to introduce a positional solution. Let $\tau \in T_h = \{t_*, t_* + h, \ldots, t^* - h\}$, $z \in R^n$. Imbed problem (1) into the family of problems

$$c'x(t^*) \to \max, \quad \dot{x} = A(t)x + b(t)u, \quad x(\tau) = z, \tag{2}$$

$$Hx(t^*) = g, \quad |u(t)| \leq 1, \quad t \in T(\tau) = [\tau, t^*],$$

depending on the position (τ, z).

Denote: $u^0(t|\tau, z)$, $t \in T(\tau)$, is an optimal open-loop control of problem (2), X_τ is a set of vectors $z \in R^n$ for which problem (2) has solutions at fixed $\tau \in T_h$.

Definition 2.4. *A function*

$$u^0(\tau, z) = u^0(\tau|\tau, z), \quad z \in X_\tau, \quad \tau \in T_h, \tag{3}$$

is said to be an optimal feedback (discrete) control of problem (1).

We close system (1) by feedback (3) and consider its behaviour under disturbance $w(t)$, $t \in T$:

$$\dot{x} = A(t)x + b(t)u^0(t,x) + w(t), \quad x(t_*) = x_0. \tag{4}$$

We define the trajectory $x(t)$, $t \in T$, of (4) as a continuous solution of the linear equation

$$\dot{x} = A(t)x + b(t)u^*(t) + w(t), \quad x(t_*) = x_0,$$

where

$$u^*(t) = u^0(t_* + kh, x(t_* + kh)),\ t \in [t_* + kh, t_* + (k+1)h[,\ k = \overline{0, N-1}.$$

The classical statement of the optimal synthesis problem assumes the construction of function (3) in an explicit form. The history of optimal control theory shows that the problem in that statement has not been solved even for the simplest problem (1).

Suppose that optimal feedback (3) has been constructed. Let us analyze how it is used in a concrete control process.

We consider the behaviour of closed system (4) where an unknown disturbance $w^*(t)$, $t \in T$, is realized. Let $x^*(t)$, $t \in T$, be the realizing trajectory of the closed system

$$\dot{x}^*(t) = A(t)x^*(t) + b(t)u^0(t, x^*(t)) + w^*(t), \quad t \in T, \quad x(t_*) = x_0^*. \tag{5}$$

From (5), one can observe that in the course of process the signals $u^*(t) = u^0(t, x^*(t))$, $t \in T$, are fed to the input of the control system. We call these signals a realization of the optimal feedback. It is clear that there is no need to know the optimal feedback (3) in all the domain of the definition, it is sufficient to get its values along the isolated trajectory $x^*(t)$, $t \in T$. Moreover, it is sufficient at each current instant $\tau \in T_h$ to calculate only its current values $u^*(\tau) = u^0(\tau, x^*(\tau))$ for the time which does not exceed h, i.e. it is sufficient to calculate this current value in real time mode.

A device which is able to fulfill this work is called Optimal Controller.

Thus, the optimal synthesis problem is reduced to constructing an algorithm for Optimal Controller.

Before describing an algorithm for Optimal Controller let us explain how to construct optimal open-loop solutions. In the class of discrete controls problem (1) is equivalent to the linear programming problem

$$\sum_{t \in T_h} c(t)u(t) \to \max, \quad \sum_{t \in T_h} d(t)u(t) = g_0, \quad |u(t)| \leq 1, \quad t \in T_h. \tag{6}$$

Here

$$c(t) = \int_t^{t+h} \psi_c'(\vartheta)b(\vartheta)d\vartheta, \quad d(t) = \int_t^{t+h} G(\vartheta)b(\vartheta)d\vartheta, \quad g_0 = g - Hx_0(t^*);$$

$x_0(t)$, $t \in T$, is a trajectory of (1) with $u(t) = 0$, $t \in T$; $\psi_c(t)$, $t \in T$, is a solution to the adjoint equation

$$\dot{\psi} = -A'(t)\psi, \quad \psi(t^*) = c; \tag{7}$$

$G(t)$, $t \in T$, is and $m \times n$-matrix function which is a solution of the equation

$$\dot{G} = -GA(t), \quad G(t^*) = H. \tag{8}$$

At small h the matrix of the condition of problem (6) $(d(t), t \in T_h)$ is large and has a high density. For that reason traditional methods of linear programming for (6) are not quite effective if h is small.

To solve problem (6) effectively a special dynamic realization of adaptive method is suggested.

The main tool of the method is a support T_{sup}. This is a totality $\{t_1, \dots, t_m\}$ consisting of m moments of the discrete interval T_h.

Definition 2.5. *A totality T_{sup} is said to be a support if the matrix $D_{sup} = \Big(d(t), t \in T_{sup}\Big)$ is nonsingular. The matrix D_{sup} is called the support matrix.*

The support matrix can be constructed using one integration of primal (1) or adjoint (7) system if m processors are used.

Every support T_{sup} is accompanied by

1) the vector of Lagrange multipliers $\nu = \nu(I)$, $I = \{1, 2, \dots, m\}$ that is a solution of the equation

$$\nu' D_{sup} = c'_{sup}, \quad c_{sup} = (c(t), \ t \in T_{sup}),$$

2) co-trajectory $\psi(t)$, $t \in T$, that is a solution of

$$\dot{\psi} = -A'(t)\psi, \quad \psi(t^*) = c - H'\nu, \tag{9}$$

3) co-control

$$\Delta(t) = \int_t^{t+h} \psi'(\vartheta) b(\vartheta) d\vartheta, \quad t \in T_h, \tag{10}$$

4) pseudocontrol $\omega(t)$, $t \in T$. Nonsupport values $\omega(t)$, $t \in T_{nonsup} = T_h \backslash T_{sup}$, are defined as

$$\omega(t) = -1 \text{ at } \Delta(t) < 0; \quad \omega(t) = 1 \text{ at } \Delta(t) > 0;$$
$$\omega(t) \in [-1, 1] \text{ at } \Delta(t) = 0, \quad t \in T_{nonsup}.$$

Support values $\omega(t)$, $t \in T_{sup}$, of pseudocontrol satisfy the equation

$$\sum_{t \in T_{sup}} d(t)\omega(t) + \sum_{t \in T_{nonsup}} d(t)\omega(t) = g_0.$$

5) quasicontrol $\tilde{\omega}(t)$, $t \in T$: $\tilde{\omega}(t) = \omega(t)$, $t \in T_{nonsup}$; $\tilde{\omega}(t) = \text{sat}\,\omega(t)$, $t \in T_{sup}$.

Any quasicontrol $\tilde{\omega}(t)$, $t \in T$, produces a discrepancy of endpoint constraints $\tilde{g}(T_{sup}) = \|g - H\tilde{x}(t^*)\|$ where $\tilde{x}(t)$, $t \in T$, is a trajectory of system (1) generated by $\tilde{\omega}(t)$, $t \in T$.

The support T_{sup} is said to be regular if $\Delta(t) \neq 0$, $t \in T_{nonsup}$.

The support T_{sup} is called optimal if it is accompanied by some pseudocontrol which satisfies the condition

$$|\omega(t)| \leq 1, \quad t \in T_{sup}. \tag{11}$$

In case (11), $\omega(t)$, $t \in T$, is the optimal open-loop control.

On the base of the introduced concepts several methods of constructing optimal open-loop controls for problem (1) are created . We describe a dual method of the solution to problem (6). This method is based on constructing the optimal support T^0_{sup}.

Suppose that a support T_{sup} does not satisfy condition (11). An iteration of the dual method is a change of an "old" support T_{sup} by a "new" support $\bar{T}_{sup}$ so that

$$\tilde{g}(\bar{T}_{sup}) \leq \tilde{g}(T_{sup}).$$

An instant $t^0 \in T_{sup}$ satisfied the relation

$$|\omega(t^0)| = \max_{t \in T_{sup}} |\omega(t)|$$

is eliminated from the support.

In the procedure of changing the support, the following information is used:

1. a set of nonsupport zeroes

$$T_{n0} = \{t \in T_{nonsup} \backslash t_* : \Delta(t-h)\Delta(t) < 0\};$$

2. a set

$$T_{sn} = T_{sup} \cup T_{n0} \cup \{t_*, t^*\} = \{t_k,\ k \in K \cup k^* + 1\}, \quad K = \{0, 1, \dots, k^*\};$$

and intervals of constant sign of a co-control:

$$T_k = \{t_{*k} = t_k, t_k + h, \dots, t^*_k = t_{k+1} - h\}, \text{ if } t_k \notin T_{sup};$$
$$T_k = \{t_{*k} = t_k + h, t_k + 2h, \dots, t^*_k = t_{k+1} - h\}, \text{ if } t_k \in T_{sup};$$

3. a value

$$\gamma = \begin{cases} \operatorname{sign} \Delta(t_*), & \text{if } t_* \notin T_{sup}; \\ \operatorname{sign} \Delta(t_* + h), & \text{if } t_* \in T_{sup}; \end{cases}$$

4. a vector

$$p = \gamma \sum_{k=0}^{k^*} (-1)^k \sum_{t \in T_k} d(t) = \gamma \sum_{k=0}^{k^*} (-1)^k \int_{t_{*k}}^{t^*_k + h} G(\vartheta) b(\vartheta) d\vartheta.$$

In the course of an iteration the following information is stored and transformed:

1. a support T_{sup};
2. a set of nonsupport zeroes T_{n0};
3. a support matrix D_{sup};
4. values $G(t)$, $\psi_c(t)$, $t \in T_{sn} \backslash t^*$;
5. a vector p;
6. a number γ, support values $\omega(t)$, $t \in T_{sup}$;
7. the Lagrange vector ν.

A general scheme of an algorithm of constructing a new support $\bar{T}_{sup}$ consists in the following:

1. A variation $\Delta\nu$ of the Lagrange vector is calculated;
2. A rate of change of the dual control criterion is calculated;

3. A short step along the direction $\Delta\nu$ providing a new zero of the varied co-control is calculated. This step causes a positive jump of the dual control criterion;
4. Transform the information stored in the computer memory;
5. Repeat operations 2—4 till the value of the rate of the change of the dual criterion reaches a nonnegative value;
6. The element t^0 of the support T_{sup} is changed by the last obtained zero $\tilde{t}^{K_0}$ of the varied co-control.

The main idea of the proposed realization concerns with operations 3, 4 of the above scheme.

The variation (direction of changing) $\Delta\nu$ of the Lagrange vector is calculated as a solution to the system of linear equations

$$-D'_{sup}\Delta\nu = (\Delta\delta(t), t \in T_{sup}), \tag{12}$$

where $\Delta\delta(t^0) = \operatorname{sign}\omega(t^0)$; $\Delta\delta(t) = 0$, $t \in T_{sup}\backslash t^0$. Set

$$\Delta\delta(t) = -\Delta\nu' d(t) =$$

$$= -\int_t^{t+h} \Delta\nu' G(\vartheta) b(\vartheta) d\vartheta, \ t \in T_n \text{ (variation of the co-control)};$$

$$\Delta(t,\sigma) = \Delta(t) + \sigma\Delta\delta(t), \quad t \in T_u; \quad \sigma \geq 0 \text{ (varied co-control)}. \tag{13}$$

Further, we assume that $\Delta\delta(t_{*k})\Delta\delta(t^*_{k-1}) > 0$, if $t_{*k} \in T_{n0}$, $k = \overline{1, k^*}$.

Denote by $\varphi(\sigma)$ a value of the control criterion of the problem dual to (6) at function (13). According to [10], in a small right-side neighborhood of the point $\sigma = 0$ the function $\varphi(\sigma)$, $\sigma \geq 0$, is linear and decreases with the rate $\alpha^1 = 1 - |\omega(t^0)| < 0$. Such behaviour remains while σ increases from $\sigma = 0$ till the value $\sigma^1 > 0$ when a new zero $t(\sigma^1)$ of function (13) arises: $\Delta(t(\sigma^1), \sigma^1) = 0$. When $\sigma = \sigma^1$ the rate of the change of the dual control criterion makes a positive jump. For changing a support, a sequence of short steps σ^k, $k = \overline{1, K_0}$, when the rate makes jumps, is formed. We calculate a long step $\sigma^* = \sigma^{K_0}$, after which the rate of the change of the dual control criterion becomes nonnegative. With the direct realization of the adaptive method [10] for detecting σ^k, $k = \overline{1, K_0}$, it is necessary to calculate steps $\sigma(t)$ for each $t \in T_n$. This requires enormous computational resources. In the proposed realization it is taken into account that with increasing σ a new zero of the varied co-control can appear only by one of the following ways:
1. At the point $t^0 + h$ or $t^0 - h$ as a result of movement of $t^0(\sigma)$, $\sigma \geq 0$: $\Delta(t^0(\sigma), \sigma) = 0$, $t^0(0) = t^0$.
2. At a point t or $t - h$ as a result of movement of a point $t(\sigma)$, $\sigma \geq 0$, $t \in T_{n0}$.
3. At the point t_* as a result of arising at the left end of the interval T.
4. At the point $t^* - h$ as a result of arising at the right end of the interval T.
5. Inside the interval T from a stationary point of the function

$$\tilde{\Delta}(t) = (\psi'_c(t) - \nu' G(t)) b(t), t \in T.$$

Thus, if we omit from consideration the case 5, which is studied according to the below scheme only not for the function $\tilde{\Delta}(t)$, $t \in T$, but for its derivative, it is sufficient to investigate the behaviour of the function $\Delta(t,\sigma)$ not on the whole set T_n but only for points 1–4.

Details of an iteration are given in the paper [12].

If the supports are regular (nondegenerate case) the optimal support is constructed for the finite number of iterations.

The problem of effectiveness of methods is principal in constructive theory of optimal control. It is not always reasonable to estimate the effectiveness of methods by the amount of iterations because different methods consist of different types of iterations. It is more natural to use as the measure of effectiveness the quantity of full integrations of the primal and the adjoint systems which are used to construct the optimal solution [13]. In this case, as a unit of complexity, one can take one integration on the whole segment T.

It is impossible to find formulae to estimate the complexity but computer experiments can give a certain representation on characteristics of the methods and on their properties.

3. Example. Open-loop controls

Here, we consider results of computer experiments fulfilled by Dr. N. V. Balashevich (Institute of Mathematics, NASB, Minsk). The optimization problem in question deals with the 4th-order dynamic system:

$$\int_0^{25} u(t)dt \longrightarrow \min, \quad \dot{x}_1 = x_3, \quad \dot{x}_2 = x_4,$$

$$\dot{x}_3 = -x_1 + x_2 + u, \quad \dot{x}_4 = 0.1x_1 - 1.02x_2 \tag{14}$$

$$x_1(0) = x_2(0) = 0,\ x_3(0) = 2,\ x_4(0) = 1,$$

$$x_1(25) = x_2(25) = x_3(25) = x_4(25) = 0, \quad 0 \le u(t) \le 1,\ t \in [0, 25[.$$

If we interpret $u(t)$ as a fuel consumption per second, then problem (14) is to stop oscillations of a two-mass system with the minimal fuel consumption (Figure 1). The problem was solved with the quantization period $h = 25/1000 = 0.025$.

As an initial support a set $T_{sup} = \{5, 10, 15, 20\}$ was taken. The optimal value of the cost function was equal to 6.331252. Movements of moments $t \in T_0 = T_{n0} \cup t^0$ in the course of iterations are pictured in Figure 2. Numbers near each jump of curves show movements of the corresponding moments during the iteration in h units. The described method admits two ways of realization [12] (economical and fast).

The complexity of the economical variant equals 2.433. One integration was used to construct D_{sup}, one to test optimality conditions and only 0.433 integration to obtain the optimal open-loop control. The complexity of the fast way in the example equals 2.231.

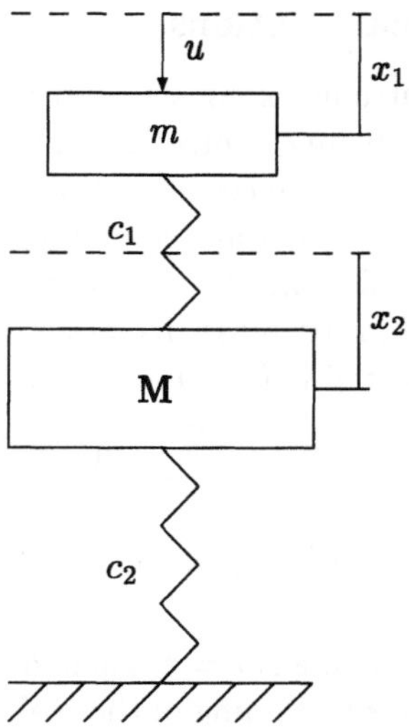

Fig. 1.

The table contains some information on solution to problem (14) for other quantization periods.

h	number of iterations	value of the control criterion	complexity	
			economical way	fast way
0.25	9	6.353339	2.64	2.41
0.0025	15	6.330941	2.4149	2.2018
0.001	20	6.330938	2.3912	2.23564

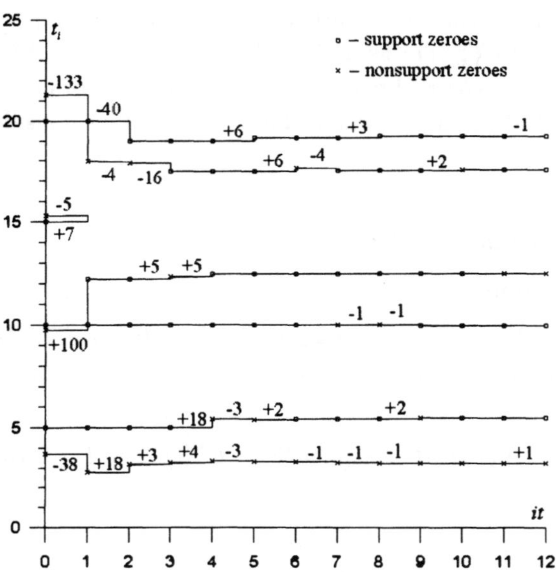

4. The optimal positional solutions

Suppose that Optimal Controller is constructed and has been acting during the time $\{t_*, t_* + h, \ldots, \tau\}$ producing control signals $u^*(t_*)$, $u^*(t_* + h)$, ..., $u^*(\tau)$. These signals and realized disturbances $w^*(t_*)$, $w^*(t_* + h)$, ..., $w^*(\tau)$ transfer the dynamic system under optimization at the moment $\tau + h$ in the state $x^*(\tau + h)$. The task of Optimal Controller at the moment $\tau + h$ is to calculate a current signal $u^*(\tau + h) = u^0(\tau + h, x^*(\tau + h))$. By assumption, at the previous moment τ Optimal Controller calculated $u^*(\tau)$, i.e. it solved the problem

$$\sum_{t \in T_h(\tau)} c(t)u(t) \to \max, \quad \sum_{t \in T_h(\tau)} d(t)u(t) = g - G(\tau)x^*(\tau), \tag{15}$$

$$|u(t)| \leq 1, \ t \in T_h(\tau),$$

and it knows $T^0_{sup}(\tau)$.

Problem (15) at initial moment $\tau = t_*$ coincides with (1) and can be solved by Optimal Controller before the beginning of the real control process to get $T^0_{sup}(t_*)$, $u^*(t_*)$. The current state $x^*(\tau + h)$ is connected with the previous state $x^*(\tau)$ by the formula

$$x^*(\tau + h) = F(\tau + h)F^{-1}(\tau)x^*(\tau) + \int_{\tau}^{\tau+h} F(\tau + h)F^{-1}(t)b(t)dtu^*(\tau)+$$

$$+\int_{\tau}^{\tau+h} F(\tau + h)F^{-1}(t)w^*(t)dt$$

Because the value $w^*(\tau)$ is bounded, the difference between two states $x^*(\tau)$ and $x^*(\tau + h)$ is small. Therefore the problem

$$\sum_{t \in T_h(\tau+h)} c(t)u(t) \to \max, \quad \sum_{t \in T_h(\tau+h)} d(t)u(t) = g - G(\tau + h)x^*(\tau + h),$$

$$|u(t)| \leq 1, \ t \in T_h(\tau + h),$$

that Optimal Controller have to solve at $\tau + h$ differs from (15) a little if h is small. At such situation the dual method under consideration is very effective. Optimal Controller uses $T^0_{sup}(\tau)$ as an initial support $T_{sup}(\tau + h)$, constructs the optimal support $T^0_{sup}(\tau + h)$ and calculates $u^*(\tau + h)$. To obtain a representation about possibility to fulfill calculations in the real time mode consider the previous example.

5. Example. Positional solutions

Let the perturbed system

$$\dot{x}_1 = x_3, \quad \dot{x}_2 = x_4, \quad \dot{x}_3 = -x_1 + x_2 + u, \quad \dot{x}_4 = 0.1x_1 - 1.02x_2 + w \tag{16}$$

be closed by the optimal feedback $u^* = u^0(x_1, \ldots, x_n)$ and be acting under a disturbance

$$w^*(t) = 0.3 \sin 4t, \ t \in [0, 9.75[; \quad w^*(t) \equiv 0, \ t \geq 9.75.$$

Optimal Controller does not know this disturbance but at each moment $\tau \in T_h$ it can measure a current state $x^*(\tau)$. Figure 3 a, b presents projections on the planes

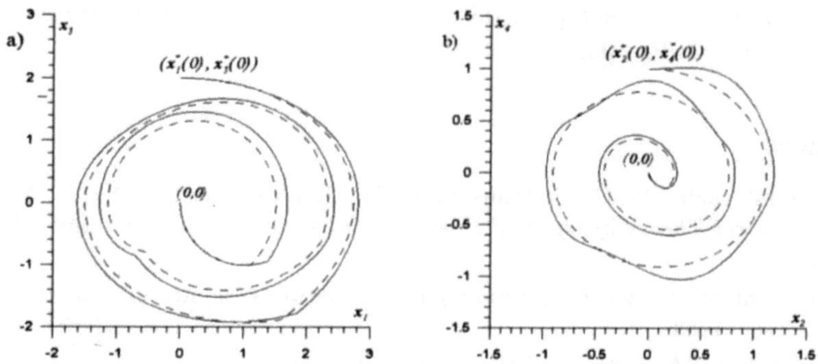

Fig. 3

x_1x_3, x_2x_4 of trajectories of system (14) closed by the optimal feedback. Dash lines are optimal open-loop trajectories, solid lines are trajectories of the system with the disturbance $w^*(t)$, $t \in T$. Movements of zeros $t_i(\tau) \in T_0(\tau)$, $\tau \in T_h$, are presented in Figure 4. It is seen from Figure 5 that in the course of control process new elements of the set $T_0(\tau)$ appear and disappear several times. The realization $u^*(t)$, $t \in T$, is given in Figure 5. Complexity of correction of current supports by

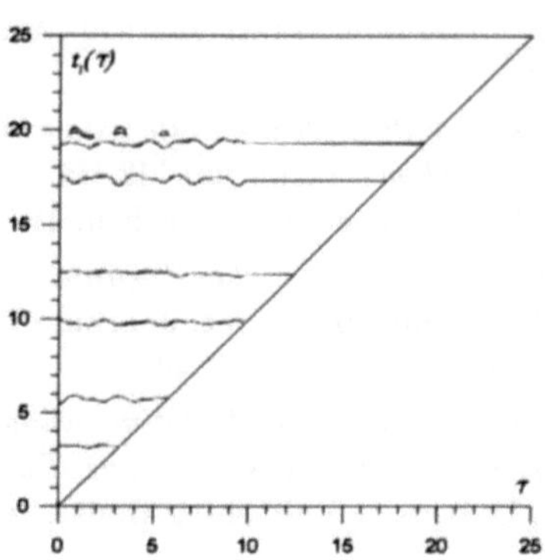

Fig. 4

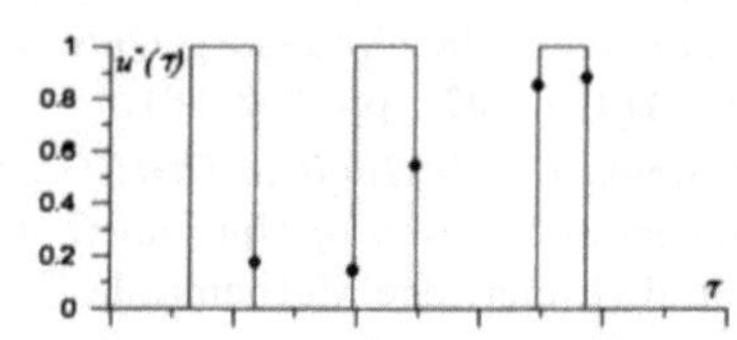

Fig. 5

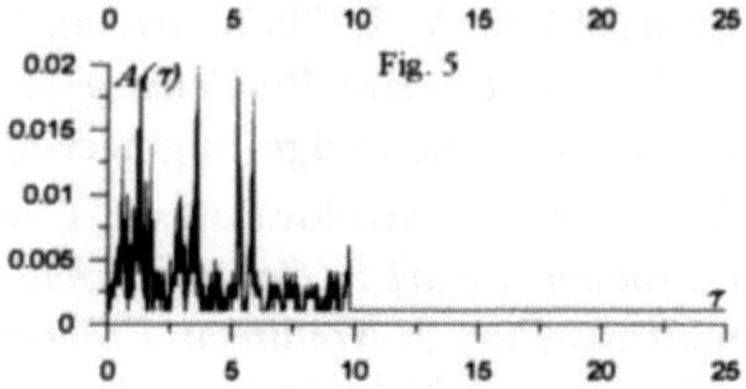

Fig. 6

the dual method is given in Figure 6. As it is seen from Figure 6 during the control process the complexity does not exceed 0.02. It means that computer calculated $u^*(\tau)$ for the time less than 0.02α where α is the computer time for one integration of system (16) on T. For many modern microprocessors the time 0.02α is less than the quantization period h if real time is measured, for example, in seconds.

The authors wish to thank Dr. N. V. Balashevich and Dr. N. M. Dmitruk for the help in preparing the paper.

This paper is partly supported by Belarusian Republican Foundation of Basic Research (grant F99R – 002).

References

[1] L. S. Pontryagin, V. G. Boltyanskii, R. V. Gamkrelidze, and Ye. F. Mischenko, *Mathematical Theory of Optimal Processes*, Interscience Publishers Inc., New York, 1962.

[2] G. B. Dantzig, *Linear Programming and Extentions*, Princeton University Press, New Jersey, 1963.

[3] R. Bulirsch, F. Montrone, and H. J. Pesch, *Abort Landing in the Presence of Windshear as a Minimax Optimal Control Problem, Part 1: Necessary Conditions*, Journal of Optimization Theory and Applications, **70(1)** (1991), pp. 1–23.

[4] R. Bulirsch, F. Montrone, and H. J. Pesch, *Abort Landing in the Presence of Windshear as a Minimax Optimal Control Problem, Part 2: Multiple Shooting and Homotopy*, Journal of Optimization Theory and Applications, **70(2)** (1991), pp. 223–254.

[5] H. J. Pesch, *Real-time Computation of Feedback Controls for Constrained Optimal Control Problems. Part 1: Neighbouring Extremals*, Optimal Control Applications and Methods, **10** (1989), pp. 129–145.

[6] H. J. Pesch, *Real-time Computation of Feedback Controls for Constrained Optimal Control Problems. Part 2: A Correction Method Based on Multiple Shooting*, Optimal Control Applications and Methods, **10** (1989), pp. 147–171.

[7] K. Chudej, *Realistic Modelled Optimal Control Problems in Aerospace Engineering – A Challenge to the Necessary Optimality Conditions*, Mathematical Modelling of Systems, **2(4)** (1996), pp. 252–261.

[8] O. von Stryk, *User's Guide to DIRCOL Version 2.0: A Direct Collocation Method for the Numerical Solution of Optimal Control Problems*, Lehrstuhl M2 Höhere Mathematik und Numerische Mathematik, Technische Universität München, 1999, 128 p.

[9] A. I. Tyatushkin, A. I. Zholudev, and N. M. Erinchek, *The Program System for Solving Optimal Control Problems with Phase Constraints*, Int. Journal of Software Engineering and Knowledge Engineering, **3(4)** (1993), pp. 487–497.

[10] R. Gabasov, F. M. Kirillova et al., *Constructive Methods of Optimization. Part 1: Linear Problems. Part 2: Control Problems. Part 3: Network Problems. Part 4: Convex Problems. Part 5: Nonlinear Problems*, University Publishing House, Minsk, Belarus, 1984, 1984, 1986, 1987, 1998.

[11] R. Gabasov, F. M. Kirillova, and S. V. Prischepova, *Optimal Feedback Control*, Lecture Notes in Control and Information Sciences, M. Thoma, ed., Springer-Verlag, **207**, 1995.

[12] R. Gabasov, F. M. Kirillova, and N. V. Balashevich, *Open-loop and Closed-loop Optimization of Linear Control Systems*, Asian Journal of Control, **2(3)** (2000), pp. 155–168.

[13] R. P. Fedorenko, *Approximate Solution of Optimal Control Problems*, Moscow, Nauka, 1973.

Optimal Sizing Using Automatic Differentiation

Gundolf Haase, Ulrich Langer, Ewald Lindner,
and Wolfram Mühlhuber

Abstract. This paper is devoted to optimal sizing. As a model example we consider the minimization of the mass of the frame of an injection moulding machine. The deformation of the frame is described by a generalized plane stress state with an elasticity modulus scaled by the thickness. This constrained nonlinear optimization problem is solved by sequential quadratic programming (SQP) which requires gradients of the objective and the constraints with respect to the design parameters.

As long as the number of design parameters is small, finite differences may be used. In order to handle also several hundreds of varying thickness parameters, we use the reverse mode of automatic differentiation for differentiating the function evaluation. This approach works fine but requires huge memory and disk capabilities. Furthermore, the use of iterative solvers for the governing state equations is limited. Therefore, we combine it with the adjoint method to get a fast and flexible gradient evaluation procedure.

The presented numerical results show the potential of this approach and imply that this method can also be used for finding an initial guess for a shape optimization.

1. Introduction

The design of a machine or of a mechanical structure has to fulfill various constraints in many industrial applications. In most cases, an optimal design subject to several restrictions is desired. These restrictions are mainly induced by technological limits, requirements of the individual customer or indirectly by competition on the world market. Unfortunately, due to lack of time, engineers designing a machine component have to stop their design process after a few iterations – in most cases only two or three. Then they take the best design obtained so far, because no more time is left for drafts that would possibly meet the requirements to a larger extent.

Therefore, tools accelerating such a design process have to fulfill mainly two goals:

This work was supported by the Austrian Science Fund – 'Fonds zur Förderung der wissenschaftlichen Forschung (FWF)' – SFB F013 'Numerical and Symbolic Scientific Computing'.

- They have to be flexible enough to handle the various requirements. Nevertheless, they also have to be robust to produce reliable results. Especially, it is desirable to spend only little work on modifying the code when the requirements change.
- On the other hand, these tools have to be fast. The faster the tool, the more design drafts can be optimized.

An extensive review of the various methods for structural optimization using finite elements is given in the monograph of [8]. A monograph specializing more on topology optimization is [1]. Both contain many references on the various aspects of shape optimization in general. The publications [10] and [11] focus on an optimal sizing approach similar to the one in this paper, but they use approximate representations of the objective and of the constraints to reduce the overall costs of the method. In [17] a generalization of that approach to a class of nonlinearly elastic materials is presented. Some stability results with respect to an inexact solution of the state equation are given in [16].

In [13] connections between stable discretizations for a material distribution problem and stable ones for Stokes' problem are presented. An efficient solution strategy for this problem was presented in [9]. For approaches using a topology optimization for getting an initial guess of the topology used in a shape optimization afterwards see, e.g., [12] or [15].

This paper deals with minimizing the mass of the frame of an injection moulding machine as an example for a typical optimal sizing problem. After modeling the problem, we show how to transform it into an equivalent one which can be managed by the optimizer in an easier way. Furthermore, a short sketch on the optimization method used is presented. The main problem in applying standard optimization methods is to calculate gradients of the given objective and the restrictions. Since implementing analytic derivatives is an improper approach, various alternatives are presented and compared to each other. Besides finite differences, a very flexible approach using automatic differentiation is presented. But in order to get also an efficient and fast method, automatic differentiation has to be coupled with a well-known approach from shape optimization – the so-called adjoint method. Numerical results show the strength of this approach.

The paper is organized as follows: The problem is modeled and the arising optimization problem is stated in Section 2. Section 3 deals with the handling of the problem from the optimization's point of view. Section 4 presents some methods for calculating gradients of the given objective or the restrictions. In Section 5, numerical results are presented.

2. The modeling of the design of the frame of an injection moulding machine

The frame of an injection moulding machine is briefly sketched by its 2D-cut Ω given in Figure 1. For a frame of homogeneous thickness, typical dimensions are:

FIGURE 1. Cross section of the original shape

- thickness of one plate: 180 mm
- mass of one plate: 3.8 tons
- clumping force (surface force): 300 tons $\approxeq$ 16 N/mm^2
- length: 2.8 m
- height: 1.7 m
- 2 supporting areas

The primary goal of the design phase is to minimize the mass of the frame of the injection moulding machine. Several other requirements have to be fulfilled in addition, e.g.

- maximal v. Mises stress: $\sigma^{\mathrm{vM}} \leq \sigma^{\mathrm{vM}}_{\mathbf{max}}$,
- maximal tensile stress: $\sigma^{\mathrm{ten}} \leq \sigma^{\mathrm{ten}}_{\mathbf{max}}$,
- shrinking angle of the clumping unit (vertical edges on top, called *wings*): $\alpha \leq \alpha_{\mathbf{max}}$,
- handling of the machine or the feeding mechanism,
- easy and cost efficient manufacturing.

Some of these constraints can be integrated into the optimization procedure directly (e.g., restrictions on the stresses), whereas others like the easy manufacturing have to be considered in a post-processing step.

In order to evaluate the stresses σ, the displacement field u of the frame under some load F has to be known.

For a fixed thickness $\rho(x)$, the displacement field $u(x), x = (x_1, x_2) \in \Omega$, fulfills

$$a(\rho; u, v) = F(v) \qquad \text{for all } v \in V_0 \tag{1}$$

with

$$a(\rho; u, v) = \int_\Omega \rho \frac{\partial u_i}{\partial x_j} E_{ijkl} \frac{\partial v_k}{\partial x_l}\, dx, \quad F(v) = \int_\Omega \langle f, v\rangle\, dx + \int_{\Gamma_N} g\, v\, ds,$$

where E_{ijkl} denotes the elasticity tensor, f the volume force density and g the surface force density on a part Γ_N of the boundary. The displacement field u and v are assumed to be in $V_0 = \left\{v \in H^1(\Omega) \mid v = 0 \text{ on } \Gamma_D, \text{meas } \Gamma_D > 0\right\}$ (set of admissible displacements), where $\partial\Omega = \Gamma_D \cup \Gamma_N$, $\Gamma_D \cap \Gamma_N = \emptyset$.

The elasticity tensor E_{ijkl} is given by

$$E_{ijkl} = \lambda\delta_{ij}\delta_{kl} + \mu(\delta_{ik}\delta_{jl} + \delta_{il}\delta_{jk}), \tag{2}$$

where δ_{ij} denotes the Kronecker - Delta and λ, μ denote Lame's constants. These can be calculated from Young's modulus E and Poisson's ratio ν by

$$\lambda = \frac{E\,\nu}{(1+\nu)(1-\nu)}, \qquad \mu = \frac{E}{2(1+\nu)}. \tag{3}$$

In equation (1) we have assumed that

- a generalized plane stress problem is considered, i.e., we consider a body Ω as a plate that is thin in x_3-direction (compared to the other coordinate directions) and that can carry stresses only in parallel to the x_1-x_2-plane, and
- the applied surface tractions and the body forces are independent of x_3, and therefore, there is no displacement in x_3-direction and the other two displacement components are also independent of x_3.

The design problem can be stated as follows:

$$\begin{aligned}
& \int_\Omega \rho\, dx \to \min_{u,\rho} \\
\text{subject to} \quad & a(\rho; u, v) = F(v) && \text{for all } v \in V_0, \\
& \underline{\rho} \le \rho \le \overline{\rho} && \text{a.e. in } \Omega, \\
& \sigma^{\text{vM}}(u) \le \sigma^{\text{vM}}_{\max} && \text{a.e. in } \Omega, \\
& \sigma^{\text{ten}}(u) \le \sigma^{\text{ten}}_{\max} && \text{a.e. in } \Omega, \\
& \alpha(u) \le \alpha_{\max}.
\end{aligned} \tag{4}$$

Here$\sigma^{\text{vM}}(u)$ denotes the v. Mises stress, $\sigma^{\text{ten}}(u)$ the tensile stress in the frame. The change in the shrinking angle of the clumping unit (vertical edges on top, called *wings*) is denoted by $\alpha(u)$.

For discretizing the problem, we use triangular finite elements with piece-wise constant shape functions for approximating ρ and piece-wise quadratic ones for approximating u. In many situations it is additionally assumed that ρ is constant in certain non overlapping subregions Ω_i. We denote the discrete approximation of ρ and u again by ρ and u.

Summarizing all those considerations, the discretized optimization problem can be formulated as follows:

$$
\begin{aligned}
\text{mass}(\rho) = \int_{\Omega_h} \rho(x)dx \to \min_{\rho} & \\
\text{s.t.} \qquad K(\rho)\,u &= F(\rho), \\
\sigma^{\text{vM}}(u) &\leq \sigma^{\text{vM}}_{\max}, \\
\sigma^{\text{ten}}(u) &\leq \sigma^{\text{ten}}_{\max}, \\
\alpha(u) &\leq \alpha_{\max}, \\
\underline{\rho} \leq \rho &\leq \overline{\rho},
\end{aligned}
\tag{5}
$$

with the symmetric positive definite, large, sparse stiffness matrix K. The symbol Ω_h denotes the discretized domain. For the discretized problem, the constraints on σ^{vM}, σ^{ten} and ρ have to be understood component-wise.

In our application, the upper limits on the angle and the stresses are either treated as constraints or as soft limits, which can be violated to some extent, if the mass would be severely smaller then. Furthermore, the pointwise constraints on σ^{vM} and σ^{ten} are replaced by using a higher order ℓ^p norm. Treating the upper limits as soft constraints leads to the following reformulation:

$$
\begin{aligned}
&\text{mass}(\rho) + \omega_1 \big(\max\,(\|\sigma^{\text{vM}}\|_p - \sigma^{\text{vM}}_{\max}, 0)\big)^2 \\
&\quad + \omega_2 \big(\max\,(\|\sigma^{\text{ten}}\|_p - \sigma^{\text{ten}}_{\max}, 0)\big)^2 \\
&\quad + \omega_3 \big(\max\,(\alpha - \alpha_{\max}, 0)\big)^2 \quad \to \min_{u,\rho}, \\
\text{s.t.} \qquad & K(\rho)\,u = F(\rho), \\
& \underline{\rho} \leq \rho \leq \overline{\rho},
\end{aligned}
\tag{6}
$$

where ω_i denote user-chosen factors of influence. Note that this modification leads to an objective in C^1. It looks similar to a penalty formulation of the constraints, but the weights ω_i are kept fixed during the iteration and do not tend to infinity like in penalty methods. Therefore, the problem of ill-conditioning of the Hessian is avoided.

3. A review from the optimizer's point of view

From the optimization's point of view the problem (6) reads as

$$
\begin{aligned}
J(u,\rho) &\to \min_{u,\rho} \\
\text{subject to} \qquad K(\rho)\,u &= F(\rho), \\
\underline{\rho} \leq \rho &\leq \overline{\rho},
\end{aligned}
\tag{7}
$$

where ρ denotes the vector of design parameters and u the solution of the governing finite element (FE) state equation. The splitting of the parameter vector into design parameters ρ and the solution of the FE equation u is typical for problems in shape optimization. From the optimization's point of view the discretized state equation can be interpreted as equality constraints. In our case, it is linear with respect to u and $K(\rho)$ is symmetric and positive definite for all admissible parameters ρ. Therefore, eliminating u formally, we arrive at

$$\begin{aligned} \tilde{J}(\rho) = J(K^{-1}(\rho)\, F(\rho),\, \rho) &\to \min_{\rho} \\ \text{subject to} \qquad \underline{\rho} \leq \rho &\leq \overline{\rho}. \end{aligned} \tag{8}$$

Since we want to use a standard SQP method for optimizing, the formulation (8) is advantageous compared to (7) as it has much fewer parameters. This relies on the fact, that our implementation of the SQP method is mainly based on linear algebra with dense matrices. A short sketch onto a model SQP algorithm for solving

$$\begin{aligned} \tilde{J}(\rho) &\to \min_{\rho} \\ \text{subject to} \qquad c(\rho) &\leq 0 \end{aligned} \tag{9}$$

is given in Algorithm 1. Details can be found, e.g., in [3] or [4].

Algorithm 1 SQP model algorithm

$B_0 = I$
$g_0 = \operatorname{grad} \tilde{J}(\rho_0)$,
/* *linearize constraints,* $\doteq$ *denotes a first order approximation* */
$c(\rho_0 + \delta) \doteq A_0\delta + b_0$
$k = 1$
while not converged **do**
  /* *calculate search direction* s_k */
  Solve
    $\frac{1}{2} s^T B_k\, s + g_k^T s \to \min$
  under the constraints $A_k s \leq -b_k - A_k \rho_k$
  /* *line search procedure* */
  Calculate $\alpha_k \in (0, 1]$ as large as possible such that
    $\Phi_k(\rho_k) + \mu_1 \alpha_k \Phi'_k(\rho_k) \leq \Phi_k(\rho_k + \alpha_k s_k) \leq \Phi_k(\rho_k) + \mu_2 \alpha_k \Phi'_k(\rho_k)$
    with $0 < \mu_1 < \frac{1}{2} < \mu_2 < 1$ with a suitable merit function Φ_k
  /* *Update several quantities* */
  $\rho_{k+1} = \rho_k + \alpha_k s_k$
  $g_{k+1} = \operatorname{grad} \tilde{J}(\rho_{k+1})$
  $c(\rho_{k+1} + \delta) \doteq A_{k+1}\delta + b_{k+1}$
  Update Hessian matrix $\to B_{k+1}$
  $k = k + 1$
end while

The optimizer used in our code is based on a Quasi-Newton approximation of the Hessian using a modified BFGS update formula following [14] in order to avoid the need for Hessian information of the objective. The quadratic subproblem is solved by a range space based QP method combined with an active set strategy. The line search procedure uses an exact penalty function

$$\Phi_k(\rho) = \tilde{J}(\rho) + \sum_j \overline{\sigma}_j^k \max(\rho_j - \overline{\rho}, 0) + \underline{\sigma}_j^k \max(\underline{\rho} - \rho_j, 0) \tag{10}$$

with suitable chosen penalty parameters $\overline{\sigma}_j^k$, $\underline{\sigma}_j^k$ as merit function (c.f. [7]).

4. Calculating gradients

Using a Quasi-Newton strategy and update formulas within the SQP method as proposed in the section before, the remaining main problem is the calculation of gradients for the objective and the constraints. As in this paper, the problem was transformed into a problem with box constraints only, routines providing analytic gradients for these constraints can be implemented easily. But for the objective, the implementation of an analytic derivative is by far too complicated and time consuming. Furthermore, it would not be well suited for the use in a design process, as we would loose the flexibility of the code completely. That is the reason why we have to think of alternative methods for calculating the gradients.

Several methods are presented and compared to each other in this section. On the one hand we have black box methods like finite differences or automatic differentiation (c.f. [5]), on the other hand, methods exploiting the special structure of the state equation are available, e.g., the direct method or the adjoint method (c.f. [8]).

As none of these methods is really well suited for our problem, a hybrid method combining automatic differentiation and the adjoint method is developed.

4.1. Finite differences

If no analytic derivatives of a function can be implemented due to the high complexity, an approximation by finite differences is often the first idea. One approximation is the central difference quotient

$$\mathrm{D}\, f(x) = \frac{f(x+h) - f(x-h)}{2h}. \tag{11}$$

The choice of the increment h is rather critical for getting good results and depends on estimates of the third derivative of f.

In order to improve the accuracy of finite differences, extrapolation methods can be used. For an initial increment H the sequence

$$\mathrm{D}\, f(H), \mathrm{D}\, f(\frac{H}{2}), \mathrm{D}\, f(\frac{H}{4}), \ldots, \mathrm{D}\, f(\frac{H}{2^i}), \ldots \tag{12}$$

is calculated and extrapolated for $i \to \infty$. These methods return not only a value for the derivative, but also an estimate on the precision of that value which can be used for controlling the order of the extrapolation scheme (c.f. [18]).

The main properties are summarized as follows:

- Two function evaluations are needed per difference quotient and in most cases several difference quotients are needed in order to reach the desired precision of the derivative. Furthermore, the number of function evaluations is proportional to the number of design parameters. Due to the high effort finite differences are only well suited for problems with few design parameters.
- Finite differences can easily be used for very complex functions, as they do not rely on any special properties. On the other hand they neither can use any special properties which makes the use of finite differences rather inefficient in certain cases.
- From the user's point of view finite differences are very flexible. Changes in the desired objective imply only a re-implementation of the objective function. Time consuming changes of the gradient routine do not appear, which is especially important for the acceptance of such a method in an industrial design process.
- The possible use of iterative methods for solving the state equation is very important for our problem as the number of parameters in the state equation may be rather large. As finite differences do not rely on any special properties of the function, the coupling with iterative solvers can be done without any problems.

4.2. Automatic differentiation

Compared to finite differences, automatic differentiation (AD) follows a completely different approach. Finite differences try to approximate the derivative and therefore do not provide accurate results, whereas AD methods incur no truncation error at all and usually yield derivatives with working accuracy. Starting point is a computer program that calculates numerical values for a function. First, a symbolic evaluation graph mapping the design parameters to the function values is built. Like symbolic differentiation, AD operates by systematic application of the chain rule, familiar from elementary differential calculus. However, in the case of AD, the chain rule is applied not to symbolic expressions, but to actual numerical values. By using all the intermediate results generated by the function evaluation, the exponential growth of the evaluation complexity of symbolic differentiation can be avoided, as many common subexpressions can be used repeatedly when the gradient is evaluated at any particular point. Furthermore, optimizations made for the function evaluation also pay off for its derivative. Details on how the AD technique works, as well as many related issues on calculating higher order derivatives can be found in [5].

Two different kinds of tools are known in the AD community: The first group is based on source code transformation, e.g., ADIFOR written for FORTRAN codes, c.f. [2]. The other group is based on evaluation graphs generated at runtime, e.g., ADOL-C written for C and C++ codes, c.f. [6]. As our finite element code is completely written in C++ and uses heavily virtual inheritance, source code transformation tools can not be used. It has to be mentioned, that some of the properties of AD listed in the following rely on the use of runtime tools.

- ADOL-C needs a file containing the evaluation graph in a symbolic form for evaluating the function and its gradient. This file is generated at runtime. For structural optimization problems, huge memory and disk capabilities are required for that purpose. Due to the need of an evaluation graph, ADOL-C can only be applied to functions of moderate complexity. The limiting factor is not the inherent complexity of the function itself, but the size of the generated files and the time needed for reading and writing the data. To give an example, the files storing the evaluation graph for a problem with about 450 design parameters and about 7500 DOFs in the FE state equation need about 1 GB of disk space.
- The flexibility of AD with respect to changes in the objective is similar to finite differences. Changes in the objective need only a re-implementation of the objective function, but no changes in the gradient routine when runtime tools are used. Sometimes, special care has to be taken for a correct generation of the evaluation graph, especially in the context of conditional statements.
- AD using ADOL-C is a black box method. The use of the evaluation graph is a drawback of the method, especially when debugging is needed. This is compensated to some extent by the good runtime behaviour of the method. For the so-called reverse mode, the calculation time of the gradient is independent of the number of design parameters and takes the time of about 15 - 20 native C++ function evaluations as long as the evaluation graph can be stored in the main memory of the computer. Compared to the use of finite differences, this is a tremendous speedup, even for problems with only 10 - 20 design parameters.
- The coupling of AD with iterative solvers is a problem of current research (see, e.g., [5] and references therein). As the use of iterative methods (e.g., multilevel methods) is important for solving fine FE discretizations of the state equation efficiently, the applicability is limited to problems, where direct solvers can be used.

4.3. Direct and Adjoint Method

The direct and the adjoint method are both well-known in the shape optimization community (see, e.g., [8]) and take into account the special structure of the state equation. They differentiate the state equation with respect to a design parameter ρ_i leading to

$$K \frac{\partial u}{\partial \rho_i} = \frac{\partial F}{\partial \rho_i} - \frac{\partial K}{\partial \rho_i} u. \tag{13}$$

For the direct method, (13) is solved numerically using the same methods as for the state problem itself. Then the gradient of the objective can be calculated by

$$\frac{\mathrm{d}\tilde{J}}{\mathrm{d}\rho_i} = \frac{\partial J}{\partial \rho_i} + \langle \frac{\partial J}{\partial u}, \frac{\partial u}{\partial \rho_i} \rangle. \tag{14}$$

On the contrary to the direct method, the adjoint method solves (13) formally and inserts the result in (14) which leads to

$$\frac{\mathrm{d}\tilde{J}}{\mathrm{d}\rho_i} = \frac{\partial J}{\partial \rho_i} + \langle K^{-T} \frac{\partial J}{\partial u}, \frac{\partial F}{\partial \rho_i} - \frac{\partial K}{\partial \rho_i} u \rangle \tag{15}$$

In the following, the main properties are summarized:

- The solution of one state equation per design parameter is needed for the direct method, whereas the adjoint method needs the solution of one adjoint problem for the objective and in principle for each constraint. As K is symmetric in our case, the effort for solving the adjoint problem is the same as for solving the state equation itself. Depending on the number of design parameter and constraints, the better suited method can be chosen.
- As analytic partial derivatives of J with respect to ρ and u are needed ($\frac{\partial J}{\partial \rho}, \frac{\partial J}{\partial u}, \frac{\partial K}{\partial \rho}, \frac{\partial F}{\partial \rho}$), both methods can only be applied to simple objectives, where this can easily be done. Furthermore, the flexibility of the method suffers from the need of hand-coded gradient routines.
- Compared to finite differences or the use of AD for the whole function, this approach is much faster. Finite differences need much more solutions of the design problem, compared to AD the huge evaluation graph which originates mainly from the solution of the state equation is avoided.
- Any solver can be used for solving the state problem, especially the use of iterative solvers like conjugate gradient methods with multilevel preconditioning is no problem.

4.4. Hybrid Method

Comparing the last two sections, it can be seen that the strengths of these methods lie in completely different areas. AD provides very high flexibility with respect to the objective used, but has drawbacks with respect to the necessary computer requirements, the use of iterative solvers for the state equation and with respect to longer runtime. On the other hand the direct and the adjoint method can easily be combined with iterative solvers and provide a fast way for calculating the necessary gradients, but they lack from the flexibility needed. However, both approaches can be combined to a new hybrid method combining their strengths in the following way:

The main drawback of the direct or the adjoint method is the need of analytic partial derivatives of the objective and the constraints with respect to ρ and u. But these derivatives can easily be provided by using AD tools. Then only $\frac{\partial K}{\partial \rho_i}$ and $\frac{\partial f}{\partial \rho_i}$ remain, for which hand-coded routines have to be implemented or AD can be used. For optimal sizing problems, these routines can be hand-coded easily. Furthermore,

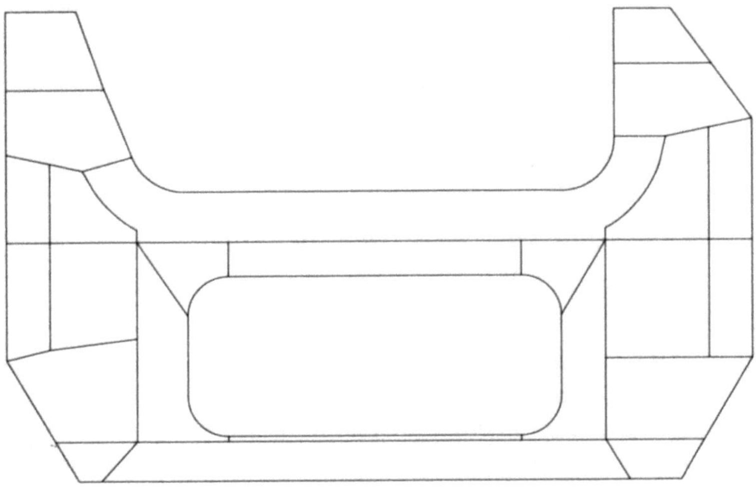

FIGURE 2. Frame with 24 sub-domains

they do not depend on the specific problem which justifies the additional effort for coding even for more complex problems.

5. Numerical results

In the following, some numerical results for the problem stated in Section 2 are presented. They were calculated on an SGI Origin 2000 with 300 MHz.

At the beginning, we tried to use only few design parameters. Therefore, we divided our domain into a number of sub-domains (see Figure 2) and approximated the thickness with a constant function per sub-domain. The state equation was discretized using triangular finite elements with quadratic FE functions.

For evaluating the gradient, either finite differences, a pure AD approach or the hybrid method were used. For a better comparison, the calculation was terminated after a fixed number of steps (The run using finite differences terminated earlier because the search direction was no descent direction anymore). Detailed results can be found in Table 1. All three methods lead to a similar design with about 5 % reduction of the mass compared to the starting configuration (which is the current design of the frame). The optimized thickness distribution can be seen in Figure 3 (the darker the color, the thicker the frame), Figure 4 shows the distribution of the van Mises stresses in the optimized frame (the lighter the color, the higher the stresses).

It can be seen in Table 1 that for a few design parameters the main effort consists in solving the FE state equation, respectively calculating the gradient of the objective. Compared to finite differences and the pure AD approach, the

		Finite Diff.	Pure AD	Hybrid M.
Problem dim.	Nr. of design params	24	24	24
	Nr. of elements (quad.)	3 981	3 981	3 981
	DOFs of state equ.	16 690	16 690	16 690
Optimizer statistics	Iterations	83	100	100
	Function evaluations	19 752	315	236
	Gradient evaluations	84	101	101
Runtime	Total CPU time	12.4 h	4.88 h	0.39 h
	Total elapsed time	12.6 h	8.42 h	0.40 h
Elapsed time	Optimizer	0.01 h	0.03 h	0.01 h
	Function evaluation	0.23 h	2.36 h	0.20 h
	Gradient evaluation	12.40 h	6.00 h	0.18 h

TABLE 1. Comparison of the runtime for various differentiation strategies

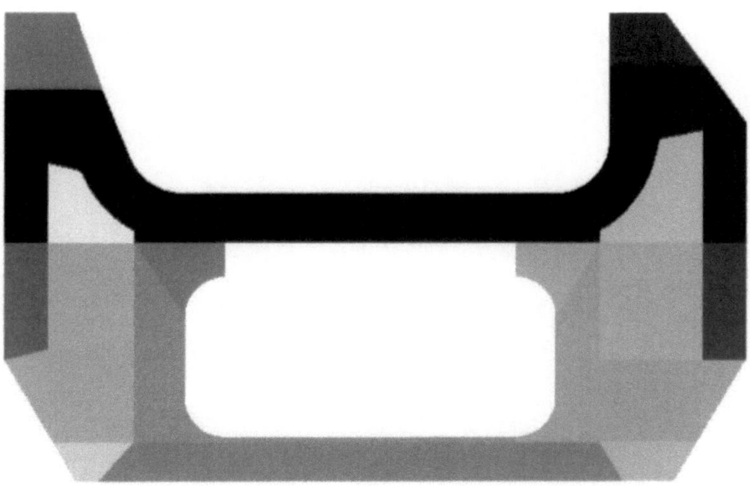

FIGURE 3. Optimized thickness distribution for 24 sub-domains

hybrid method is severely faster, as it combines a fast function evaluation and a fast gradient evaluation. The gradient evaluation is the main drawback for finite differences. For the pure AD approach, we had to implement additional safeguards. In order to detect when a regeneration of the evaluation graph was necessary, we compared the value of the objective using the evaluation graph and the value using a native C++ implementation which explains the longer runtime of the function evaluation.

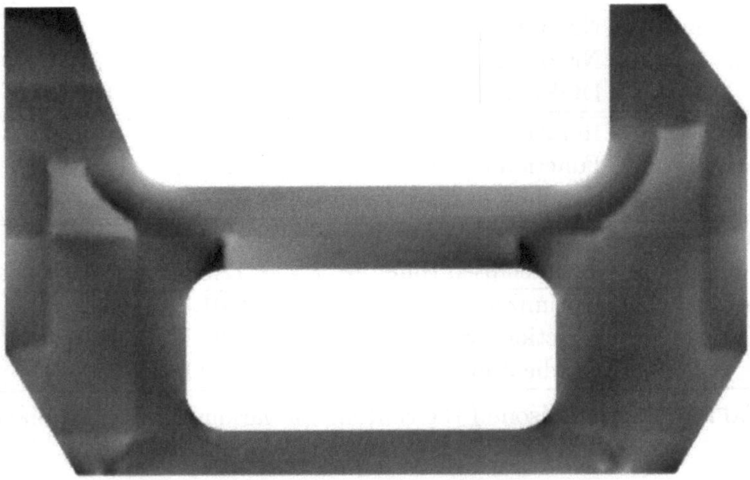

FIGURE 4. Van Mises stresses for 24 sub-domains

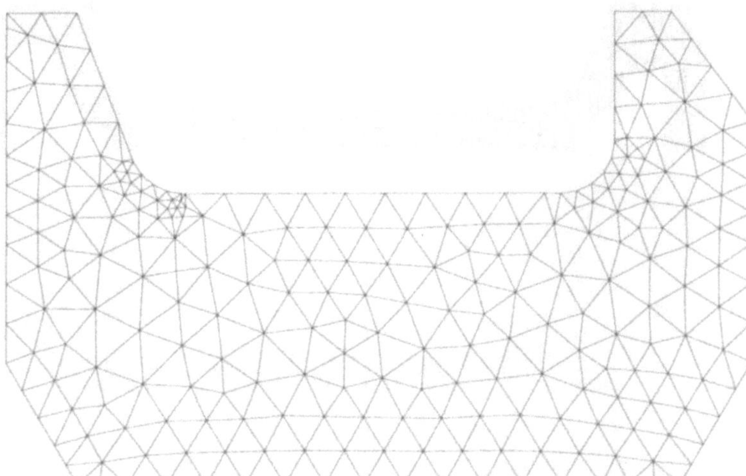

FIGURE 5. Frame with 449 sub-domains

In order to find a better suited splitting of the domain in few sub-domains we increased the number of sub-domains to about 450. We used the coarsest grid of our FE triangulation also for discretizing the thickness distribution (c.f. Figure 5). For solving the design problem, each coarse grid finite element was subdivided into

		Pure AD	Hybrid M.	Hybrid M.
Problem dim.	Nr. of design params	449	449	449
	Nr. of elements (quad.)	1 796	1 796	7 184
	DOFs of state equ.	7 518	7 518	29 402
Evaluation graph	Independents	449	9 314	36 586
	Dependents	1	1	1
	Operations	45 521 797	1 399 910	5 578 270
	Maxlive	540 302	28 140	110 122
	Valstacksize	51 995 116	1 644 461	6 552 653
	Total file size	953 MB	32.4 MB	129.2 MB
Optimizer Statistics	Iterations	800	800	800
	Function evaluations	5 811	3 744	3 745
	Gradient evaluations	801	801	801
Runtime	Total CPU time	32.3 h	3.73 h	14.01 h
	Total runtime	38.5 h	3.76 h	14.12 h
Elapsed time	Optimizer	4.0 h	1.93 h	2.64 h
	Function evaluation	16.5 h	1.29 h	8.13 h
	Gradient evaluation	18.0 h	0.54 h	3.35 h

TABLE 2. Comparison of the runtime for many design parameters

16 elements using 2 levels of uniform refinement. On this refined triangulation the state equation was discretized using finite elements with quadratic FE functions. As finite differences are no more suitable for this number of design parameters, Table 2 contains only results for the pure AD approach and the hybrid method. Additionally, we present also results for a finer discretization of the state equation.

The values reported for the evaluation graph have the following meaning: *Independents* is the number of independent variables of the function differentiated automatically, *dependents* the number of dependent ones (only one dependent variable, as only the objective is differentiated using ADOL-C). *Operations* gives us the number of arithmetic operations in the evaluation graph, *maxlive* the maximal number of live active variables (maximal number of variables allocated at one moment of time during the evaluation of the objective), *valstacksize* the size of the value stack (number of intermediate results).

The optimized thickness distribution can be seen in Figure 6, the corresponding stress distribution in Figure 7. Analyzing the runtime behaviour of the two methods in Table 2, it can be seen that the pure AD approach is no more competitive due to the large file containing the evaluation graph. Furthermore, it can be seen that for the hybrid approach the optimizer needs already a considerable amount of the total runtime. Its relative amount of the runtime even grows when using more design parameters as the complexity of one optimization step is proportional to $(\dim \rho)^3$ (due to the use of dense matrix linear algebra), whereas the

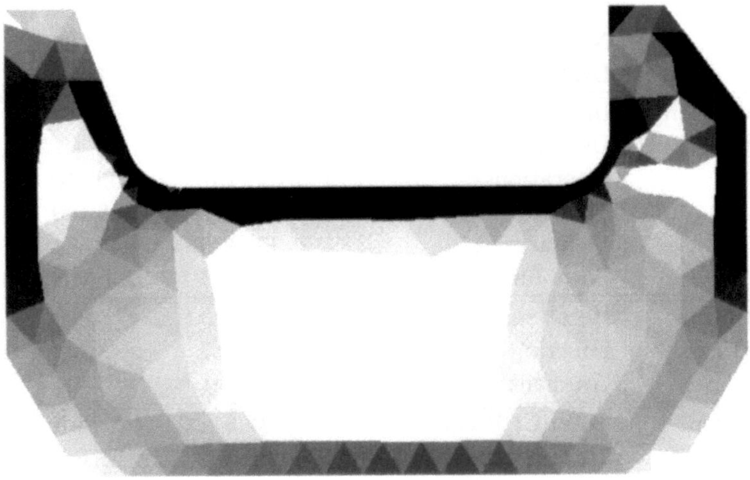

FIGURE 6. Optimized thickness distribution for 449 sub-domains

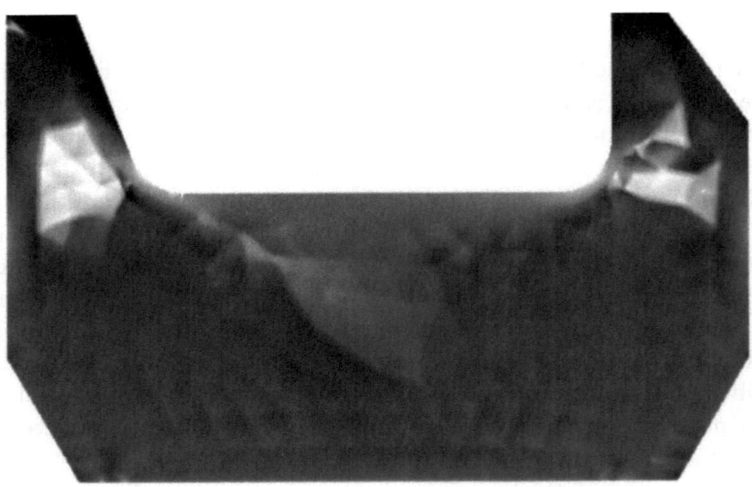

FIGURE 7. Van Mises stresses for 449 sub-domains

FIGURE 8. Optimized thickness distribution for 1078 sub-domains

complexity of solving one FE state equation is proportional to $\dim u$ (if solvers with optimal complexity, e.g., conjugated gradients with multigrid or multilevel preconditioning are used).

Just for comparison, the thickness distribution was discretized using about 1100 design parameters. The optimized thickness distribution and the corresponding distribution of the v. Mises stresses can be found in Figure 8 and Figure 9 respectively. The main problem using that much design parameters is the runtime needed by our optimization module, which dominates the time needed by all function and gradient evaluations completely (about 90 % of the runtime, about 18000 DOFs of the FE state equation).

6. Remarks and Conclusions

In this paper various differentiation strategies needed for solving a real life optimal sizing problem were presented. During the comparison, we focussed our attention on the flexibility of the gradient routine and on the possibility to combine the gradient module with fast iterative solvers for the FE state equations. We compared finite differences and the use of automatic differentiation with the adjoint method and the newly developed hybrid approach.

Finite differences provide a flexible way of getting an approximation of the gradient and can be combined easily with iterative solvers for the FE state equations. On the other hand, extrapolation schemes have to be used to increase their

FIGURE 9. Van Mises stresses for 1078 sub-domains

accuracy, which results in a method with rather high computational effort. Furthermore, finite differences are badly-suited for problems with many design parameters.

Automatic differentiation needs a symbolic evaluation graph and operates on this graph by systematic application of the chain rule. The evaluation time for the gradient is independent of the number of design parameters, but this approach needs huge memory and disk requirements for storing the evaluation graph. Moreover it is not clear how to combine automatic differentiation with iterative solvers for the state equation in a flexible way.

In contrast to finite differences and automatic differentiation, which are black box methods to some extent, the adjoint method is very efficient from the computational point of view if only few constraints are imposed. On the other hand, we loose flexibility, as some parts of the gradient routine have to be re-coded when the objective changes. Moreover, it can easily be combined with iterative solvers for the state equation.

The hybrid method avoids this drawback as it combines the strengths of automatic differentiation and the adjoint method. This method preserves the flexibility of the pure AD approach more or less at the costs of a completely hand-coded gradient routine. Furthermore, the huge memory and disk requirements of the pure AD approach are reduced severely.

Coming back to the optimization routine itself, it must be noted that our current implementation of the optimizer is based on dense matrix linear algebra and therefore, it is only well suited for small to medium size optimization problems.

But in order to close the gap to topology optimization, which is of high practical importance, new optimization methods for large scale problems have to be developed. An approach using multigrid methods also for solving the optimization problem was proposed by [9].

References

[1] M. P. Bendsøe, *Optimization of structural topology, shape and material*, Springer, Berlin, 1995.

[2] C. H. Bischof, A. Carle, P. M. Khademi, and A. Mauer, *The ADIFOR 2.0 system for the automatic differentiation of Fortran 77 programs*, Comput. Sci. Engrg., **3**, No. 3 (1996).

[3] R. Flechter, *Practical methods for optimization*, Constrained Optimization, Vol. **2**, John Wiley & Sons Ltd., Chinchester, 1981.

[4] P. E. Gill, W. Murray, and M. H. Wright, *Practical optimization*, Academic Press, London, San Diego, New York, 1981.

[5] A. Griewank, *Evaluating derivatives, principles and techniques of algorithmic differentiation*, Frontiers in Applied Mathematics, Vol. **19**, SIAM, Philadelphia, 2000.

[6] A. Griewank, D. Juedes, and J. Utke, *ADOL-C, a package for the automatic differentiation of algorithms written in C/C++*, ACM Trans. Math. Software, **22**, No. 3 (1996), pp. 131–167.

[7] S. P. Han, *A globally convergent method for nonlinear programming*, J. Optimization Theory Appl., **22** (1977), pp. 297–309.

[8] J. Haslinger and P. Neittaanmäki, *Finite element approximation for optimal shape design: Theory and applications*, John Wiley & Sons Ltd., Chinchester, 1988.

[9] B. Maar and V. Schulz, *Interior point multigrid methods for topology optimization*, Struct. Multidisc. Optim., **19** (2000), pp. 214–224.

[10] K. G. Mahmoud, *Approximations in optimum structural design*, in: Advanced in Structural Optimization, B. H. V. Topping and M. Papadrakakis, eds., Civil-Comp Press, Edinburgh, 1994, pp. 57–67.

[11] K. G. Mahmoud, H. Engl, and L. Holzleitner, *Optimum structural design using MSC/NASTRAN and sequential quadratic programming*, Computers & Structures, **52** (1994), pp. 437–447.

[12] K. Maute and E. Ramm, *Adaptive topology optimization*, Structural Optim., **10** (1995), pp. 100–112.

[13] J. Petersson, *Finite element analyses of topology optimization of elastic continua*, in: ENUMATH 97 – 2nd European Conference on Numerical Mathematics and Advanced Applications, H. G. Bock, G. Kanschat, R. Rannacher, F. Brezzi, R. Glowinski, Y. A. Kuznetsov, and J. Périaux, eds., World Scientific Publishing, Singapore, 1998, pp. 503–510.

[14] M. J. D. Powell, *A fast algorithm for nonlinearly constrained optimization calculations*, in: Numerical Analysis, G. A. Watson, ed., Lecture Notes in Math., **630**, Springer, Berlin, 1978, pp. 144–157.

[15] E. Ramm, K.-U. Bletzinger, R. Reitinger, and K. Maute, *The challenge of structural optimization*, in: Advanced in Structural Optimization, B. H. V. Topping and M. Papadrakakis, eds., Civil-Comp Press, Edinburgh, 1994, pp. 27–52.

[16] C. Stangl, *Stability of the optimal design with respect to inaccurate solution of the nonlinear state problem*, submitted.

[17] ———, *Optimal sizing for a class of nonlinearly elastic materials*, SIAM J. Optim., **9**, No. 2 (1999), pp. 414–443.

[18] J. Stoer, *Numerische Mathematik 1*, Springer, Berlin, 1994.

On a Globalized Augmented Lagrangian-SQP Algorithm for Nonlinear Optimal Control Problems with Box Constraints

Michael Hintermüller

Abstract. A globalized augmented Lagrangian-SQP algorithm for nonlinear optimal control problems with box constraints on the control is presented. For the solution of the quadratic subproblems of the SQP-type algorithm a very efficient primal-dual method yielding exact subproblem solutions is utilized. The globalization strategy is based on a suitable modification of the Hessian of the Lagrangian and on a line search utilizing an exact penalty functional. Finally, a report on numerical test runs is given.

1. Introduction

Nonlinear optimal control problems for partial differential equations arise in many different areas of science. Typical instances include control problems in fluid dynamics [7, 9], superconductivity [5, 14], phase field modelling [4], and many more. Frequently, due to technically motivated reasons the control must satisfy certain constraints such as non-negativity (– more general – boundedness from below) and/or boundedness from above. The numerical treatment of these problems still represents a significant challenge.

In this note we study a class of nonlinear optimal control problems where the underlying state equation is semilinear, and the control has to satisfy pointwise bound (box) constraints. The algorithm we shall propose is based on an augmented Lagrangian-SQP method [11] which is extended to inequality constraints. Typically, in an inner iteration of the new method a linear-quadratic optimal control problem subject to bound constraints has to be solved. Frequently this part is the most time consuming one and is more involved as in the case where no (inequality) constraints are present.

After discretization these subproblems belong to the class of convex quadratic programs. From a black-box point of view primal-dual path following interior point algorithms are very efficient in solving large convex quadratic programming problems; see [15, 16, 17] and the references therein. However, there is evidence – see [3] – that the recently developed primal-dual active set strategies [2, 8] outperform interior point methods when applied to discretized optimal control problems with control constraints. Therefore the numerical solution of the discretized subproblems arising in this paper is computed by using the efficient primal-dual active set method of [8].

We also address the aspect of globalization, i.e. convergence of the proposed algorithm for arbitrary starting values. It is well-known that compared to pure SQP-algorithms the augmentation in the above mentioned method increases the local convergence radius, and very good numerical behavior is observed [11]. However, there still remains the aspect of locality and the fact that for certain problems and/or initializations no convergence is observed. We propose a simple strategy such that the resulting algorithm converges globally while the fast local convergence rates are preserved. The key ingredients are a modification of the Hessian of the Lagrangian and a line search.

The paper is organized as follows. In section 2 we state the optimal control problem under consideration and establish the corresponding first order necessary conditions. The adaptation of the augmented Lagrangian-SQP method to problems with bound (box) constraints is subject of section 3. Moreover we point out the differences to problems without inequality constraints. The globalization strategy is discussed in section 4, and in section 5 numerical results illustrating the efficiency of the new method are given.

2. Problem statement and first order necessary conditions

The nonlinear optimal control problem under consideration is given by

$$\begin{aligned}
&\text{minimize} && \tfrac{1}{2}\|y-z\|_\Omega^2 + \tfrac{\alpha}{2}\|u-u_d\|_{\Gamma_1}^2 =: J(y,u) && \text{(1a)}\\
&\text{subject to} && -\Delta y + f(y) = g \quad \text{in } \Omega, && \text{(1b)}\\
& && \frac{\partial y}{\partial n} = u \quad \text{on } \Gamma_1, \qquad \frac{\partial y}{\partial n} = h \quad \text{on } \Gamma_2, && \text{(1c)}\\
& && u \in U_{ad} \subset L^2(\Gamma_1)\,, && \text{(1d)}
\end{aligned}$$

where $\alpha > 0$, $z, g \in L^2(\Omega)$, $u_d \in L^2(\Gamma_1)$, $h \in L^2(\Gamma_2)$ are fixed, and $u \in L^2(\Gamma_1)$ is the control variable. The domain $\Omega \subset \mathbb{R}^n$ is bounded, and its boundary $\Gamma = \Gamma_1 \cup \Gamma_2$, with $\Gamma_1 \cap \Gamma_2 = \emptyset$ and Γ_2 possibly empty, is assumed to be sufficiently smooth. The L^2-norm on a set S is denoted by $\|\cdot\|_S$. Moreover, we suppose that $f \in C^2(\mathbb{R})$ with $f(H^1(\Omega)) \subset L^{1+\epsilon}(\Omega)$ for some $\epsilon > 0$ if $n = 2$, and $f(H^1(\Omega)) \subset L^{\frac{2n}{n+2}}(\Omega)$ if $n \geq 3$. The set of feasible controls, i.e. U_{ad}, is given by

$$U_{ad} = \{w \in L^2(\Gamma_1) | a \leq w \leq b \text{ a.e. on } \Gamma_1\} \subset L^2(\Gamma_1)\,,$$

where $a, b \in L^\infty(\Gamma_1)$, with

$$b(x) - a(x) > 0 \quad \text{for almost all } x \in \Gamma_1.$$

The state equation (1b)-(1c) is understood in the weak sense, i.e.

$$\langle \tilde{e}(y,u), \phi \rangle_{(H^1)^*,H^1} = \langle \nabla y, \nabla \phi \rangle_\Omega + \langle f(y) - g, \phi \rangle_\Omega - \langle \tilde{h}, \tau\phi \rangle_\Gamma \tag{2}$$

for all $\phi \in H^1(\Omega)$, with $\tilde{h} = u$ on Γ_1 and $\tilde{h} = h$ on Γ_2, and $\tau : H^1(\Omega) \to L^2(\Gamma)$ the zero order trace operator. Above $\langle \cdot, \cdot \rangle_\Omega$ denotes the duality pairing between functions in $L^p(\Omega)$ and $L^q(\Omega)$ with $p^{-1} + q^{-1} = 1$, and $\langle \cdot, \cdot \rangle_\Gamma$ is the usual $L^2(\Gamma)$-inner product.

In a more abstract way (1) can be written as

$$\text{minimize} \quad J(y,u) \tag{3a}$$

$$\text{subject to} \quad e(y,u) = 0, \tag{3b}$$

$$u \in U_{ad}, \tag{3c}$$

where $e : H^1(\Omega) \times L^2(\Gamma_1) \to H^1(\Omega)$, with $e = \mathcal{N}\tilde{e}$ and $\mathcal{N} : (H^1(\Omega))^* \to H^1(\Omega)$ the Neumann solution operator associated with

$$\langle \nabla v, \nabla \phi \rangle_\Omega + \langle v, \phi \rangle_\Omega = \langle s, \phi \rangle_\Omega \quad \text{for all } \phi \in H^1(\Omega),$$

with $s \in (H^1(\Omega))^*$.

We invoke the following assumptions:

(H1) There exists a (local) solution $(y^*, u^*) \in H^1(\Omega) \times L^2(\Gamma_1)$ of (3), and e is twice continuously Fréchet differentiable with Lipschitz continuous second Fréchet derivative in a neighborhood U^* of (y^*, u^*).

(H2) The Fréchet derivative $e'(y^*, u^*)$ is surjective.

Remark 2.1. *The regularity requirements for e are implied by (h0) of* [11]. *Assumption (H2) holds, if* $f'(y^*) \geq \underline{\epsilon} > 0$.

The a-active set, b-active set, and the inactive set at u^* are defined by

$$\mathcal{A}_a^* = \{x \in \Gamma_1 | u^* = a\},\ \mathcal{A}_b^* = \{x \in \Gamma_1 | u^* = b\},\ \mathcal{I}^* = \Gamma_1 \setminus (\mathcal{A}_a^* \cup \mathcal{A}_b^*), \tag{4}$$

where the equalities are understood in the pointwise almost everywhere sense.

Moreover, for $(y,u,p) \in H^1(\Omega) \times L^2(\Gamma_1) \times H^1(\Omega)$ we introduce the Lagrangian functional of the unconstrained problem, i.e. (3) without the constraint $u \in U_{ad}$, as

$$\tilde{\mathcal{L}}(y,u,p) = J(y,u) + \langle p, e(y,u) \rangle_{H^1}.$$

The first order necessary conditions are specified next,

Theorem 2.2. (**First order necessary conditions**) *Let (H1) and (H2) hold, and suppose that* $f'(y^*)\phi \in L^2(\Omega)$ *for all* $\phi \in H^1(\Omega)$. *Then an optimal solution* (y^*, u^*) *is characterized by the existence of (Lagrange) multipliers* $p^* \in H^1(\Omega)$ *and* $\lambda^* \in L^2(\Gamma_1)$ *satisfying (in the weak sense)*

$$-\Delta p^* + f'(y^*)p^* = -(y^* - z) \text{ in } \Omega, \quad \frac{\partial p^*}{\partial n} = 0 \text{ on } \Gamma, \tag{5a}$$

$$\tau_{|\Gamma_1} p^* - \lambda^* = \alpha(u^* - u_d), \tag{5b}$$

$$\lambda^*_{|\mathcal{A}_a^*} \leq 0, \quad \lambda^*_{|\mathcal{A}_b^*} \geq 0, \quad \lambda^*_{|\mathcal{I}^*} = 0, \tag{5c}$$

together with $e(y^*, u^*) = 0$ *and* $u^* \in U_{ad}$.

Proof. The existence of $p^* \in H^1(\Omega)$ follows from (H2). For every $v \in H^1(\Omega)$ we have

$$\tilde{\mathcal{L}}_y(y^*,u^*,p^*)(v) = \langle y^* - z, v\rangle_\Omega + \langle \nabla p^*, \nabla v\rangle_\Omega + \langle p^*, f'(y^*)v\rangle_\Omega$$

which yields (5a). On the other hand, for every $w - u^* \in U_{ad}$ we obtain

$$\tilde{\mathcal{L}}_u(y^*,u^*,p^*)(w-u^*) = \alpha\langle u^* - u_d, w - u^*\rangle_{\Gamma_1} - \langle \tau_{|\Gamma_1} p^*, w - u^*\rangle_{\Gamma_1} \geq 0\,.$$

Define $\lambda^* = \tau_{|\Gamma_1} p^* - \alpha(u^* - u_d)$. Then similar arguments as in the proof of [8, Thm. 2.3] yield (5b)-(5c). □

If we decompose λ^* into its a- and b-part, i.e.

$$\lambda^* = \lambda_b^* - \lambda_a^*, \quad \text{with} \quad \lambda_a^* \geq 0,\ \lambda_b^* \geq 0, \quad \text{and} \tag{6a}$$

$$\lambda^*_{a|\mathcal{I}^*\cup\mathcal{A}_b^*} = 0\,, \quad \lambda^*_{b|\mathcal{I}^*\cup\mathcal{A}_a^*} = 0\,, \tag{6b}$$

then the following result can be proved.

Proposition 2.3. *Let λ^* be decomposed corresponding to (6). Then*

(i) $\lambda^*_{a|\mathcal{A}_a^*} = -\lambda^*_{|\mathcal{A}_a^*}$ *and* $\lambda^*_{b|\mathcal{A}_b^*} = \lambda^*_{|\mathcal{A}_b^*}$;

(ii)

$$\lambda_a^* = \max\{\lambda_a^* + c(a - u^*), 0\}\,, \tag{7}$$

$$\lambda_b^* = \max\{\lambda_b^* + c(u^* - b), 0\}\,, \tag{8}$$

for arbitrarily fixed $c > 0$.

Proof. (i) The first assertion directly follows from (5c).

(ii) The second result follows from a result from convex analysis. In fact, we utilize the generalized Moreau-Yosida regularization of the indicator function χ_C of a convex set $C \subset L^2(\Gamma_1)$, i.e.

$$\varphi(\tilde{w}) = \inf_{w\in L^2(\Omega)} \{\chi_C(\tilde{w} - w) + \langle\lambda, w\rangle_{\Gamma_1} + \tfrac{c}{2}\|w\|^2_{\Gamma_1}\}$$

with $c > 0$ arbitrarily fixed. Note that the infimum above is attained at

$$w^*(\tilde{w},\lambda) = \tilde{w} - P_C(\tilde{w} + c^{-1}\lambda)\,,$$

where P_C denotes the Hilbert space projection onto C. Further observe that in the case of $C = C_{a,b} = \{v | a \leq v \leq b\}$ the projection P_C is defined by

$$P_C(w)(x) = \begin{cases} a(x) & \text{if } w(x) < a(x), \\ w(x) & \text{if } a(x) \leq w(x) \leq b(x), \\ b(x) & \text{if } w(x) > b(x), \end{cases}$$

for almost all $x \in \Gamma_1$. Due to [12, Thm. 2.1], $\varphi(\tilde{w})$ is continuously Fréchet differentiable with derivative $\varphi'(\tilde{w}) = \lambda + cw^*(\tilde{w},\lambda)$. Moreover [12, Thm. 2.4] yields

$$\lambda \in \partial\chi_{C_{a,b}}(\tilde{w}) \quad \Leftrightarrow \quad \lambda = \varphi'(\tilde{w}).$$

This implies $cw^*(\tilde{w},\lambda) = 0$, and therefore

$$\tilde{w} = P_{C_{a,b}}(\tilde{w} + c^{-1}\lambda).$$

Using $\tilde{w} = u^*$, $\lambda = \lambda^*$, and (i) the result follows easily. □

Remark 2.4. *From the proof of Proposition 2.3, and* [12, Thm 2.1, Thm. 2.4] *we have the alternative characterizations*

$$u^* \in U_{ad}, (5c) \quad \Leftrightarrow \quad \lambda^* \in \partial\chi_{C_{a,b}}(u^*) \quad \Leftrightarrow \quad u^* = P_{C_{a,b}}(u^* + c^{-1}\lambda^*)\,. \tag{9}$$

3. Augmented Lagrangian-SQP algorithm

In this section we adapt the augmented Lagrangian-SQP (AL-SQP) algorithm in [11] to inequality constraints.

3.1. Basic algorithm

The basic tool of the AL-SQP algorithm is the augmented Lagrangian functional

$$\mathcal{L}_\nu(y,u,p,\lambda_a,\lambda_b) = \mathcal{L}(y,u,p,\lambda_a,\lambda_b) + \tfrac{\nu}{2}\|e(y,u)\|^2_{H^1}\,,$$

where $\nu > 0$ is a penalty parameter, and $\mathcal{L} : H^1(\Omega) \times L^2(\Gamma_1) \times H^1(\Omega) \times L^2(\Gamma_1) \times L^2(\Gamma_1)$ denotes the Lagrangian functional associated with (1), i.e.

$$\mathcal{L}(y,u,p,\lambda_a,\lambda_b) = \tilde{\mathcal{L}}(y,u,p) + \langle\lambda_a, a-u\rangle_{\Gamma_1} + \langle\lambda_b, u-b\rangle_{\Gamma_1}\,.$$

Subsequently, we denote by primes the derivative with respect to (y,u). Note that the first order conditions (5) imply

$$\begin{aligned}
\mathcal{L}'_\nu(y^*,u^*,p^*,\lambda_a^*,\lambda_b^*) &= 0, \quad \text{for all } \nu \geq 0, && \text{(10a)}\\
e(y^*,u^*) &= 0, && \text{(10b)}\\
\lambda_a^* - \max\{\lambda_a^* + c(a-u^*),0\} &= 0, && \text{(10c)}\\
\lambda_b^* - \max\{\lambda_b^* + c(u^*-b),0\} &= 0 && \text{(10d)}
\end{aligned}$$

for all $c > 0$. We take a closer look at (10a) and find

$$\begin{aligned}
\mathcal{L}'_\nu(y^*,u^*,p^*,\lambda_a^*,\lambda_b^*) &= J'(y^*,u^*) + e'(y^*,u^*)^*(p^* + \nu e(y^*,u^*)) + \\
&\quad \begin{pmatrix}0\\ \lambda_b^*\end{pmatrix} - \begin{pmatrix}0\\ \lambda_a^*\end{pmatrix}\\
&= J'(y^*,u^*) + e'(y^*,u^*)^*(p^* + \nu e(y^*,u^*)) + \begin{pmatrix}0\\ \lambda^*\end{pmatrix}
\end{aligned}$$

due to (6). Here $e'(y^*,u^*)^*$ denotes the adjoint of $e'(y^*,u^*)$. Replacing λ_a^* and λ_b^* by λ^*, in view of (9) the conditions (10c)-(10d) are equivalent to

$$\lambda^* = c(u^* + c^{-1}\lambda^*) - P_{C_{a,b}}(u^* + c^{-1}\lambda^*).$$

Thus to find (y^*,u^*) numerically observe that (y^*,u^*,p^*,λ^*) is a solution to

$$\begin{aligned}
J'(y,u) + e'(y,u)^*\tilde{p} + \begin{pmatrix}0\\ \lambda\end{pmatrix} &= 0, && \text{(11a)}\\
e(y,u) &= 0, && \text{(11b)}\\
c(u + c^{-1}\lambda) - P_{C_{a,b}}(u + c^{-1}\lambda) &= \lambda, && \text{(11c)}
\end{aligned}$$

with $\tilde{p} = p + \nu e(y,u)$. For the iterative solution of (11) we apply the following technique. Suppose that $(y_n, u_n, p_n, \lambda_n)$ is the current iterate. Then the next iterate $(y_{n+1}, u_{n+1}, p_{n+1}, \lambda_{n+1})$ is the solution to

$$\begin{pmatrix} \tilde{\mathcal{L}}''_\nu(y_n,u_n,p_n) & e'(y_n,u_n)^\star \\ e'(y_n,u_n) & 0 \end{pmatrix} \begin{pmatrix} \binom{\delta y_n}{\delta u_n} \\ \delta p_n \end{pmatrix} + \begin{pmatrix} \binom{0}{\hat{\lambda}_{n+1}} \\ 0 \end{pmatrix} = -\begin{pmatrix} \tilde{\mathcal{L}}'_\nu(y_n,u_n,p_n) \\ e(y_n,u_n) \end{pmatrix}, \tag{12a}$$

$$c(u_{n+1} + c^{-1}\hat{\lambda}_{n+1}) - P_{C_{a,b}}(u_{n+1} + c^{-1}\hat{\lambda}_{n+1}) = \hat{\lambda}_{n+1}, \tag{12b}$$

where $\delta y_n = y_{n+1} - y_n$, $\delta u_n = u_{n+1} - u_n$, and $\delta p_n = p_{n+1} - p_n$. Moreover

$$\tilde{\mathcal{L}}_\nu(y_n,u_n,p_n) = \tilde{\mathcal{L}}(y_n,u_n,p_n) + \tfrac{\nu}{2}\|e(y_n,u_n)\|^2_{H^1}$$

is the augmented Lagrangian of the unconstrained problem. Note that the system (12) represents the first order necessary and sufficient conditions of

$$\begin{aligned} &\text{minimize} && J'_\nu(y_n,u_n)\begin{pmatrix}\delta y\\ \delta u\end{pmatrix} + \tfrac{1}{2}\tilde{\mathcal{L}}''_\nu(y_n,u_n,p_n)\begin{pmatrix}\delta y\\ \delta u\end{pmatrix}^2 \\ &\text{subject to} && e(y_n,u_n) + e'(y_n,u_n)\begin{pmatrix}\delta y\\ \delta u\end{pmatrix} = 0, \\ & && u_n + \delta u \in U_{ad} \end{aligned} \tag{13}$$

provided that $\tilde{\mathcal{L}}''_\nu(y_n,u_n,p_n)$ is positive definite. In (13) we use

$$J_\nu(y,u) = J(y,u) + \tfrac{\nu}{2}\|e(y,u)\|^2_{H^1}.$$

Essentially following the arguments in [1] and [10] we can avoid evaluating $\tilde{\mathcal{L}}'_\nu(y,u,p)$ and $\tilde{\mathcal{L}}''_\nu(y,u,p)$. In fact, observe that

$$\begin{aligned} \tilde{\mathcal{L}}_\nu(y,u,p) &= \tilde{\mathcal{L}}(y,u,p + \tfrac{\nu}{2}e(y,u)), \\ \tilde{\mathcal{L}}'_\nu(y,u,p) &= \tilde{\mathcal{L}}'(y,u,p + \nu e(y,u)), \\ \tilde{\mathcal{L}}''_\nu(y,u,p) &= \tilde{\mathcal{L}}''(y,u,p + \nu e(y,u)) + \nu\langle e'(y,u)(\cdot), e'(y,u)(\cdot)\rangle_{\Omega\times\Gamma_1}. \end{aligned}$$

Thus (12) is equivalent to

$$\begin{pmatrix} \tilde{\mathcal{L}}''(y_n,u_n,\tilde{p}_n) & e'(y_n,u_n)^\star \\ e'(y_n,u_n) & 0 \end{pmatrix} \begin{pmatrix} \binom{\delta y_n}{\delta u_n} \\ \delta\tilde{p}_n \end{pmatrix} + \begin{pmatrix} \binom{0}{\hat{\lambda}_{n+1}} \\ 0 \end{pmatrix} = -\begin{pmatrix} \tilde{\mathcal{L}}'(y_n,u_n,\tilde{p}_n) \\ e(y_n,u_n) \end{pmatrix}, \tag{14a}$$

$$c(u_{n+1} + c^{-1}\hat{\lambda}_{n+1}) - P_{C_{a,b}}(u_{n+1} + c^{-1}\hat{\lambda}_{n+1}) = \hat{\lambda}_{n+1}, \tag{14b}$$

where $\delta\tilde{p}_n = p_{n+1} - \tilde{p}_n$, with $\tilde{p}_n = p_n + \nu e(y_n,u_n)$.

Next we specify the AL-SQP algorithm for the box constrained nonlinear optimal control problem considered here.

AL-SQP Algorithm

(1) *(Initialization). Choose* $(y_o,u_o,p_o,\lambda_o) \in H^1(\Omega) \times U_{ad} \times H^1(\Omega) \times L^2(\Gamma_1)$, $\nu > 0$, *and set* $n = 0$.
(2) *(Augmentation). Set* $\tilde{p}_n = p_n + \nu e(y_n,u_n)$.
(3) *(QP-problem). Solve (14) for* $(\delta y_n, \delta u_n, \delta\tilde{p}_n, \hat{\lambda}_{n+1})$.

(4) *(Step). Set* $(y_{n+1}, u_{n+1}, p_{n+1}, \lambda_{n+1}) = (y_n + \delta y_n, u_n + \delta u_n, \tilde{p}_n + \delta\tilde{p}_n, \hat{\lambda}_{n+1})$, *and* $n = n+1$. *Go to step (2).*

Formally, the AL-SQP algorithm corresponds to the algorithm in [11]. Here, however, step (3) is more involved. Indeed, while for the unconstrained problems considered in [11] the analogue of step (3) amounts to computing the solution of a system of linear equalities, for the present step (3) the additional nonlinear (and nonsmooth) equation (14b) must be taken into account. Therefore, compared to the unconstrained case a more sophisticated QP-solver in step (3) is needed.

3.2. QP-solver

In [2] a primal-dual active set algorithm for control constrained optimal control problems governed by linear elliptic partial differential equations with distributed control was introduced and analyzed. The efficiency of this method was confirmed by the comparison with primal-dual path following interior point methods carried out in [3]. In [8] the basic method of [2] was carried over to the case where the control only acts on the boundary of the domain, and where bilateral constraints on the control are present.

Recall that the first order condition of (13) is given by (12) which, in turn, is equivalent to (14). Moreover, the primal-dual algorithm in [8] can readily be adapted to (13). Here, we only give a brief outline of the algorithm and refer to [8] for details and proofs. We assume that $(y_n, u_n, \tilde{p}_n, \lambda_n)$ is the actual iterate, and that all QP-subproblems admit a solution. The running index of the algorithm below is denoted by superscript k.

Primal-dual active set algorithm (QP-solver)

(1) *(Initialization). Put* $\delta y^0 = 0$, $\delta u^0 = 0$, $\delta\tilde{p}^0 = 0$, $\hat{\lambda}^0 = 0$, *and* $k = 1$.

(2) *(Active-inactive sets). Determine the a-active, b-active, and inactive sets, i.e.*

$$\begin{aligned} \mathcal{A}_a^k &= \{x \in \Gamma_1 | u_n + \delta u^{k-1} + c^{-1}\hat{\lambda}^{k-1} < a\}, \\ \mathcal{A}_b^k &= \{x \in \Gamma_1 | u_n + \delta u^{k-1} + c^{-1}\hat{\lambda}^{k-1} > b\}, \\ \mathcal{A}^k &= \mathcal{A}_a^k \cup \mathcal{A}_b^k, \quad \mathcal{I}^k = \Gamma_1 \setminus \mathcal{A}^k. \end{aligned}$$

Put $\delta u^k = a - u_n$ *on* $\mathcal{A}_a^k$, $\delta u^k = b - u_n$ *on* $\mathcal{A}_b^k$, *and* $\hat{\lambda}^k = 0$ *on* $\mathcal{I}^k$.

(3) *If* $k > 2$ *and* $\mathcal{A}^{k-1} = \mathcal{A}^k$, *then put* $\delta y_n = \delta y^k$, $\delta u_n = \delta u^k$, $\delta\tilde{p}_n = \delta\tilde{p}^k$, $\hat{\lambda}_{n+1} = \hat{\lambda}^k$, *and* EXIT; *otherwise go to (4).*

(4) *(System solve). Solve (14a) for* δy^k, $\delta u^k_{|\mathcal{I}^k}$, $\hat{\lambda}^k_{|\mathcal{A}^k}$, *and* $\delta\tilde{p}^k$.

(5) *Put* $k = k+1$, *and go to (2).*

Note that the identification strategy of step (2) is based on (7) and (8).

4. Globalization

Usually, the AL-SQP algorithm converges only for starting values sufficiently close to a local minimum of (1); see [1, 10, 11]. In this section we introduce a numerically inexpensive globalization strategy. As a result the globalized AL-SQP (gAL-SQP) algorithm converges for arbitrary start up choices.

4.1. Positive definiteness

So far we introduced all the concepts in quite general terms. However, from now on we exploit the specific structure of the present problem class which is typical for a large number of optimal control problems arising in applications [4, 5, 7, 9, 14].

As we have already mentioned in section 3 the positive definiteness of the second order term $\tilde{\mathcal{L}}''(y_n, u_n, \tilde{p}_n)$ is essential for the AL-SQP algorithm to be well defined. Therefore, let us study the structure of $\mathcal{L}''(y,u,p)$ the crucial component of the augmented Lagrangian. We invoke the following assumptions (see also [11]):

$$\left.\begin{array}{lcl} f'(y) & \in & L^{2+\epsilon}(\Omega) \text{ for some } \epsilon > 0 \text{ if } n = 2, \\ f'(y) & \in & L^{n}(\Omega) \text{ if } n \geq 3. \end{array}\right\} \tag{A1}$$

$$\left.\begin{array}{lcl} f''(y) & \in & L^{1+\epsilon}(\Omega) \text{ for some } \epsilon > 0 \text{ if } n = 2, \\ f''(y) & \in & L^{2n/(6-n)}(\Omega) \text{ if } 3 \leq n \leq 6. \end{array}\right\} \tag{A2}$$

Under (H1), (H2), (A1) and (A2) the following characterization of $\mathcal{L}''(y,u,p)$ holds true (for the proof we refer to [11, Prop. 3.4]):

$$\mathcal{L}''(y,u,p)(\delta y, \delta u)^2 = \|\delta y\|^2_\Omega + \alpha\|\delta u\|^2_{\Gamma_1} + \langle p, f''(y)\delta y^2\rangle_\Omega, \tag{15}$$

for all $(\delta y, \delta u) \in H^1(\Omega) \times L^2(\Gamma_1)$.

In view of the AL-SQP algorithm we observe that whenever the expression $\mathcal{L}''(y_n, u_n, \tilde{p}_n)(\delta y_n, \delta u_n)^2$ becomes negative, then the (obviously negative) term $\langle \tilde{p}_n, f''(y_n)\delta y_n^2\rangle$ dominates. This guides us to the following strategy: For $0 \leq \gamma \leq 1$ define

$$L''_\gamma(y,u,p)(\delta y, \delta u)^2 = J''(y,u)(\delta y, \delta u)^2 + \gamma\langle p, f''(y)\delta y^2\rangle_\Omega,$$

and put $\gamma = 1$. Then the system (14) is solved for δy_n, δu_n and $\delta\tilde{p}_n$, and the test

$$L''_\gamma(y_n, u_n, \tilde{p}_n)(\delta y_n, \delta u_n)^2 \geq \kappa(\|\delta y_n\|^2_\Omega + \alpha\|\delta u_n\|^2_{\Gamma_1}) \tag{16}$$

with $0 < \kappa \ll 1$, is performed. If (16) fails to hold then γ is reduced by

$$\tilde{\gamma} = \beta\frac{(\kappa - 1)(\|\delta y_n\|^2_\Omega + \alpha\|\delta u_n\|^2_{\Gamma_1})}{\langle \tilde{p}_n, f''(y_n)\delta y_n^2\rangle}, \tag{17}$$

where $0 < \beta < 1$ is fixed, and $\gamma = \tilde{\gamma}$ is set. Then the system (14) is solved again, the test (16) is performed, and, if necessary, γ is reduced again. This procedure is repeated until (16) is satisfied.

This adjustment scheme is justified by the non-negativity and convexity properties of J. In fact, assume that there exist constants $M, \epsilon > 0$ such that $|\langle \tilde{p}_n, f''(y_n)\delta y_n^2\rangle| \leq M$ and $\min\{\|\delta y_n\|_\Omega, \|\delta u_n\|_{\Gamma_1}\} \geq \epsilon$ for all trial values $(\delta y_n, \delta u_n)$. Since the right hand side in (16) is equal to $\kappa J''(y_n, u_n)(\delta y_n, \delta u_n)^2$, due to the definition of L''_γ the test (16) is satisfied for sufficiently small γ.

In conclusion, step (3) of the AL-SQP algorithm is replaced by

(3) *Put* $\gamma = 1$, *and* $j = 1$.
  (a) *Solve (14) with* $\mathcal{L}''(y_n, u_n, \tilde{p}_n)$ *replaced by* $L''_\gamma(y_n, u_n, \tilde{p}_n)$ *for* δy_n, δu_n, $\delta\tilde{p}_n$, *and* $\hat{\lambda}_{n+1}$.
  (b) *If (16) fails to hold, then compute*

$$\tilde{\gamma} = \min\left\{\beta^j, \beta\frac{(\kappa-1)(\|\delta y_n\|^2_\Omega + \alpha\|\delta u_n\|^2_{\Gamma_1})}{\langle\tilde{p}_n, f''(y_n)\delta y_n^2\rangle}\right\}. \tag{18}$$

  Put $\gamma = \tilde{\gamma}$ *and* $j = j+1$. *Go to (a).*

Note that β^j in (18) represents a safeguard that shall guarantee that γ decreases sufficiently fast. Of course, other (more progressive) choices are possible: for instance, consider the switching technique, where $\gamma := 0$ if (16) is not satisfied.

4.2. Line search

Even if $\mathcal{L}''(y_n, u_n, \tilde{p}_n)$ is positive definite along the iteration sequence, the algorithm may fail to converge. One option to overcome this situation is to include a line search.

We introduce the exact penalty function

$$S(y,u) = J(y,u) + \mu\|e(y,u)\|_{H^1},$$

with $\mu > 0$ denoting a suitable penalty parameter. Its linearization at some $(\bar{y}, \bar{u})$ is defined by

$$\bar{S}(\bar{y},\bar{u})(\delta y, \delta u) = J(\bar{y},\bar{u}) + J'(\bar{y},\bar{u})(\delta y,\delta u) + \mu\|e(\bar{y},\bar{u}) + e'(\bar{y},\bar{u})(\delta y,\delta u)\|_{H^1}.$$

We use $\bar{S}^n(\tau)$ for $\bar{S}(y_n,u_n)(\tau\delta y_n, \tau\delta u_n)$, and $S^n(\tau)$ for $S(y_n+\tau\delta y_n, u_n+\tau\delta u_n)$ with $\tau\in[0,1]$, respectively.

A suitable descent test is given by the well-known *Armijo*-rule

$$S^n(\tau) - S^n(0) \le \sigma\tau(\bar{S}(1) - \bar{S}(0)), \quad \text{with } 0<\sigma<\tfrac{1}{2} \text{ (fixed)}. \tag{19}$$

From the fact that (12) represents the first order conditions for (13), it can easily be derived that the choice $\mu = \mu_n \ge \|\tilde{p}_n + \delta\tilde{p}_n\|_{H^1} + \varepsilon_S$, with $\varepsilon_S > 0$, yields

$$\bar{S}^n(1) - \bar{S}^n(0) \le -\kappa(\|\delta y_n\|^2_\Omega + \alpha\|\delta u_n\|^2_{\Gamma_1}) - \varepsilon_S\|e(y_n,u_n)\|_{H^1} < 0,$$

where we have also used (16).

Note further that

$$\begin{aligned}
S^n(\tau) \;-\; S^n(0) &\le \tau J'(y_n,u_n)(\delta y_n,\delta u_n) + \tfrac{\tau^2}{2}J''(y_n,u_n)(\delta y_n,\delta u_n)^2 + \\
&\quad \mu_n\left(\|e(y_n,u_n) + \tau e'(y_n,u_n)(\delta y_n,\delta u_n)\|_{H^1} - \|e(y_n,u_n)\|_{H^1}\right) + o(\tau) \\
&\le -\tau\left(J(y_n,u_n) + \mu_n\|e(y_n,u_n)\|_{H^1}\right) + \tau J(y_n,u_n) + \\
&\quad \tau\left(J'(y_n,u_n)(\delta y_n,\delta u_n) + \mu_n\|e(y_n,u_n) + e'(y_n,u_n)(\delta y_n,\delta u_n)\|_{H^1}\right) + \\
&\quad \tfrac{\tau^2}{2}\left(\|\delta y_n\|^2_\Omega + \alpha\|\delta u_n\|^2_{\Gamma_1}\right) + o(\tau) \\
&= \tau(\bar{S}^n(1) - \bar{S}^n(0)) + \tfrac{\tau^2}{2}\left(\|\delta y_n\|^2_\Omega + \alpha\|\delta u_n\|^2_{\Gamma_1}\right) + o(\tau)
\end{aligned}$$

Next consider

$$\tau(\bar{S}^n(1)-\bar{S}^n(0))+\tfrac{\tau^2}{2}\left(\|\delta y_n\|_\Omega^2+\alpha\|\delta u_n\|_{\Gamma_1}^2\right)+o(\tau)\le\sigma\,\tau(\bar{S}^n(1)-\bar{S}^n(0))$$

which is equivalent to

$$\tfrac{\tau}{2}\left(\|\delta y_n\|_\Omega^2+\alpha\|\delta u_n\|_{\Gamma_1}^2\right)+\frac{o(\tau)}{\tau}\le(\sigma-1)(\bar{S}^n(1)-\bar{S}^n(0)).$$

Since $o(\tau)/\tau\to 0$ as $\tau\to 0$ and $(\sigma-1)(\bar{S}^n(1)-\bar{S}^n(0))>0$, there exists $\bar{\tau}>0$ such that (19) is fulfilled for all $0\le\tau\le\bar{\tau}$.

This proves the following

Lemma 4.1. *Suppose that* $\max\{\|\delta y_n\|_\Omega,\|\delta u_n\|_{\Gamma_1},\mu_n\}\le C_1$ *for some* $C_1\in\mathbb{R}_+$. *Then there exists* $\bar{\tau}>0$ *such that the Armijo test (19) is satisfied for all* $\tau\le\bar{\tau}$.

5. Numerical results

We shall now report on preliminary numerical results attained by the globalized augmented Lagrangian SQP algorithm (gAL-SQP). We pick one two dimensional example and discuss it in detail. A comprehensive numerical study of the gAL-SQP algorithm and a comparison with other techniques will be presented elsewhere.

The problem data of the test example are as follows: $\Omega=(0,1)^2$, $\Gamma_1=\Gamma$, $\Gamma_2=\emptyset$. The desired state z is given by $z=\sin(1.5\pi x_1)+\cos(1.5\pi x_2)$, and $u_d\equiv 0$. The cost parameter is $\alpha=10^{-3}$. Moreover, $f(y)=75(y^3-y)$, and $g=-10(x_1^2+x_2^2)$. The lower and upper bounds on the control are $a\equiv-\infty$ and $b=10((x_1-\frac{1}{2})^2+(x_2-\frac{1}{2})^2)$, respectively. All variables are set to zero initially.

For discretizing (1b) and (1c) finite differences with mesh-size $h=2^{-5}$ are used. The Laplacian is approximated by the well-known five point stencil. For the normal derivatives we use symmetric differences; see [6].

The algorithm is stopped when the respective norms of the relative residuals are smaller than 10^{-8} or the maximum of the respective norms of the search directions is smaller than 10^{-8}. In the sequel the quantities $\bar{y}$, $\bar{u}$, $\bar{\lambda}$, and $\bar{p}$ denote the solution $(\bar{y},\bar{u})$ with corresponding multipliers $\bar{\lambda}$ and adjoint state $\bar{p}$ obtained from the gAL-SQP algorithm.

5.1. General remarks

Figure 1 shows the optimal state $\bar{y}$ and the optimal control $\bar{u}$ upon termination of the algorithm. The dotted curve in the right graph corresponds to the upper bound b.

The active set $\bar{\mathcal{A}}=\bar{\mathcal{A}}_b$ can be identified form the right graph of Figure 1. The Lagrange multiplier $\bar{\lambda}$ is displayed in the left graph of Figure 2.

First note that the complementarity system

$$\bar{u}-b_h\le 0,\quad \bar{\lambda}\ge 0,\quad (\bar{u}-b_h)^t\bar{\lambda}=0$$

is satisfied. Here subscript h refers to discretized quantities. For the active parts on $\Gamma_l=\{(x_1,x_2)|x_1=0\}$ and $\Gamma_r=\{(x_1,x_2)|x_1=1\}$ we detect that $\bar{\lambda}$ is zero or very

TABLE 1. γ_n in the course of the iteration for $\nu = 0$.

#it	1	2	3	4	5	6	7	8	9
γ_n	1.000	0.239	0.049	0.044	0.051	0.064	0.078	0.085	0.096
#it	10	11	12	13	14	15	16	17	
γ_n	0.070	0.078	0.065	1.000	1.000	1.000	1.000	1.000	

close to zero, i.e. in the former case $\bar{u}$ is weakly active. This fact is also referred to as degeneracy of the problem and frequently causes numerical difficulties. However, we do not observe any numerical problems in the course of the iteration of the gAL-SQP algorithm.

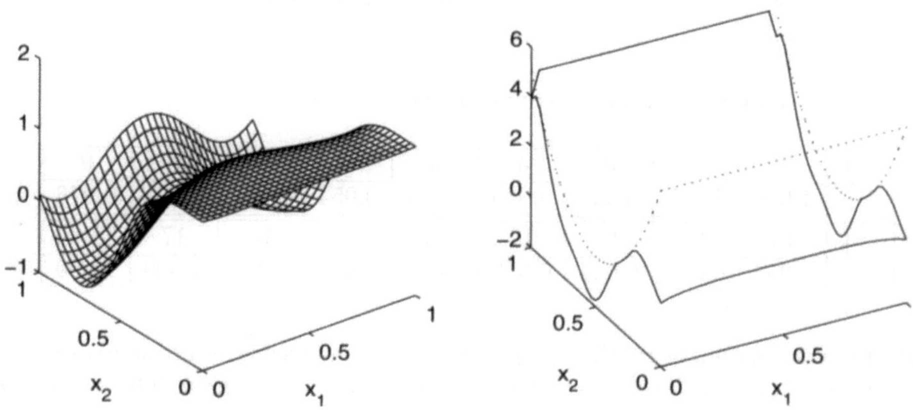

FIGURE 1. Left graph: optimal state $\bar{y}$; right graph: optimal control $\bar{u}$ (solid) and upper bound b (dotted).

5.2. Modification of the Hessian

For the first run we report on, $\nu = 0$ is chosen, i.e. no augmentation is taken into account. Since the augmentation, i.e. $\nu > 0$, yields an increase of the radius of local convergence, we expect in the case of $\nu = 0$ that many modifications of the Hessian of the Lagrangian must take place. The algorithm terminated after iteration 17. Table 1 displays γ_n in the course of the iteration. We note that in iterations 2-12 the Hessian is significantly modified in order to satisfy the test (16). Next we report the test run for $\nu = 0.01$. The algorithm stopped after 20 iterations. Table 2

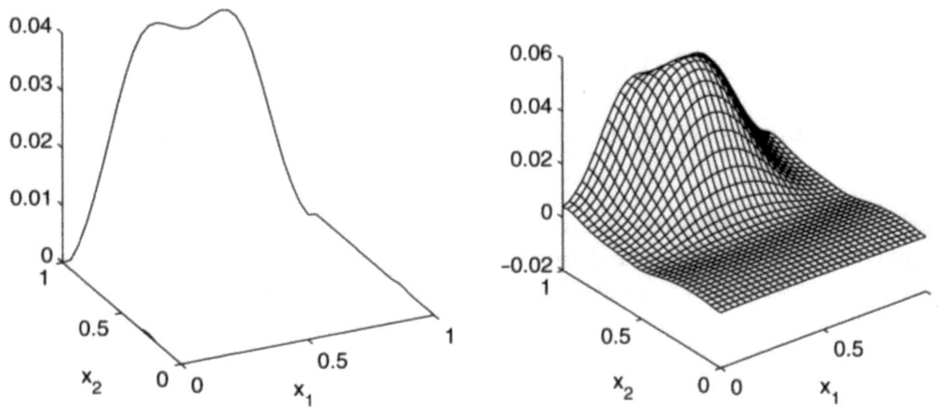

FIGURE 2. Left graph: optimal multiplier $\bar{\lambda}$; right graph: adjoint state $\bar{p}$.

TABLE 2. γ_n in the course of the iteration for $\nu = 0.01$.

#it	1	2	3	4	5	6	7	8	9
γ_n	1.000	0.117	0.058	0.085	0.044	0.054	0.061	0.061	0.076
#it	10	11	12	13	14	15	16	17	18-20
γ_n	0.064	0.058	0.044	0.059	0.024	1.000	1.000	1.000	1.000

gives the corresponding results for γ_n. Obviously, increasing ν also increased the number of iteration. This behavior of the augmented Lagrangian SQP-method is sometimes observed. Moreover, no tendency concerning γ_n can be deduced.

We have also tried larger values for ν. The number of iterations increased only slightly. The number of iterations where γ_n is modified and the magnitude of γ_n is comparable to the results of Table 2.

Finally, we remark that in the case where we skipped the modification of the Hessian of the Lagrangian, the algorithm typically did not converge.

5.3. Line search

Next we discuss the second ingredient of our globalization strategy, i.e. the line search. Tables 3 and 4 again correspond to test runs for $\nu = 0$ and $\nu = 0.01$. We can observe that the sequences of step-sizes for both runs exhibit a high degree of coincidence. This behavior is confirmed for test runs with larger ν-values. Typically, τ_n becomes (and then remains) 1 at iteration 13 or 14. This behavior of the step-length τ is expected from theoretical results [13]. Moreover, a step-length equal to 1 in the final iterations is necessary for fast local convergence. We demonstrate this fast convergence rate in Figure 3, where the behavior of the

TABLE 3. τ_n in the course of the iteration for $\nu = 0$.

#it	1	2	3	4	5	6	7	8	9
τ_n	0.500	0.500	0.250	0.125	0.125	0.125	0.125	0.125	0.125
#it	10	11	12	13	14	15	16	17	
τ_n	0.250	0.250	0.500	1.000	1.000	1.000	1.000	1.000	

TABLE 4. τ_n in the course of the iteration for $\nu = 0.01$.

#it	1	2	3	4	5	6	7	8	9
τ_n	0.500	0.500	0.250	0.125	0.063	0.125	0.125	0.125	0.125
#it	10	11	12	13	14	15	16	17	18-20
τ_n	0.125	0.125	0.250	0.500	1.000	1.000	1.000	1.000	1.000

norm of the relative maximal violation of the first order necessary conditions is documented (for $\nu = 0$). Note that the vertical axis has a logarithmic scale.

Notice that after iteration 13, i.e. the iteration after which the step-length τ remains 1, a very fast rate of convergence is observed.

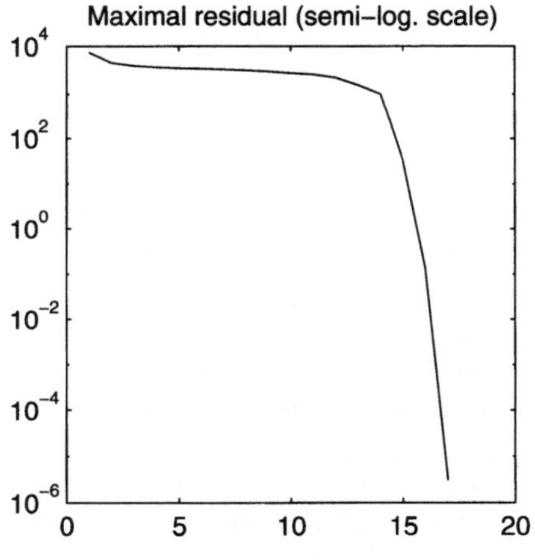

FIGURE 3. Decrease of the maximal violation of the first order system.

TABLE 5. Number of inner iterations for $\nu = 0$.

#it (outer)	1	2	3	4	5	6	7	8	9
#it (inner)	5	5	5	5	6	6	5	6	5
#it (outer)	10	11	12	13	14	15	16	17	
#it (inner)	6	6	7	5	2	2	2	2	

TABLE 6. Number of inner iterations for $\nu = 0.01$.

#it (outer)	1	2	3	4	5	6	7	8	9
#it (inner)	5	4	5	6	5	3	3	7	6
#it (outer)	10	11	12	13	14	15	16	17	18-20
#it (inner)	5	6	5	6	5	4	4	3	2

5.4. QP-solver

Finally, we address the efficient solution of the QP subproblems by the method introduced in section 3.2. The Tables 5 and 6 display the number of (inner) QP-iterations per outer iteration for $\nu = 0$ and $\nu = 0.01$, respectively.

First we can observe that at most 7 inner iterations are needed per outer iteration. Typically, 5-6 inner iterations per outer iteration are performed. Further, one can observe that at the final stages of the (outer) iteration only 2-3 inner iterations are needed. These results are typical within a range of test examples and prove the efficiency of the primal dual active set method for solving the QP-subproblems.

References

[1] D. P. Bertsekas, *Nonlinear Programming*, Athena Scientific, 1995.

[2] M. Bergounioux, K. Ito, and K. Kunisch, *Primal-dual strategy for constrained optimal control problems*, SIAM J. Control Optim., **37** (1999), pp. 1176–1194.

[3] M. Bergounioux, M. Haddou, M. Hintermüller, and K. Kunisch, *A comparison of a Moreau-Yosida-based active set strategy and interior point methods for constrained optimal control problems*, SIAM J. Optim., **11**, No. 2 (2000), pp. 495–521.

[4] Z. Chen and K. H. Hoffmann, *Numerical solutions of the optimal control problem governed by a phase field model,* in: Control and Estimation of Distributed Parameter Systems, Internat. Ser. Numer. Math., **100** (1991), pp. 79–97.

[5] M. D. Gunzburger, L. Hou, and T. P. Svobodny, *Finite element approximations of an optimal control problem associated with the scalar Ginzburg-Landau equation*, Comput. Math. Appl. (1991), pp. 123–131.

[6] W. Hackbusch, *Theorie und Numerik elliptischer Differentialgleichungen*, Teubner Verlag, Stuttgart, Germany, 1996.

[7] M. Heinkenschloss, *Formulation and analysis of a sequential quadratic programming method for the optimal Dirichlet boundary control of the Navier-Stokes flow,* in: Optimal control: theory, algorithms and applications, W. W. Hager and P. M. Pardalos, eds., Kluwer Academic Publishers, 1998, pp. 178–203.

[8] M. Hintermüller, *A primal-dual active set algorithm for bilaterally control constrained optimal control problems*, SFB-report No. 146, University of Graz, Austria, 1998.

[9] M. Hinze, *Optimal and instantaneous control of the instationary Navier-Stokes equations*, habilitation thesis, TU Berlin, Germany, 1999.

[10] K. Ito and K. Kunisch, *Augmented Lagrangian-SQP-methods in Hilbert spaces and application to control in the coefficient problems,* SIAM J. Optimization, **6** (1996), pp. 96–125.

[11] K. Ito and K. Kunisch, *Augmented Lagrangian-SQP methods for nonlinear optimal control problems of tracking type*, SIAM J. Control Optim., **34** (1996), pp. 874–891.

[12] K. Ito and K. Kunisch, *Augmented Lagrangian methods for nonsmooth convex optimization in Hilbert spaces*, Nonlinear analysis, theory, methods and applications, to appear.

[13] M. J. D. Powell, *Variable metric methods for constrained optimization*, in: Mathematical programming: the state of art, Bachem, Grötschel and Korte, eds., Springer-Verlag, Berlin, 1982.

[14] M. Tinkham, *Introduction to superconductivity*, McGraw-Hill, NY, 1975.

[15] R. J. Vanderbei, *LOQO: An interior point code for quadratic programming*, technical report SOR94-15, Princeton University, Princeton, NJ, 1994 (revised 1998).

[16] S. J. Wright, *Primal-dual interior point methods*, SIAM, Philadelphia, 1997.

[17] Y. Ye, *Interior point algorithms: theory and analysis*, Wiley-Intersci. Ser. Discrete Math. Optim., John Wiley, New York, 1997.

Recent Advances in Mesh Adaptivity for Optimal Control Problems

WenBin Liu

Abstract. In this work, we present a brief review on recent developments on adaptive finite element method for optimal control problems. We review some current approaches commonly used in adaptive finite element method, and further discuss the main obstacles in applying these approaches to finite element approximation of optimal control problems. We then discuss some recent progress in this area and possible future research. In particular, we report upon the recent advances made by the research groups in Kent and Heidelberg.

1. Introduction

Optimal control has been a very important research area for the last fifty years. Finite element approximation of optimal control problems plays a very important role in the numerical methods for these problems. The literature on this aspect is huge. For instance, there have been extensive studies in convergence of finite element approximation of optimal control problems, see, for example, [3], [14], [15], [16], [20], [21] , [25], [32], [34], and [39]. However, in our opinion, there is not yet enough research on design of the meshes tailored for finite element approximation of optimal control problems.

Only recently, has the importance of this issue been recognized. In this review, we will report upon some recent developments in mesh adaptivity for finite element approximation of optimal control problems, especially upon these made by the research groups in Kent and Heidelberg. Due to space limits, we can only examine the issue around some model control problems.

Let Ω and Ω_U be two bounded open sets in R^n $(1 \leq n \leq 3)$ with Lipschitz boundaries $\partial\Omega$ and $\partial\Omega_U$.

In this review we adopt the standard notation $W^{m,q}(\Omega)$ for Sobolev spaces on Ω with norm $\|\cdot\|_{W^{m,q}(\Omega)}$ and seminorm $|\cdot|_{W^{m,q}(\Omega)}$. We set $W_0^{m,q}(\Omega) \equiv \{w \in W^{m,q}(\Omega) : w|_{\partial\Omega} = 0\}$. We denote $W^{m,2}(\Omega)$ $(W_0^{m,2}(\Omega))$ by $H^m(\Omega)$ $(H_0^m(\Omega))$. In addition c or C denotes a general positive constant independent of h.

Let the state space $Y = H_0^1(\Omega)$ and the control space $U = L^2(\Omega_U)$. Let U_{ad} be a close convex subset in U. Let B be a continuous linear operator from U to

$L^2(\Omega)$. Let g and h be smooth strictly convex functionals on Y and U respectively. In applications, very often $g(y) = \int_\Omega |y - y_0|^2/2$ and $h(u) = \int_{\Omega_U} u^2/2$.

A typical model elliptic optimal control problem reads as follows: Given $f \in L^2(\Omega)$, find $(y, u) \in Y \times U_{ad}$ such that

$$\min_{u \in U_{ad}} \{g(y) + h(u)\}, \tag{1}$$

$$-div(A\nabla y) = f + Bu, \quad in\ \Omega, \qquad y|_{\partial\Omega} = 0,$$

where

$$A(\cdot) = (a_{i,j}(\cdot))_{n\times n} \in (L^\infty(\Omega))^{n\times n},$$

such that there is a constant $c > 0$ for any vector $X \in R^n$

$$X^t AX \geq c\|X\|^2_{R^n}.$$

The weak formulation of the above control problem is as follows: Let

$$a(y, w) = \int_\Omega (A\nabla y) \cdot \nabla w, \quad \forall y, w \in Y,$$

$$(f_1, f_2) = \int_\Omega f_1 f_2, \quad \forall (f_1, f_2) \in L^2(\Omega) \times L^2(\Omega).$$

Then the control problem (1) can be restated as (QCP):

$$\min_{u \in U_{ad}} \{g(y) + h(u)\}\ , \tag{2}$$

$$a(y(u), w) = (f + Bu, w), \qquad \forall w \in Y = H^1_0(\Omega). \tag{3}$$

It is well known (see [23]) that under some mild conditions for y_0, h, g, there exists a unique solution for this problem with the following optimality conditions (QCP-OPT): there is a co-state $p^* \in Y$ such that the triplet (y^*, p^*, u^*) satisfies

$$a(y^*, w) = (f + Bu^*, w) \qquad \forall w \in Y = H^1_0(\Omega), \tag{4}$$

$$a(q, p^*) = (g'(y^*), q) \qquad \forall q \in Y = H^1_0(\Omega), \tag{5}$$

$$(h'(u^*) + B^*p^*, v - u^*)_U \geq 0 \qquad \forall v \in U_{ad} \subset U = L^2(\Omega_U), \tag{6}$$

where B^* is the adjoint operator of B, and h', g' are the derivatives.

It is clear that the optimality conditions consist of a variational inequality coupled with two elliptic equations. When $U_{ad} = U$, the above control problem is called a unconstrained control problem. Sometimes, there are further constraints for the state y. Then the control problem is referred to as an optimal control problem with state constraints, and is much more difficult to study, see, e.g., [38]. When A also depends on the control u, the problem is called a coefficient control problem, see [6]. In general a coefficient control problem is very complicated.

For time-dependent optimal control, we consider the following model parabolic optimal control problem:

$$\min_{u(t) \in U_{ad}} \left\{ \int_0^T (g(y(t)) + h(u(t)))dt \right\}$$

$$y_t - div(A\nabla y) = f + Bu, \quad x \in \Omega, \quad t \in [0, T],$$

$$y(x,t)=0,\quad x\in\partial\Omega,\quad t\in[0,T],\quad y(x,0)=y_0(x),\quad x\in\Omega.$$

Its weak formulation is quite similar to that of the model elliptic control problem, see [23]. Again this control problem has a unique solution under some mild conditions for the objective functions. see [23]. Like the elliptic case above, one can also derive optimality conditions similar to (4)-(6), which are again equivalent to the original control problem.

There are many other types of important optimal control problems, such as boundary control problems, where the control appears in the boundary conditions of the state equation; the nonlinear control problems, where the state equation is nonlinear; optimal shape design, where the control is some geometric quantity; and flow control problems, etc. However we will not be able to examine them in detail.

2. Finite Element Approximation of Optimal Control Problems

For ease of exposition here we only consider n-simplex elements as they are among the most widely used ones. Also we only consider conforming Lagrange elements.

Let Ω^h be a polygonal approximation to Ω with the boundary $\partial\Omega^h$. Let T^h be a partitioning of Ω^h into disjoint open regular n-simplices τ, so that $\bar{\Omega}^h=\bigcup_{\tau\in T^h}\bar{\tau}$. Each element has at most one face on $\partial\Omega^h$, and $\bar{\tau}$ and $\bar{\tau}'$ have either only one common vertex, or a whole edge, or a whole face if τ and $\tau'\in T^h$.

We further require that $P_i\in\partial\Omega^h\Rightarrow P_i\in\partial\Omega$ where $\{P_i\}(i=1....J)$ is the vertex set associated with the partitioning T^h. Associated with T^h is a finite dimensional subspace S^h of $C^0(\bar{\Omega}^h)$, such that $\chi_{|\tau}$ are polynomials of order k ($k\geq 1$) for all $\chi\in S^h$ and $\tau\in T^h$.

For ease of exposition we will assume that $\Omega^h=\Omega$, though all the results can be extended to the more general case where $\Omega^h\subset\Omega$, except for boundary control problems, whose variation crime is rather complicated for analysis, see [27].

Let $V_0^h=\{\chi\in S^h:\chi(P_i)=0 \text{ for any } P_i\in\partial\Omega\}$. Let $Y^h=V_0^h$. It is easy to see that $Y^h\subset Y$. Let $\pi_h: C^0(\bar{\Omega}^h)$ to S^h, denote the interpolation operator such that for any $v\in C^0(\bar{\Omega}^h)$, $\pi_h v(P_i)=v(P_i)$ $(i=1,...,J)$.

Let Ω_U^h be a polygonal approximation to Ω_U with the boundary $\partial\Omega_U^h$. Let T_U^h be a partitioning of Ω_U^h into disjoint open regular n-simplices τ_U, so that $\bar{\Omega}_U^h=\bigcup_{\tau_U\in T_U^h}\bar{\tau}_U$. Again each element has at most one face on $\partial\Omega_U^h$, and $\bar{\tau}_U$ and $\bar{\tau}_U'$ have either only one common vertex, or a whole edge, or a whole face if τ_U and $\tau_U'\in T_U^h$.

We further require that $P_i^U\in\partial\Omega_U^h\Rightarrow P_i^U\in\partial\Omega_U$ where $\{P_i^U\}(i=1....J')$ is the vertex set associated with the partitioning T_U^h.

Associated with T_U^h is another finite dimensional subspace W_U^h of $L^2(\Omega_U^h)$, such that $\chi_{|\tau_U}$ are polynomials of k-order ($k\geq 0$) for all $\chi\in W_U^h$ and $\tau_U\in T_U^h$. Here there is no requirement for the continuity or boundary conditions. We again assume that $\Omega_U^h=\Omega_U$.

Let $U^h = W_U^h$. It is easy to see that $U^h \subset U$. Let h_τ (h_{τ_U}) denote the maximum diameter of the element τ (τ_U) in T^h (T_U^h).

In this review we only consider the piecewise constant finite element space for approximation of the control, though the resulting error estimators can be used in higher order approximations.

Let us first consider finite element approximation of the model elliptic control problem. A possible finite element approximation of the optimal control problem is as follows: $(QCP)^h$

$$\min_{u_h \in U_{ad}^h \subset L^2(\Omega_U)} g_h(y_h) + h_h(u_h) \tag{7}$$

$$a(y_h(u_h), w_h) = (f + Bu_h, w_h), \qquad \forall w_h \in Y^h, \tag{8}$$

where U_{ad}^h is a closed convex set in U^h, and g_h and h_h are some approximations of g and h respectively. It is well known that this finite dimensional convex optimization problem has a unique solution. Furthermore, the approximator (y_h, u_h) converges to the true solution under some mild conditions, see the references cited in Section 1. One can then use some mathematical programming algorithms like Projected Gradient Algorithm, Interior Point Algorithm, and SQP Algorithm, etc to solve it.

For time-dependent control problems, one has to consider what time-discretization schemes should be used. For instance, it is now quite popular to apply Discontinuous Galerkin Scheme in computing. However, here, we only consider the simplest backward Euler Scheme for time-discretization.

Let I^k be a partitioning of $(0,T)$ such that $0 = t_0 < t_1 < \cdots < t_{N-1} < t_N = T$, $k_i = t_i - t_{i-1}$, $i = 1, 2, \cdots, N$. For $i = 1, 2, \cdots, N$, assume that one has constructed the finite element spaces $Y_i^h \in H_0^1(\Omega)$ with the mesh T_i^h. Similarly, construct the finite element spaces $U_i^h \in L^2(\Omega_U)$ with the mesh $(T_U^h)_i$. Let h_{τ^i} $(h_{\tau_U^i})$ denote the maximum diameter of the element τ^i (τ_U^i) in T_i^h $((T_U^h)_i)$. Let $(U_{ad}^h)_i \subset U_i^h \cap K$. Then a possible fully discrete approximation scheme $(QCP)^{hk}$ is to find $(y_h^i, u_h^i) \in V_i^h \times K_i^h$, $i = 1, 2, \cdots, N$, such that

$$\min_{u_h^i \in (U_{ad}^h)_i} \left\{ \sum_{i=1}^N k_i (g_h(y_h^i(t_i)) + h_h(u_h^i(t_i))) \right\},$$

$$\left(\frac{y_h^i - y_h^{i-1}}{k_i}, w_h \right) + a(y_h^i, w_h) = (f(x, t_i) + Bu_h^i, w_h),$$

$$\forall w_h \in Y_i^h \subset Y = H_0^1(\Omega), \quad i = 1, \cdots, N,$$

$$y_h^0(x) = y_0^h(x), \quad x \in \Omega.$$

Again, this optimization problem has a unique solution, and can be solved by some mathematical programming algorithms, see [36].

One can similarly consider finite element discrete approximations for other control problems, see [18], [36] and [37], for instance.

Adaptive Approximation and a Posteriori Error Estimates

In order to obtain a numerical solution of acceptable accuracy for an optimal control problem, the finite element meshes have to be refined, or adjusted according to a mesh refinement scheme. Adaptive finite element approximation either refines the area where the approximation error may be larger, or adjusts the meshes so that a higher density of nodes is distributed over the area where the solution is difficult to approximate. At the heart of any adaptive finite element method is an a posteriori error indicator or monitor. The decision of whether further refinement of the meshes is necessary is based on the estimate of the discretization error. If further refinement is to be performed then the error indicator or monitor is used as a guide as to show how the refinement or adjustment might be accomplished most efficiently in reducing approximation error. The literature in this area is very extensive. It seems difficult to give even a very brief review here. Some general references can be found in [1] and [41], and for some relevant techniques, see [1], [4], [7], [8], [13], [29], and [41].

The general idea behind adaptive finite element approximation is to refine or adjust meshes such that approximation errors are "equally" distributed over the computational meshes. There are different ways to implement this idea. In practice, there are four major types of adaptive finite element methods, namely, the h-method (mesh refinement), the p-method (order enrichment), the r-method (mesh motion), and the hp-method. The following is an example of the so called h-version refinement procedure. In this method, an a posteriori error estimator η usually has the form of

$$\eta^2 = \sum_{\tau_i} \eta^2_{\tau_i},$$

where τ_i is a finite element and η_{τ_i} can be calculated via finite element approximators.

At each iteration, an average quantity of all $\eta^2_{\tau_i}$ is calculated, and each $\eta^2_{\tau_i}$ is then compared with this quantity. The element τ_i is to be refined if $\eta^2_{\tau_i}$ is larger than this quantity. As $\eta^2_{\tau_i}$ represents the distribution of total approximation error over τ_i, this strategy makes sure that a higher density of nodes is distributed over the area where the error is higher, see [1] and [41] for the details.

Another method is to use $\eta^2_{\tau_i}$ to adjust meshes - the moving mesh method. In this method, the above error estimators are usually referred to as error monitors. The meshes are not further refined but only adjusted to equi-distribute the error, while the total number of the elements is fixed in the procedure. An initial numerical solution is firstly solved on a uniform mesh, and then an error monitor is calculated in order to decide the motion of each node in order to generate a new mesh. More details can be found in [22] and [35].

Clearly efficiency and reliability of adaptive finite element approximation very much rely on how accurate the a posteriori estimators or monitors used are. For finite element approximation of the following elliptic equation:

$$a(y,v) = (f,v), \forall v \in H_0^1(\Omega)$$

a posteriori error estimates normally provide upper or lower bounds for an error estimator η. Let y_h be the finite element approximator of the equation. The estimator η is said to be reliable if there is $C > 0$ such that

$$\|y - y_h\| \leq C\eta, \tag{9}$$

the estimator is said to be efficient if there is $c > 0$ such that

$$\|y - y_h\| \geq c\eta, \tag{10}$$

or the estimator is said to be equivalent if,

$$c\|y - y_h\| \leq \eta \leq C\|y - y_h\|. \tag{11}$$

From our computational experience, an equivalent error estimator normally leads to efficient mesh refinement or adjustment schemes.

For finite element approximation of the linear elliptic *state equation* (of the elliptic model control problem):

$$a(y,v) = (f + Bu, v), \forall v \in H_0^1(\Omega)$$

there are several ways to construct an error indicator or monitor if the control u is assumed to be known. One popular way in engineering computations is to use *gradient* recovery techniques. For the elliptic state equation, it is possible to construct a gradient recovery operator G: such that $|Gy_h - \nabla y|_{L^2(\Omega)}$ is much smaller than $|\nabla y_h - \nabla y|_{L^2(\Omega)}$. It can be shown that then the a posteriori quantity $|Gy_h - \nabla y_h|_{L^2(\Omega)}$ may be a good error indicator for the approximation error of the state y. Another widely used method is to first solve auxiliary local boundary problems numerically, which are simpler than the original elliptic equation, and then to construct an error estimator using the numerical solutions similarly as above.

A posteriori estimation of residual $a(y_h, v) - (f + Bu_h, v)$ can also derive a posteriori error estimators or indicators. These estimators are commonly referred to as a posteriori error estimators of residual type. One can use the energy norm or $L^2(\Omega)$ norm to bound the residual to derive similar estimates as in (9)-(11), but using the latter norm requires to apply the so called duality technique (see [1] and [41]) to the dual equation.

One of the major limitations of this type indicators is that in general the constants c, C in (9)-(11) are unknown. Thus they cannot be used to estimate the actual approximation error. Nevertheless, error estimators of residual type have been successfully applied to mesh adaptivity in many practical problems.

However, optimal control problems require to solve the coupled (y, u) which satisfies a complicated nonlinear system (4)-(6), consisting of the state and co-state equations and a variational inequality. It is not obvious how to apply the known approaches examined above to optimal control problems. This will be discussed in the next section.

3. Some recent advances in mesh adaptivity of optimal control problems

Although adaptive finite element approximation has been widely used in numerical simulations, it has not been *fully* utilised in optimal design problems yet. Initial attempts in this aspect have only been reported recently for some design problems, see, e.g., [2], [5], [12], [33], and [40]. However a posteriori error indicators of a heuristic nature are widely used in most applications. For instance, in some existing work on adaptive finite element approximation of optimal design, the mesh refinement is guided by a posteriori error estimators based on a posteriori error estimates *solely* from the state equation (3) for a fixed u_h. Thus error information from approximation of the control (design) is not utilised. This strategy was found to be inefficient in recent numerical experiments (see [10]). From our numerical experience, mesh refinements based on such error indicators may lead to the meshes which give even larger error of *control* approximation than the uniform meshes. As one would be more interested in computing the optimal control than the state, these indicators should not be used in general. In other work (see [12]), a preassigned mesh refinement scheme is applied around the possible singularity points of the state equation. Although these methods may work well in some particular applications, they cannot be applied confidently in general. It is unlikely that the potential power of adaptive finite element approximation has been fully utilised due to the lack of more sophisticated a posteriori error indicators.

It is not straightforward to derive suitable a posteriori error estimators or monitors for general optimal control problems. For instance, it seems difficult to apply the gradient recovery techniques as ∇u is not well defined; the control u is only in $L^2(\Omega_U)$. Recovering approximation in function value is in general difficult. For a similar reason, it seems also difficult to apply the local solution strategy.

Very recently, some error indicators of residual type were developed by two research groups in Heidelberg and in Kent, see, [10, 11], and [27, 28, 30, 31].

When there is no constraint in a control problem (i.e. $U_{ad} = U$), the optimality conditions (4)-(6) actually consist of coupled elliptic equations only. Consequently one should be able to write down the dual system, and then to apply the weighted a posteriori error estimation technique to obtain a posteriori estimators for *objective functional* approximation error of the control problem, see [10, 11]. Such estimators have indeed been derived for some unconstrained elliptic control problems, and have proved quite efficient in the numerical tests carried out in [10].

However in general, there are always some constraints for the control in applications. In such cases, the optimality conditions often contain a variational inequality, and then have some very different properties. For example, the dual system is unknown. Thus it seems very difficult to apply the techniques used in [10, 11] to constrained control problems.

In our work, more general cases are studied via residual estimation using the norms of energy type. A posteriori error estimators are derived for some constrained control problems where U_{ad} is convex. Here we report some details for the model problems.

Let us first consider the following finite element approximation of the model elliptic control problem (QCP^h):

$$\min_{u_h \in U_{ad}^h} \{g(y_h) + h(u_h)\} , \tag{12}$$

$$a(y_h, w_h) = (f + Bu_h, w_h), \qquad \forall w_h \in Y^h \subset H_0^1(\Omega). \tag{13}$$

Similarly we have the following optimality conditions $(QCP - OPT^h)$: there is a co-state $p_h^* \in Y_h$ such that the triplet (y_h^*, p_h^*, u_h^*) satisfies

$$a(y_h^*, w_h) = (f + Bu_h^*, w_h) \quad \forall w_h \in Y^h \subset H_0^1(\Omega), \tag{14}$$

$$a(q_h, p_h^*) = (g'(y_h^*), q_h) \quad \forall q_h \in Y^h \subset H_0^1(\Omega), \tag{15}$$

$$(h'(u_h^*) + B^* p_h^*, v_h - u_h^*)_U \geq 0 \quad \forall v_h \in U_{ad}^h \subset U^h \subset L^2(\Omega_U). \tag{16}$$

For the above finite element approximation scheme we have the following a posteriori error estimates (see [28] and [31]):
Let (y^*, p^*, u^*) and (y_h^*, p_h^*, u_h^*) be the solutions of (4)-(6) and (14)-(16) respectively. Then

$$\|u_h^* - u^*\|_{0,\Omega_U}^2 + \|y_h^* - y^*\|_{0,\Omega}^2 + \|p_h^* - p^*\|_{0,\Omega}^2$$

$$\leq C\eta^2 = C(\eta_1^2 + \eta_2^2) = C\left(\sum_{\tau_U \subset T_U^h} h_{\tau_U}^2 |B^* p_h^* + h'(u_h^*)|_{H^1(\tau_U)}^2 + \eta_2^2\right), \tag{17}$$

with

$$\eta_2^2 =$$

$$\sum_\tau h_\tau^4 \int_\tau (g'(y_h^*) + div(A^* \nabla p_h^*))^2 + \sum_l h_l^3 \int_l [A^* \nabla p_h^* \cdot n]^2$$

$$+ \sum_\tau h_\tau^4 \int_\tau (f + Bu_h^* + div(A \nabla y_h^*))^2 + \sum_l h_l^3 \int_l [A \nabla y_h^* \cdot n]^2,$$

where $[(A\nabla y_h^* \cdot n)]_l$ and $[(A^* \nabla p_h^* \cdot n)]_l$ are the A-normal derivative jump over the interior face l, defined by

$$[A\nabla y_h^* \cdot n]_l = (A\nabla y_h^*|_{\tau_l^1} - A\nabla y_h^*|_{\tau_l^2}) \cdot n,$$

and

$$[A^* \nabla p_h^* \cdot n]_l = (A^* \nabla p_h^*|_{\tau_l^1} - A^* \nabla p_h^*|_{\tau_l^2}) \cdot n,$$

where n is the unit normal vector on $l = \bar{\tau}_l^1 \cap \bar{\tau}_l^2$ outwards τ_l^1. For simplicity, let $[A\nabla y_h^* \cdot n]_l = 0$ and $[A^* \nabla p_h^* \cdot n]_l = 0$ when $l \subset \partial\Omega$. Let h_τ, h_l be the maximum diameters of the elements τ and the face l. If we use the energy norm, we then have

$$\|u_h^* - u^*\|_{0,\Omega_U}^2 + \|y_h^* - y^*\|_{1,\Omega}^2 + \|p_h^* - p^*\|_{1,\Omega}^2$$

$$\leq C \sum_{\tau_U \subset T_U^h} h_{\tau_U}^2 |B^* p_h^* + h'(u_h^*)|_{H^1(\tau_U)}^2 + C\eta_3^2, \tag{18}$$

where

$$\eta_3^2 =$$

$$\sum_\tau h_\tau^2 \int_\tau (g'(y_h^*) + div(A^* \nabla p_h^*))^2 + \sum_l h_l \int_l [A^* \nabla p_h^* \cdot n]^2$$

$$+ \sum_\tau h_\tau^2 \int_\tau (f + Bu_h^* + div(A \nabla y_h^*))^2 + \sum_l h_l \int_l [A \nabla y_h^* \cdot n]^2.$$

Error estimates of this type have been extended to some nonlinear elliptic control problems in [31]. It is clear that there are three parts in the above estimates: the first term is contributed from control approximation error and the other terms come from the state and co-state approximation error. The first term η_1 is indeed the main part of the estimators if one is mostly interested in computing the values of the control and state, and this has been confirmed by numerical experiments. If using the state equation solely to guide mesh refining, only higher order terms will be used and thus the mesh refinement is unlikely to be correct for the control approximation.

If one is mainly interested in computing the optimal control, then it may not be efficient to use the same mesh to approximate the control and state: - as one only needs to use much coarser meshes for the state and co-state approximation.

For Neumann boundary control problems, an error estimator very similar to $\eta_1^2 + \eta_3^2$ was derived in [27], though the main part

$$\hat{\eta}_1^2 = \sum_{s \subset T_U^h} h_s^2 |h'(u_h^*) + B^* p_h^*|_{H^1(s)}^2$$

is now defined on the boundary of Ω, see [26] and [27] for the details. It seems possible to further improve these estimates.

For the model parabolic control problem, the obtained error estimator for the backward Euler Scheme is much more complicated, see [30]. However again the main part has the following simpler form:

$$\bar{\eta}_1^2 = \sum_1^N k_i \sum_{\tau_U \subset T_U^{hk}} h_{\tau_U}^2 |B^* p_h^{k-1} + h'(u_h^k)|_{H^1(\tau_U)}^2,$$

where p^{k-1} is the finite element approximation of the adjoint equation of the state equation, see [30] for the details.

Error indicators for other types of optimal control problems are unknown yet in the literature, but are being investigated. Also there is only very limited work on how to implement these new error estimators in adaptive finite element schemes. Thus there is much future research work to do.

4. Future Research

This is an emerging research area, and there are many open problems.

Computational issues

It is much needed to investigate implementation issues for these error estimators developed recently. For instance, currently the same meshes are used for the control and the state approximation in numerical computation of optimal control problems. It follows from the estimates (17) that one could use much coarser meshes for the state and co-state approximation. However this strategy seems to cause some extra implementation complexity. Also there exist many new challenging issues in applying the new error estimators to mesh refinement for time-dependent optimal control problems.

It is clear that mesh refinement, PDE solvers, and finite dimensional optimization techniques all play very important roles in the numerical methods for optimal control problems. They should not be considered separately.

Theoretical issues

It seems possible to further improve these new error estimators, at least in some model problems. For instance, in solving the model elliptic control problem with $U_{ad} = \{u : u \geq 0\}$ and $U_{ad}^h = \{u_h : u_h \geq 0\}$, we actually used

$$\hat{\eta}_1^2 = \sum_{\tau_U \subset T_U^h} h_{\tau_U}^2 |(B^* p_h^* + h'(u_h^*))\chi_{u_h^*}|_{H^1(\tau_U)}^2,$$

where

$$\chi_{u_h^*} = u_h^*/(u_h^* + h^{0.5}).$$

The motivation is that inside the area $\{x : u(x) = 0\}$, the meshes need not be much refined, see [29]. This idea can be generalized to other cases. This error estimator worked very well in our numerical tests, although theoretical justification has not been established yet. It seems possible to follow the approach used in [29], where an equivalent a posteriori error estimator is derived for some classic variational inequalities. Although the variational inequalities studied in [29] are quite different from those in (4)-(6), it seems possible to extend the techniques used there to this case. It will also be very interesting to extend such sharper error estimators to more control problems.

It is important to extend the existing results on simpler convex control problems to more complicated control problems where the state equations may be nonlinear, or state constraints may be presented. It does not seem straightforward to extend the techniques in [27, 28] to such cases. In [25], a priori error estimates have been derived for the finite element approximation of a class of nonlinear optimal control problem with state and control constraints. One of the key steps in that paper is to reformulate the optimal control problem into an optimization problem in some Banach spaces by using the solution operator of the state equation. It turns out that the format of the optimality conditions for the abstract optimization problem is very suitable for deriving a posteriori error estimates. This idea works for the following weaker nonlinear state equation (see [31]):

$$-div(\nabla y) + a(y) = f + Bu$$

where the function $s = a(t)$ is smooth and monotone. However this is not so clear for the strongly nonlinear state equation:

$$-div(k(|\nabla y|)\nabla y) = f + Bu,$$

where k is smooth, and satisfies some non-degeneracy conditions, see [24].

Flow control is another very important case where the state equations may be nonlinear. It is important to develop suitable a posteriori error estimators for the optimal control problems involving fluid control. One may start with the case where the state equation is the Stokes system, It seems that the techniques used in [28, 31] can be applied to the Stokes system. Then one may be able to extend the results to the control problem governed by some quasi-Newtonian flows, see, e.g., [9]. Although a priori error estimates have been much studied on the fluid control problems, see [17], there seems no a posteriori error estimate available in the literature at all.

It is also interesting to consider Discontinuous Galerkin Scheme in time discretization of parabolic optimal control problems.

It is challenging to derive a posteriori error estimators for optimal design problems in structure optimization, see [19]. For instance, it seems very difficult to obtain such estimators for optimal shape design, where the *optimal control* is normally discontinuous.

References

[1] M. Ainsworth and J. T. Oden, *A posteriori error estimators in finite element analysis*, Comput. Methods Appl. Mech. Engrg., **142** (1997), pp. 1–88.

[2] P. Alotto, etc, *Mesh adaption and optimisation techniques in magnet design*, IEEE Trans. on Magnetics, **32**, 1996.

[3] W. Alt, *On the approximation of infinite optimisation problems with an application to optimal control problems*, Appl. Math. Optim., **12** (1984), pp. 15–27.

[4] I. Babuska and W. C. Rheinboldt, *Error estimates for adaptive finite element computations*, SIAM. J. Numer. Anal., **5** (1978), pp. 736–754.

[5] N. V. Banichuk, etc, *Mesh refinement for shape optimisation*, Structural optimisation, **9** (1995), pp. 45–51.

[6] H. T. Bank and K. Kunisch, *Estimation techniques for distributed parameter systems*, Birkhäuser, Boston, 1989.

[7] R. E. Bank and A. Weiser, *Some a posteriori error estimators for elliptic partial differential equations*, Math. Comp., **44** (1985), pp. 283–301.

[8] J. Baranger and H. E. Amri, *A posteriori error estimators in finite element approximation of quasi-Newtonian flows*, M^2AN, **25** (1991), pp. 31–48.

[9] J. W. Barrett and W. B. Liu, *Finite element approximation of quasi-Newtonian flows*, Numer. Math., **68** (1994), pp. 437–456.

[10] R. Becker and H. Kapp, *Optimization in PDE models with adaptive finite element discretization*, Report 98-20 (SFB 359), University of Heidelberg.

[11] R. Becker, H. Kapp, and R. Rannacher, *Adaptive finite element methods for optimal control of partial differential equations: basic concept*, SFB 359, University Heidelberg, 1998.

[12] Z. H. Ding, L. Ji, and J. X. Zhou, *Constrained LQR problems in elliptic distributed control systems with point observations*, SIAM J. Control Optim., **34** (1996), pp. 264–294.

[13] K. Eriksson and C. Johnson, *Adaptive finite element methods for parabolic problems I: a linear model problem*, SIAM J. Numer. Anal., **28**, No. 1 (1991), pp. 43–77.

[14] F. S. Falk, *Approximation of a class of optimal control problems with order of convergence estimates*, J. Math. Anal. Appl., **44** (1973), pp. 28–47.

[15] D. A. French and J. T. King, *Approximation of an elliptic control problem by the finite element method*, Numer. Funct. Anal. Appl., **12** (1991), pp. 299–315.

[16] T. Geveci, *On the approximation of the solution of an optimal control problem governed by an elliptic equation*, RAIRO Anal. Numer., **13** (1979), pp. 313–328.

[17] M. D. Gunzburger, L. Hou, and Th. Svobodny, *Analysis and finite element approximation of optimal control problems for stationary Navier-Stokes equations with Dirichlet controls*, RAIRO Model. Math. Anal. Numer., **25** (1991), pp. 711–748.

[18] J. Haslinger and P. Neittaanmaki, *Finite element approximation for optimal shape design*, John Wiley and Sons, Chichester, 1989.

[19] E. J. Haug and J. Cea (eds.), *Optimization of distributed parameter structures*, Sijthoff & Noordhoff, Alphen aan den Rijn, Netherlands, 1980.

[20] G. Knowles, *Finite element approximation of parabolic time optimal control problems*, SIAM J. Control Optim., **20** (1982), pp. 414–427.

[21] I. Lasiecka, *Ritz-Galerkin approximation of the time optimal boundary control problem for parabolic systems with Dirichlet boundary conditions*, SIAM J. Control Optim., **22** (1984), pp. 744–500.

[22] R. Li, T. Tang, and P. W. Zhang, *Moving mesh methods in multiple dimensions based on harmonic maps*, to appear.

[23] J. L. Lions, *Optimal control of systems governed by partial differential equations*, Springer-Verlag, Berlin, 1971.

[24] W. B. Liu and J. W. Barrett, *Finite element approximation of some degenerate monotone quasi-linear elliptic systems*, SIAM. J. Numer. Anal., **33** (1996), pp. 88–106.

[25] W. B. Liu and D. Tiba, *Error estimates in the approximation of optimal control problems*, to be published in Numer. Funct. Anal. Optim., 2000.

[26] W. B. Liu and N. Yan, *A posteriori error estimates for a model boundary optimal control problem*, J. Comput. Appl. Math., **120**, No. 1-2 (2000) (special issue), pp. 159–173.

[27] W. B. Liu and N. N. Yan, *A posteriori error analysis for convex boundary control problems*, SIAM. J. Numer. Anal., **39**, No. 1 (2001), pp. 73–99.

[28] W. B. Liu and N. N. Yan, *A posteriori error analysis for convex distributed optimal control problems*, an invited submission to a special issue on Adaptive Finite Element Method, Adv. Comp. Math., 2000.

[29] W. B. Liu and N. N. Yan, *A posteriori error estimators for a class of variational inequalities*, to appear.

[30] W. B. Liu and N. Yan, *A posteriori error estimates for parabolic optimal control problems*, submitted to Numer. Math., 2000.

[31] W. B. Liu and N. N. Yan, *A posteriori error estimates for a model nonlinear optimal control problem*, to be published in Proceedings of 3rd European Conference on Numerical Mathematics and Advanced Applications, 2000.

[32] K. Malanowski, *Convergence of approximations vs. regularity of solutions for convex, control constrained, optimal control systems*, Appl. Math. Optim., **8** (1982).

[33] K. Maute, S. Schwarz, and E. Ramm, *Adaptive topology optimization of elastoplastic structures*, Structural Optimization, **15**, No. 2 (1998), pp. 81–91.

[34] R. S. McKnight and Jr. Borsarge, *The Ritz-Galerkin procedure for parabolic control problems*, SIAM J. Control Optim., **11** (1973), 510–524.

[35] K. Miller and R. N. Miller, *Moving finite element methods I*, SIAM J. Numer. Anal., **18** (1981), 1019–1032.

[36] P. Neittaanmaki and D. Tiba, *Optimal control of nonlinear parabolic systems*, Theory, algorithms and applications, M. Dekker, New York, 1994.

[37] O. Pironneau, *Optimal shape design for elliptic systems*, Springer-Verlag, Berlin, 1984.

[38] R. Pytlak, *Numerical Methods for Optimal Control Problems With State Constraints*, Lecture Notes in Math., **1707**, Springer-Verlag, 1999.

[39] D. Tiba and F. Tröltzsch, *Error estimates for the discretization of state constrained convex control problems*, Numer. Funct. Anal. Optim., **17** (1996), pp. 1005–1028.

[40] A. Schleupen, K. Maute, E. Ramm, *Adaptive FE-procedures in shape optimization*, Structural and Multidisciplinary Optimization, **19** (2000), pp. 282–302.

[41] R. Verfurth, *A review of posteriori error estimation and adaptive mesh refinement techniques*, Wiley-Teubner, 1996.

Optimal Geometrical Design of Bingham Parameter Measurement Devices*

Dmitriy Logashenko, Bernd Maar, Volker Schulz,
and Gabriel Wittum

Abstract. Usually, Bingham material parameters are determined in a rather cumbersome and time consuming manner. Recently, automatic numerical parameter identification techniques for Bingham fluids have been developed. Here, a method is presented to compute optimal shapes of corresponding measurement devices which lead to a high reliability of the resulting parameter estimation.

1. Introduction

In a joint effort together with Braun GmbH (Friedrichshafen, Germany), we have developed a model based measurement technique, which allows the simultaneous determination of all model parameters of certain Bingham flow models. Since analytical approaches are not sufficient, a numerical parameter identification method was developed and implemented as software based on the *ug* toolbox [1]. This was reported on in detail in [20]. It turned out that existing measurement devices are not suitable for a reliable determination of the model parameters. Therefore, in this paper a shape optimization technique is investigated that allows to increase the reliability of the estimation by changes of the geometry of the measurement device.

Pastes are used, e.g., in the production of bricks from clay or bodies of catalytic converters from ceramic pastes. Usually they are extruded, where the quality of the extrusion product depends on the velocity distribution of the flow within the extrusion device. Recently, substantial progress has been achieved in the development of numerical simulation techniques for paste extrusion based on Bingham models [12]. However, in practice these numerical techniques can only be used, if certain parameters of the underlying flow model are known. These parameter values cannot be accessed by direct measurements, but are ususally determined up to now in a process involving a rather high empirical effort and using analytical approximation approaches. The efforts described in [20] show that numerical

*This research has been supported by the German federal ministry for research and education under grant number 03-WI7HE1-8.

methods can be constructed for the estimation of the model parameters from measurements of normal stresses, which need only approximately twice as much computing time as a pure flow simulation. Furthermore, all interesting parameters are estimated simultaneously and automatically (see also section 5). However, as will be clarified later on, the statistical reliability of this approach is rather low for existing measurement devices. Here methods of optimal experimental design are used to improve this reliability by geometry variations.

Optimal experimental design is a well known technique in many practical applications for linear system models. A good introduction can be found, e.g., in [14]. However, there are only few publications on optimal experimental design for nonlinear models and especially for models including differential equations. Most recent investigations are in the field of parameter identification and optimal control for ordinary differential equations and differential algebraic equations [9, 10, 6, 2].

This paper deals with an optimum experimental design problem in the form of a shape optimization problem for a highly nonlinear elliptic partial differential equations (PDE). It is organized in the following way: In section 2 we describe the basic flow model used for the Bingham flow description. Section 3 explaines the basic parameter identification formulation used in order to get values for the Bingham material parameters from measurements of normal flow stresses. In order to be able to solve the Bingham model equations, they have to be dicretized, which is described in section 4. Based on the discretization and the parameter identification formulation, an algorithm for the solution of a resulting finite dimensional least-squares problem is given in section 5. This algorithm is based on inexact RSQP methods. Section 6 presents the basic definitions for the formulation of the optimal experimental design problems. In section 7 we investigate a specific experiment with real measurement data and show how necessary improvements to the current measurement technology are. Optimized devices are presented in section 8.

2. Bingham Fluids

The non-Newtonian flow of ceramic pastes can be described by the following PDE system from continuum mechanics for the case of incompressible fluids and low Mach number velocities (cf. [7]): the continuity equation

$$\operatorname{div} \mathbf{u} = 0, \tag{1}$$

and the momentum equation

$$\rho \frac{\partial \mathbf{u}}{\partial t} = \operatorname{div} \mathbf{T} + \mathbf{f}, \tag{2}$$

where ρ is the (constant) density of the paste, $\mathbf{u}$ is the velocity field of the flow, $\mathbf{f}$ is a vector of the body forces and $\mathbf{T}$ is the stress tensor. According to the angular momentum conservation law, $\mathbf{T}$ is symmetric. It is usually assumed that it depends only on the pressure and the strain tensor

$$\mathbf{D} = \tfrac{1}{2}\left(\nabla \mathbf{u} + (\nabla \mathbf{u})^T\right).$$

In general form this dependence can be written as follows:

$$\mathbf{T} = -pI + \mathbf{T}^E(\mathbf{D}), \tag{3}$$

where p is the pressure and $\mathbf{T}^E$ the extra stress tensor describing the viscous forces in the fluid. The special choice of the extra stress tensor specifies the mathematical model of the ceramic pastes.

For the materials we are interested in, the following form of the extra stress tensor is used (cf. [11, 12]):

$$\mathbf{T}^E(\mathbf{D}) = 2\mu(\mathbf{D})\mathbf{D} = 2\left(\eta_B + \tau_F (2 I\!I_{\mathbf{D}})^{-\frac{1}{2}}\right)\mathbf{D}. \tag{4}$$

(Here $I\!I_{\mathbf{D}}$ is the second invariant of $\mathbf{D}$, $I\!I_{\mathbf{D}} = \frac{1}{2}(\mathrm{Tr}\,\mathbf{D}^2 - (\mathrm{Tr}\,\mathbf{D})^2)$.) But this form is assumed to be true if and only if

$$|I\!I_{\mathbf{T}^E}| > \tau_F^2. \tag{5}$$

Otherwise, i. e. if the inner stresses are small enough, the material is rigid and does not flow so that

$$\mathbf{D} = 0 \quad \text{for } |I\!I_{\mathbf{T}^E}| \le \tau_F^2. \tag{6}$$

Material whose flow is described by equations (1 – 2) with the stress tensor (3 – 4), under the condition (5), and (6) is called a *Bingham fluid*. The parameters η_B and τ_F are said to be *Bingham viscosity* and *yield stress* respectively.

We shall consider only flow of ceramic pastes and assume that condition (5) is true respectively. Besides we shall consider only stationary flow, so that the time derivative in (2) is zero. The gravitation, being the only body force, does not play any essential role in our situation as well, because it is negligible in comparison with the viscous forces. Thus, further we omit also $\mathbf{f}$. This simplification leads to the following system describing our situation:

$$\begin{aligned} \operatorname{div} \mathbf{u} &= 0, \\ -\operatorname{div}\,(2\mu(\mathbf{D}) \cdot \mathbf{D}) + \nabla p &= 0. \end{aligned} \tag{7}$$

In this work, we consider only a two-dimensional model.

We would encounter difficulties when trying to get an approximate solution of system (7). The matter is that the extra stress tensor tends to infinity as the term $I\!I_{\mathbf{D}}$ tends to zero. In reality this is not the case for the restriction (5) "switching" to the other equation, but omitting this condition we always come to such a situation. To avoid it, system (7) should be regularized. This means that we involve a small parameter δ in the generalized viscosity function μ so that

$$\mu(\mathbf{D}) = \eta_B + \tau_F(\delta + 2 I\!I_{\mathbf{D}})^{-\frac{1}{2}}. \tag{8}$$

In this way we get an approximation of the system (7) at which μ is bounded. The solution of the regularized system deviates from the unregularized Bingham solution. An exploration of the regularization error in a model problem is carried out in [12]. There, it is shown that this error tends to zero like $O(\sqrt{\delta})$.

Remark: As for incompressible fluids $\mathrm{Tr}\,\mathbf{D} = 0$, $I\!I_{\mathbf{D}} = \frac{1}{2}\mathrm{Tr}\,\mathbf{D}^2$. Thus (8) attains the form

$$\mu(\mathbf{D}) = \eta_B + \tau_F(\delta + \mathrm{Tr}\,\mathbf{D}^2)^{-\frac{1}{2}}. \tag{9}$$

The stress tensor (3) attains the form

$$\mathbf{T} = -pI + 2\mu(\mathbf{D})\mathbf{D}. \tag{10}$$

In addition to these equations, the model of the ceramic pastes requires special boundary conditions. The reason for this is a phenomenon called wall sliding. Microscopically we have a two-phase flow on the boundary yielding a lubrication effect. Macroscopically this is described by sliding, resulting in a Navier-Type boundary condition that has in 2D the form

$$\begin{gathered} \mathbf{n}^T\mathbf{T}\mathbf{t} = \mathrm{k}\mathbf{u}^T\mathbf{t} + \tau_G, \\ \mathbf{u}^T\mathbf{n} = 0. \end{gathered} \tag{11}$$

where $\mathbf{t}$ and $\mathbf{n}$ are the unit normal and unit tangent vectors to the boundary respectively. Here we assume that the tangential stress and the tangential velocity are codirected on the boundary. These condition include two additional scalar parameters: a wall sliding factor k and a sliding limit τ_G.

Thus for the mathematical description of the flows of the ceramic pastes we have the system of PDEs (7) with the boundary conditions (11) on walls. (Besides inflow and outflow boundary conditions can be imposed on some parts of boundary). The whole model involves four parameters: η_B, τ_F, k and τ_G. The aim of the parameter estimation procedure described here is to find these values by using experimental data obtained by the device described in the next section. The nonlinearity of the PDE system together with its boundary conditions poses quite a challenge to the numerical treatment of the resulting parameter identification problem. It is necessary to note as well that the pressure p is defined by this system only up to a constant. Methods for the discretization of this system and numerical solution of the resulting nonlinear discrete equations are considered in detail in [12, 11].

3. Parameter Identification Technique

The parameters are estimated by using data obtained from a device whose scheme is shown on Figure 1. This is a conical channel with rigid walls. The paste is pressed through this channel with constant velocity in the direction of the large arrow. During this process we measure the normal stress at seven fixed points on the upper wall (further refered to as *measurement points*). The values of the normal stress, as well as the inflow velocity of the paste, are then used for the parameter identification.

For the experiments we used the device with the following sizes: $H = 30\,\mathrm{mm}$, $h = 10\,\mathrm{mm}$, $L = 244\,\mathrm{mm}$, so $\alpha = 2.35^o$. The inflow velocity was about $80\,\frac{mm}{s}$ (compare with the experimental data below). Figure 2 shows a photo of the device.

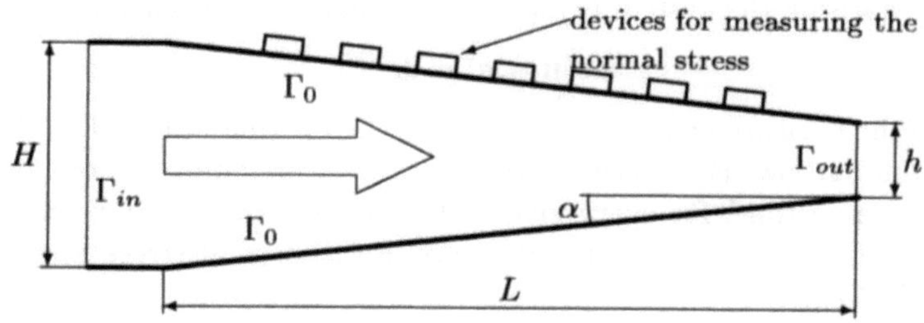

FIGURE 1. Scheme of the measurement device

FIGURE 2. Photo of the measurement device

To model the flow of the paste inside the device we consider the interior (the polygon) as a region Ω on which we impose the PDE system (7) with the generalized viscosity function (9). On the part $\Gamma_0 \subset \partial\Omega$ of the boundary corresponding to the rigid walls we assume boundary conditions (11). On the inflow boundary Γ_{in} we impose Dirichlet boundary conditions for the velocity specifying the constant inflow velocity v_0 along the whole segment. On the outflow boundary Γ_{out} we impose zero vertical velocity as boundary condition and require $\int_{\Gamma_{out}} p\, ds = \text{const.}$

For every set $q = (\eta_B, \tau_F, \mathrm{k}, \tau_G)^T$ of four parameters this system defines the velocity field $\mathbf{u}$ and the pressure field p. Further we write all the partial differential equations comprising the system (7) and the boundary conditions in the form

$$\mathbf{c}(\mathbf{u}, p, q) = 0. \tag{12}$$

Denote the measurement points by $P_1, P_2, \ldots, P_K$ (in our case $K = 7$), in the order from the outflow. For every P_i we have a measured value $\hat{\pi}_i$ of the normal stress. In the same time, for every given set q of the parameters we can get the fields $\mathbf{u}$ and p from equation (12) and compute the normal stress $\pi_P(\mathbf{u}, p, q)$ at every point $P \in \Gamma_0$:

$$\pi_P(\mathbf{u}, p, q) = \mathbf{n}_P^T \, \mathbf{T}_P(\mathbf{u}, p, q) \, \mathbf{n}_P,$$

where $\mathbf{n}_P$ and $\mathbf{T}_P(\mathbf{u}, p, q)$ respectively are the unit normal vector to the boundary and the stress tensor (3) at the point P. The "correct" parameters are then determined as the solution of the nonlinear constrained optimization problem

$$\begin{aligned} & f(\mathbf{u}, p, q) = \sum_{i=2}^{K} \frac{1}{\sigma_i^2} \Big((\pi_{P_i}(\mathbf{u}, p, q) - \pi_{P_1}(\mathbf{u}, p, q)) - (\hat{\pi}_i - \hat{\pi}_1) \Big)^2 \to \min \\ & \textbf{s. t.} \quad \mathbf{c}(\mathbf{u}, p, q) = 0, \end{aligned} \tag{13}$$

where $\sigma_i = 0.08(\hat{\pi}_i + \hat{\pi}_1)$ are the standard deviations for the difference evaulations, if all measurements are assumed to be independently normally distributed with expectation $\hat{\pi}_i$ and standard deviation $0.08\hat{\pi}_i$. Although the model defines the normal stress only up to a constant, the differences of the stresses are defined exactly and should approximate the differences of the measured normal stresses. The numerical solution is carried out in a direct approach, i.e. by discretization of the model and the objective functional. This leads to a finite dimensional nonlinearly constrained optimization problem of a very large size that requires the application of structure exploiting methods to reduce the computation time.

4. Discretization of the Model

We descretize the PDE system (7) by the finite volume method from [12]. The application of this method to Bingham equations is similar to the case of a simple Stokes equation and inherits the similar problems. The implementation is based on *ug*-modules described for Navier-Stockes equations in [15].

We discretize the velocity and pressure fields on quadrilateral elements using a collocated scheme. Such discretizations require stabilization (cf. [3]). For the stabilization the idea described in [16] is used. This technique introduces a new term into the continuity equation, yielding:

$$-\gamma^2 \operatorname{div} \nabla p + \operatorname{div} \mathbf{u} = 0$$

with a scalar stabilization parameter γ. The influence of this stabilization term to the discretization of the Bingham model is discussed in [12]. The stabilization from [16] does not require any additional stabilization parameters.

FIGURE 3. Results of the numerical simulation of a flow in the measurement device. The upper picture shows the horizontal velocity, the lower one — the pressure

Discretizing the outflow boundary conditions we use so called natural boundary conditions of our finite-volume scheme, namely $\int_{\Gamma_{out}} p\,ds = \text{const}$. That determines the pressure completely. The constant is given by the stabilization.

Further we shall denote the whole discretized system by

$$\mathbf{c}_h(\mathbf{u}, p, q) = 0. \tag{14}$$

Here $\mathbf{u}$ and p are respectively two- and one-dimensional grid functions defined on the same index sets. We underline that this is a nonlinear system for all variables. Further it will be also convenient to denote a set $(\mathbf{u}, p)$ by the single letter $\mathbf{x}$. So we shall write $\mathbf{c}_h(\mathbf{x}, q)$ instead of $\mathbf{c}_h(\mathbf{u}, p, q)$.

Since we aim at fast solution techniques we employ multigrid methods for the solution of (14). However, the direct application of multigrid methods (with an ILU smoother) to the linearized system from a Newton approach fails. Therefore we rewrite the discretized system (14) in the form

$$\mathbf{A}(\mathbf{x}, q)\,\mathbf{x} = \mathbf{f}, \tag{15}$$

where $\mathbf{A}$ is a sparse matrix. Based on this form we can easily apply a fixed-point method with inner linear multi-grid solvers.

The grid levels were obtained by uniform refinement of a coarser grid containing less then hundred quadrilaterals. The matrices on each grid level were constructed by discretization of the PDE system. In Figure 3 we show an example of numerical results obtained by this discretization.

For the discretization of the whole system the strain tensor $\mathbf{D}$ and the generalized viscosity function were also defined using finite dimensional grid functions $\mathbf{u}$. This allows to discretize the normal stress functions π_{P_i}. We denote the discrete variants by $\pi_{h,i}$ respectively. This completes the discretization of the whole optimization problem which now reads

$$
\begin{aligned}
&f_h(\mathbf{x}, q) = \sum_{i=2}^{K} \frac{1}{\sigma_i^2} \Big((\pi_{h,i}(\mathbf{x}, q) - \pi_{h,1}(\mathbf{x}, q)) - (\hat{\pi}_i - \hat{\pi}_1) \Big)^2 \to \min \\
&\textbf{s. t.} \quad \mathbf{c}_h(\mathbf{x}, q) = 0.
\end{aligned}
\tag{16}
$$

This is a nonlinearly constrained finite dimensional optimization problem.

5. The Parameter Estimation Procedure

Here we consider the optimization problem (16) in the abstract form, omitting the subscript h:

$$
\begin{aligned}
f(\mathbf{x}, q) &\to \min, \\
\text{s. t.} \quad \mathbf{c}(\mathbf{x}, q) &= 0,
\end{aligned}
\tag{17}
$$

with $f : \mathbf{R}^{n \times m} \to \mathbf{R}$ and $\mathbf{c} : \mathbf{R}^{n \times m} \to \mathbf{R}^n$ and the Jacobian, $\mathbf{J} = \dfrac{\partial \mathbf{c}}{\partial \mathbf{x}}$, which is assumed to be nonsingular. In our case the number of parameters is 4. However, n, the dimension of the grid functions, can be very large.

For the solution of the problem (17) we use a reduced SQP method. A detailed discussion of this approach can be found in [18, 17]. Here we sketch only the idea.

Reduced SQP methods are related to projected Lagrangian methods (cf. [4]) and are most advantagous in the case that the number of degrees of freedom (here the parameters) is small compared to the number of state variables. The constraints are linearized by a Taylor expansion up to first order terms, so that all steps $(\Delta\mathbf{x}, \Delta q)$ lie in the tangent space of $\mathbf{c}$ of the current approximation $(\mathbf{x}, q)$:

$$
\mathbf{c}(\mathbf{x}, q) + \mathbf{J}(\mathbf{x}, q)\, \Delta\mathbf{x} + \frac{\partial \mathbf{c}}{\partial q}(\mathbf{x}, q)\, \Delta q = 0.
$$

Then the problem is projected to this tangent space and approximated by a quadratic problem with the projected Hessian of the Lagrangian

$$
L(\mathbf{x}, q, \lambda) = f(\mathbf{x}, q) - \lambda^T \mathbf{c}(\mathbf{x}, q).
\tag{18}
$$

This leads to the following algorithm:

Algorithm: The RSQP method with an approximate Jacobian.

(0) Set $k := 0$; start at some initial guess $\mathbf{x}_0$, q_0.

(1) Compute the increment of the adjoint variables from the linear system
$\mathbf{A}^T(\mathbf{x}_k, q_k)\,\Delta\lambda_k := \nabla_{\mathbf{x}} f(\mathbf{x}_k, q_k) - \mathbf{J}^T(\mathbf{x}_k, q_k)\,\lambda_k$;
compute the reduced gradient

$$\gamma_k := \nabla_q f(\mathbf{x}_k, q_k) - \left(\frac{\partial \mathbf{c}}{\partial q}(\mathbf{x}_k, q_k)\right)^T (\lambda_k + \Delta\lambda_k);$$

determine some approximation B_k of the projected Hessian of the Lagrangian.

(2) solve $B_k \Delta q_k = -\gamma_k$.

(3) compute step on $\mathbf{x}$ form the linear system

$$\mathbf{A}(\mathbf{x}_k, q_k)\,\Delta\mathbf{x}_k := -\frac{\partial \mathbf{c}}{\partial q}(\mathbf{x}_k, q_k)\,\Delta q_k + \mathbf{c}(\mathbf{x}_k, q_k).$$

(4) Set $\mathbf{x}_{k+1} := \mathbf{x}_k + \Delta\mathbf{x}_k$, $q_{k+1} := q_k + \Delta q_k$ and $\lambda_{k+1} = \lambda_k + \Delta\lambda_k$.

(5) $k := k + 1$; go to (1) until convergence.

Here, $\mathbf{A}$ is a preconditioner for the Jacobian $\mathbf{J}$ and B_k are approximations of the Hessian of the Lagrangian at the steps. These latter matrices are usually computed using an update formula correcting the approximation on the previous step according to the first order information about the gradient of the Lagrangian. A typical choice is the BFGS update that reads

$$B_{k+1} = B_k + \mathrm{Update}(B_k, s_k, v_k), \tag{19}$$

where $s_k = q_k - q_{k-1}$, $v_k = \gamma_k - \gamma_{k-1}$ and

$$\mathrm{Update}(B, s, v) = \frac{vv^T}{v^T s} - \frac{(Bs)(Bs)^T}{s^T B s}. \tag{20}$$

As an initial guess one can take $B_0 = \alpha I$ with some real $\alpha > 0$.

A step of this method can be also interpreted as an approximate Newton step for the necessary conditions of extremum for the problem (17) since the updates of the variables are computed according to a linear system

$$\begin{pmatrix} 0 & 0 & \mathbf{A}^T \\ 0 & B_k & \left(\dfrac{\partial \mathbf{c}}{\partial q}\right)^T \\ \mathbf{A} & \dfrac{\partial \mathbf{c}}{\partial q} & 0 \end{pmatrix} \begin{pmatrix} \Delta\mathbf{x} \\ \Delta q \\ \Delta\lambda \end{pmatrix} = \begin{pmatrix} \nabla_{\mathbf{x}} L \\ \nabla_q L \\ -\mathbf{c} \end{pmatrix}. \tag{21}$$

Some notes on the convergence properties of the reduced SQP methods can be found in [18]. The positive definiteness of the approximations B_k should be preserved. The BGFS update formula (19 – 20) yields a positive definite B_{k+1} for a positive definite B_k if and only if $v_k^T s_k > 0$ (cf. [4]). In the RSQP method started with an arbitrary initial guess this property can be violated that can lead to indefinite approximations of the Hessian. As a remedy we used the Powell modification of the BGFS update formula [13].

The next important issue is that these methods converge only locally and starting with an arbitrary initial approximation Algorithm always requires some damping in step 4. The damping factor is usually chosen by a line search. It is required that the damping parameter supplies sufficient descent of a merit function that involves the constraint as well as the value of the objective function. This is done to provide also the approximate feasibility of the guess. We used the merit function

$$\psi(\mathbf{x}, q) = f(\mathbf{x}, q) + \sum_i \mu_i |c_i(\mathbf{x}, q)|, \tag{22}$$

where $\mu_i > 0$ are estimated upper bounds for the absolute values of the adjoint variables ($|\lambda_i|$) at the optimal solution. This merit function supplied good convergence properties up to the practical precision.

For the computations we implemented Algorithm with the Powell modification of the BGFS update formula. The initial approximation to the Hessian was $B_0 = I$. For the initial guess of the parameters we took some reasonable values. As the initial guess for $\mathbf{x}$ we took an approximate solution of the constraint for the initial parameters obtained by the fixed point iteration. Besides we used matrices of the fixed point iteration as preconditioners $\mathbf{A}$ since, as mentioned above, the true Jacobians are hardly invertible. For the solution of the large sparse linear systems with the matrices $\mathbf{A}$ and $\mathbf{A}^T$, we applied the multigrid method with ILU smoothing. This provided an approximate feasibility in the first iteration. The initial values of the adjoint variables were merely $\lambda_0 = 0$. In the next section we describe the results of the numerical experiments.

6. Confidence Intervals of the Computed Parameters

We estimate the precision of the represented measurement technique by computing the confidence intervals for the obtained parameters according to the linearized model. To this end we assume that the errors of the measurements at the measurement points are statistically independent and meet the normal distribution. Let $\Pi : \mathbf{R}^n \times \mathbf{R}^4 \to \mathbf{R}^6$ be a function that assignes the theoretical relative normal stresses to a given guess $\mathbf{x}$ and a given set of parameters:

$$\Pi(\mathbf{x}, q) = (\pi_2(\mathbf{x}, q) - \pi_1(\mathbf{x}, q), \dots, \pi_7(\mathbf{x}, q) - \pi_1(\mathbf{x}, q))^T.$$

As the velocity and pressure fields corresponding to the parameters q satisfy the equation $\mathbf{c}(\mathbf{x}, q) = 0$, the full differential of the function assigning the normal stresses at the measurement points to the parameters is a matrix

$$S(q) = \begin{pmatrix} \dfrac{\partial \Pi}{\partial \mathbf{x}} & \dfrac{\partial \Pi}{\partial q} \end{pmatrix} \begin{pmatrix} -J^{-1}\dfrac{\partial c}{\partial q} \\ I \end{pmatrix}$$

TABLE 1. Measured normal stresses

Measurement point	P_1	P_2	P_3	P_4	P_5	P_6	P_7
Normal stresses (bar)	2.66	5.2	7.45	9.3	11.2	12.4	13.5

(all the functions should be evaluated at $(\mathbf{x}, q)$, with $\mathbf{c}(\mathbf{x}, q) = 0$). Now the covariances corresponding to the parameters q are the diagonal enties of the matrix

$$\mathbf{Cov}(q) = \left[(S(q))^T D^{-2} S(q)\right]^{-1}$$

$$= \left[\begin{pmatrix} -J^{-1}\dfrac{\partial c}{\partial q} \\ I \end{pmatrix}^T \begin{pmatrix} \left(\dfrac{\partial \Pi}{\partial \mathbf{x}}\right)^T \\ \left(\dfrac{\partial \Pi}{\partial q}\right)^T \end{pmatrix} D^{-2} \begin{pmatrix} \dfrac{\partial \Pi}{\partial \mathbf{x}} & \dfrac{\partial \Pi}{\partial q} \end{pmatrix} \begin{pmatrix} -J^{-1}\dfrac{\partial c}{\partial q} \\ I \end{pmatrix} \right]^{-1}, \tag{23}$$

where $D = \text{diag}\,\{0.08(\hat{\pi}_{i-1} + \hat{\pi}_1)\}$ for the precision of the measurements 8%. Then for every parameter p (p can be η_B, τ_F, k or τ_G) the 95%-confidence interval is $p \pm 3.182\sqrt{(\mathbf{Cov}(q))_{pp}}$.

As we shall see, the confidence intervals of the parameters measured using the simple device shown on Figure 1 are too large and inacceptable for industrial purposes. This disadvantage can be corrected by the shape optimization of the device.

7. Experiment

Here we show results of one of the experiments. For this experiment we used the device with the following sizes: $H = 30\,mm$, $h = 10\,mm$, $L = 244\,mm$, $\alpha \approx 2.35^o$ (cf. Figure 1). The paste was pressed through the device at the average inflow velocity $v_0 = 80\,\frac{mm}{s}$. The measured normal stresses on the upper wall are listed in Table 1. The measurement points are numbered here from the outlet. The precision of the measurements is $\pm 8\%$.

Because of the comparatively low measurement precision the normal stresses obtained from the model do not coincide with the measured ones. The measured stresses do not correspond to real material parameters, so the minimum value of the objective function that can be expected is not equal to 0.

The optimization procedure was started with the initial parameters

$$\eta_B = 0.0005\,bar\cdot s, \quad \tau_F = 3\,bar, \quad \mathrm{k} = 0.5\,\frac{bar\cdot s}{m}, \quad \tau_G = 0.15\,bar$$

on a grid with 969 nodes (896 quadrilaterals) on the finest level. For the regularization constant δ we have chosen a value 0.01. The computed stresses after the parameter estimation are compared with the measurements in Figure 4.

Table 3 represents also the 95% confidence intervals computed for the standard deviation of 0.05 bar of the measured normal stresses. Computation of the

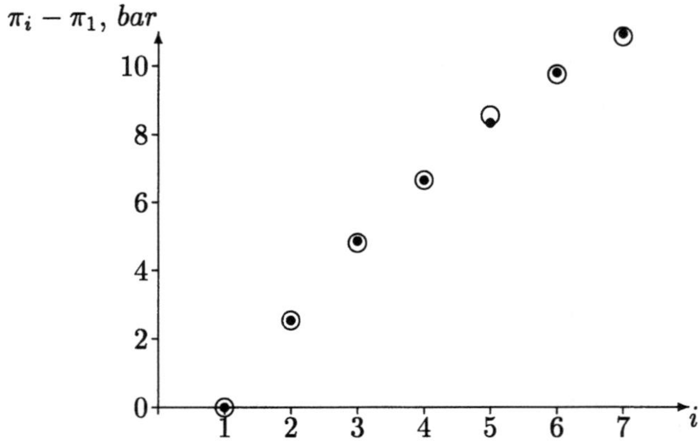

FIGURE 4. The differences of the normal stresses. Empty circles denote the relative measured stresses $(\hat{\pi}_i - \hat{\pi}_1)$, and the black circles show the stresses $\pi_{h,i}(\mathbf{x}, q) - \pi_{h,1}(\mathbf{x}, q)$ computed for the final guess $(\mathbf{x}, q)$ of the optimization procedure. The difference $\pi_{h,1}(\mathbf{x}, q) - \hat{\pi}_1$ is 13.75

covariance matrix requires inverting the Jacobian J that proved to be a very ill conditioned matrix.

As one can see the precision of the method is very poor for all parameters. This happens because in this device, the dependence of the simulated relative normal stresses at the measurement points on the parameters is very weak. This fact can be easily seen from numerical experiments: quite different sets of the parameters yield quite similar differences of the normal stresses. As a remedy one can propose to change the shape of the device. We discuss this question in the next section.

8. Optimization of the shape of the device

One can try to improve the confidence intervals by variing the shape of the device. We consider measurement devices with piecewise planar boundaries (Figure 5). There are 7 pieces and the measurement points are located at the centers of the segments. The horizontal projections of the pieces always have equal length, but the heights h_i are varied. We look for a device that for a given set q of parameters, provides the smallest (in some sence) confidence intervals for all the parameters. As the criterion for the minimization we choose the value

$$\Phi(\mathbf{Cov}) = \frac{1}{4} \operatorname{Tr} \mathbf{Cov}$$

(A-optimal design).

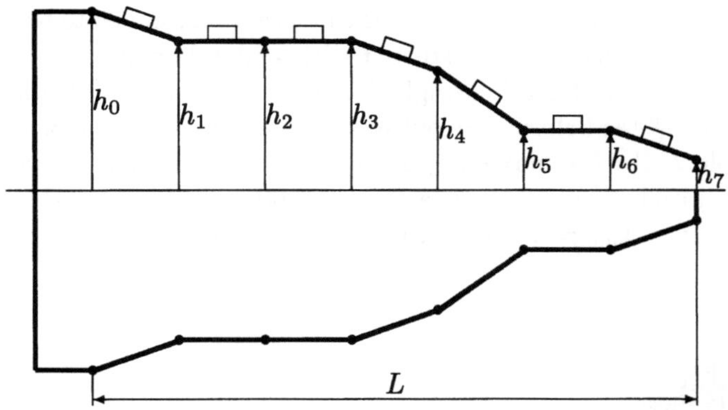

FIGURE 5. Scheme of the measurement device shape at the shape optimization

Here we need the following assumption. As the model determines the normal stresses up to a constant only, the real absolute values of these stresses are unknown, and it is impossible to determine absolute values of relative errors. So we assume that the measurements at all the measurement points are carried out with the same standard deviation $\sigma = 0.05\,bar$. Then we can compute the relative covariations according the formula (23) with $D = 2I$ (because the standard deviation of the differencies is then 2σ), the confidence intervals being $\pm 3.182\sigma\sqrt{(\mathbf{Cov}(q))_{pp}}$

We also impose some restriction on the shape of the device. First, it is reasonable to consider only devices with $h_0 \geq h_1 \geq \cdots \geq h_7$. Then, the practice shows that the heights h_i should be restricted above and below to avoid extremally large or small values. So the complete optimization problem is: for a given set q of the parameters, find $h_0, \ldots, h_7$ such that

$$\begin{array}{rl} \Phi(\mathbf{Cov}(q)) & = \frac{1}{4}\mathrm{Tr}\,\mathbf{Cov}(q) \to \min \\ \mathbf{s.\ t.} & h_0 \geq \cdots \geq h_7, \\ & h_0 \geq h_{\min}, \quad h_7 \leq h_{\max}. \end{array}$$

Note that the objective functional of this problem requires computation of $\mathbf{x}$ such that $c(\mathbf{x}, q) = 0$, i. e. simulation of the flow of the paste in the device. As the variables h_i are included in the problem implicitely (they determine the shape of the domain), it is hardly possible to differentiate the objective functional with respect to them. This implies application of a numerical optimization procedure that does not use the gradients. We applied a direct search method for nonlinear optimization [8], which resulted in h_i-values as listed in Table 2.

TABLE 2. The values of h_i for the optimized device.

i	0	1	2	3	4	5	6	7
h_i, *mm*	59.7	31.8	31.8	25.8	22.7	4.9	3.0	3.0

At the computations, the length of the device was assumed to be $L = 240\,mm$. The optimization procedure was started with a simple quadrilateral shape (like that on the Figure 1) of the followigh size: $h_0 = 30\,mm$, $h_7 = 6\,mm$. The inflow velocity was $80\,\frac{mm}{s}$ as before. The optimization was carried out for the parameter set from Table 2. For the maximal possible value of the height h_i we have chosen the doubled initial value of h_0, i. e. $h_{\max} = 60\,mm$. Analogously, the minimal possible value was $h_{\min} = 3\,mm$ — the halved initial value of h_7.

FIGURE 6. Simulation of flows in the optimized device (7 segments). The upper picture shows the horizontal velocity, the lower one — the pressure.

The shape of the device after the optimization is described by the heights h_i in the Table 2. Results of a simulation of the flows in this device are shown on Figure 6. Table 3 shows the relative covariances and the confidence intervals for the optimized case in comparison with those for the simple shape described in the previous chapters.

FIGURE 7. Simulation of flows in the optimized device with emphasis on τ_G. The upper picture shows the horizontal velocity, the lower one — the pressure.

Although the statistical reliability is now increased by a great amount, still there can be made improvements, when considereing single parameters. For instance, now the parameter determined worst is τ_G. The trace of the covariance matrix as the objective averages out the gain in reliability with respect to the four parameters. If we choose as the objective

$$\Phi(\mathbf{Cov}) = \mathbf{Cov}_{4,4}$$

instead of the trace, we emphasize the influence of τ_G and ignore the the effect of geometry changes on the statistical reliability of the remaining parameters. The resulting optimal geometry is shown in Figure 7.

9. Conclusion

In this paper, we have developed an on-line parameter estimation technique based on measuring the stresses in a flow, and comparing them with simulated ones.

TABLE 3. The parameters and the confidence intervals

	The simple shape		The optimized shape	
Parameter	Rel. covariance	Conf. interval	Rel. covariance	Conf. interval
$\eta_B = 0.302\, bar \cdot s$	27298.8	± 26.287	$9.13 \cdot 10^{-4}$	± 0.0048
$\tau_F = 3.03\, bar$	58651.2	± 38.531	0.107	± 0.052
$k = 0.497 \frac{bar \cdot s}{m}$	84305.5	± 46.195	0.188	± 0.069
$\tau_G = 0.180\, bar$	86.3207	± 1.478	0.259	± 0.081

As it is shown, this method needs a new construction of the measurement device. Its shape is suggested by means of methods influenced from the field of optimum experimental design.

References

[1] P. Bastian, K. Birken, K. Johannsen, S. Lang, N. Neuß, and C. Wieners, *UG — a flexible software toolbox for solving partial differential equations*, Comput Visual Sci., **1** (1997), pp. 27–40.

[2] I. Bauer, H. G. Bock, S. Koerkel, and J. Schlöder, *Numerical methods for optimum experimental design in DAE systems*, in [19] (2000), pp. 1–25.

[3] F. Brezzi, M. Fortin, *Mixed and Hybrid Finite Element Methods*, Springer, New York, 1991.

[4] P. E. Gill, W. Murray, and M. H. Wright, *Practical optimization*, Academic Press, London, 1981.

[5] W. Hackbusch, *Iterative Solution of Large Sparse Systems of Equations*, Springer, New York, 1994.

[6] K. D. Hilf, *Optimale Versuchsplanung zur dynamischen Roboterkalibrierung*, Fortschrittsbesrichte, **8**, VDI, Düsseldorf, 1996.

[7] L. D. Landau and E. M. Lifschitz, *Lehrbuch der Theoretischen Physik*, Band **VI**, Hydrodynamik, Akademischer Verlag, Berlin, 1991.

[8] R. M. Lewis, V. Torczon, and M. W. Trosset, *Direct search methods: then and now*, J. Comput. Appl. Math., **124**, No. 1-2 (2000), pp. 191–207.

[9] Th. Lohmann, H. G. Bock, and J. P. Schlöder, *Numerical methods for parameter identification and optimal experimental design in chemical reaction systems*, Ind. Eng. Chem. Res., **31** (1992), pp. 54–57.

[10] Th. Lohmann, *Ein numerisches Verfahren zur Berechnung optimaler Versuchspläne für beschränkte Parameteridentifizierungsprobleme*, Reihe Informatik, Verlag Shaker, Aachen, 1993.

[11] C. Lund, *Ein Verfahren zur numerischen Simulation instationärer Strömungen mit nichtlinear-viskosen Fließeigenschaften*, Fortschr.-Ber. VDI, Reihe 7 Nr. 344, VDI Verlag, Düsseldorf, 1998.

[12] B. Maar, *Nicht-Newtonsche Fluide. Mehrgitterverfahren für Bingham-Strömungen*, PhD Thesis, Universität Stuttgart, 1998.

[13] M. J. D. Powell, *A fast algorithm for nonlinearly constrained optimization calculations*, in: G. A. Watson (ed.), Numerical Analysis Proceedings, Dundee 1977, Springer, Heidelberg, 1978, pp. 144–157.

[14] F. Pukelsheim, *Optimal design of experiments*, Wiley Series in Probability and Mathematical Statistics, Wiley, New York, 1993.

[15] H. Renz-Reichert, *Robuste Mehrgitterverfahren zur Lösung der inkompressiblen Navier-Stokes Gleichung: Ein Vergleich*, PhD Thesis, Universität Stuttgart, 1996.

[16] G. E. Schneider and M. J. Raw, *Control Volume Finite-Element Method for Heat Transfer and Fluid Flow Using Colocated Variables — 1. Computational Procedure*, Numerical Heat Transfer, **11** (1987), pp. 363–390.

[17] V. H. Schulz, *Reduced SQP methods for large-scale optimal control problems in DAE with application to path planning problems for satellite mounted robots*, PhD Thesis, Universität Heidelberg, 1996.

[18] V. H. Schulz, *Solving discretized optimization problems by partially reduced SQP methods*, Comput Visual Sci, **1** (1998), pp. 83–96.

[19] V. Schulz, guest-editor, *SQP-based direct discretization methods for practical optimal control problems*, Special issue, J. Comput. Appl. Math., **120**, No. 1-2 (2000), pp. 1–254.

[20] G. Wittum, V. Schulz, B. Maar, and D. Logashenko *Numerical methods for parameter estimation in Bingham-fluids*, Mathematics – key technology for the future II, Springer, 2001, to appear.

Sufficient Optimality for Discretized Parabolic and Elliptic Control Problems

Hans D. Mittelmann

Abstract. We study optimal control problems for semilinear parabolic and elliptic equations subject to control and state constraints. We quote known second-order sufficient optimality conditions (SSC) from the literature. Both problem classes are discretized by a finite difference method. The discrete SSC are stated and numerically verified for fine discretizations with the help of sparse linear algebra techniques. This confirms initial results reported earlier for such discretized control problems. In order to relate these results to optimality for the underlying continuous problems corresponding theoretical, especially convergence, results are still unavailable at present.

1. Introduction

A wealth of literature exists on theoretical and computational aspects of control problems for ordinary differential equations. This includes in particular results on necessary and sufficient optimality conditions. Exemplarily, we mention the fundamental works [11, 15] in which it was shown that a *two-norm discrepancy* cannot be avoided in the continuous case. Other recent work on such problems includes [14], [20], and [31]. Research on control problems governed by partial differential equations has started more recently as did the investigation of optimality conditions for these problems. First, nonlinear parabolic control problems were considered, for example, in [8], then elliptic problems in [4]. Some of the most recent works are [5, 21] in which such problems with control and state constraints are treated. Second order sufficient optimality conditions (SSC) for elliptic control problems without such constraints were considered in [12, 13] while SSC were shown and partly also numerically verified for finite element (semi-) discretizations of the one dimensional Burgers equation in [28, 29]. The establishment of SSC requires a highly technical machinery including the use of a third norm as was done before in [6, 14]. The verification of the, partly restrictive, continuous optimality condition is possible only in rare circumstances, such as an analytically known solution for which all conditions can be evaluated analytically; for a parabolic example, see [1]. For the numerical solution of such control problems there are two different approaches, a direct discretization of the entire problem leading to a large finite-dimensional constrained optimization problem. This approach has also been called "all-at-once" method, see, for example, [25], "one-shot" method, "full discretization approach" etc. An iterative method can be applied in function space resulting in a series of infinite-dimensional linear-quadratic control problems which then still need to be discretized. The literature on the latter is quite extensive while the former has more recently been considered for the solution of real-life problems, when progress had been made on solution techniques for very large constrained nonlinear programming (NLP) problems. Exemplarily, we cite the special issue [24].

In [16]–[18] we have considered general semilinear elliptic control problems with control and state constraints. The first-order necessary conditions were formally derived in the continuous case. These were compared to the finite-dimensional conditions for a direct finite-difference discretization of the control problem and the latter was solved by applying state-of-the-art interior point and SQP methods.

The direct discretization will lead to huge NLP's for three-dimensional elliptic or two and three-dimensional parabolic problems. If the discretization is consistent, preferably of higher order in the discretization parameter, the numerical verification of the SSC for these problems, even for relatively coarse discretizations, may be expected to yield clues on the optimality of the approximated continuous solution. It is in this spirit that in [19] a class of both parabolic and elliptic control problems in one respectively two space dimensions was considered. The problems were discretized and the SSC were verified. However, the use of dense linear algebra techniques limited the size of the problems considerably and an argument could be made that these results need to be confirmed for the same problem sizes for which the control problems themselves were solved and for which, at least in some cases, actual error evaluations could be done showing the quality of approximation of the underlying continuous problem by its discretization.

In the following section both the parabolic and elliptic control problems are stated. Some known SSC results are quoted in the third section. In section 4 the way in which the discrete SSC are verified is described followed in section 5 by results for all parabolic cases from [19], in addition one with state constraints as well as the instationary *Burgers* equation, and one exemplary elliptic boundary control problem from [19, 18]. Concluding remarks are made in the last section.

2. Parabolic and Elliptic Control Problems

In this section we describe the continuous control problems *exemplarily* considered in the following. The first is a one-dimensional parabolic boundary control problem formulated such that it includes problems from [1, 9] and has additional pointwise state constraints.

Minimize

$$\begin{aligned} f(y,u) &= \frac{1}{2}\int_0^l (y(x,T)-y_T(x))^2dx + \frac{\alpha}{2}\int_0^T u(t)^2dt \\ &+ \int_0^T (a_y(t)y(l,t)+a_u(t)u(t))dt, \quad \alpha > 0 \end{aligned}$$

subject to (P)

$$\begin{aligned} & y_t - y_{xx} = 0 \quad \text{in } (0,l)\times(0,T) \\ & y(x,0) = a(x) \quad \text{in } (0,l) \\ & y_x(0,t) = 0 \quad \text{in } (0,T) \\ y_x(l,t) \; + \; & \beta y(l,t) = b(t) + u(t) - \varphi(y(l,t)) \quad \text{in } (0,T) \\ & \alpha_1 \le u(t) \le \alpha_2 \quad \text{in } (0,T) \\ & \alpha_3 \le y(x,t) \le \alpha_4 \quad \text{in } (0,l)\times(0,T) \end{aligned}$$

The notation above is that of [1] and the corresponding data will be defined in section 5 below. The problem does, however, include also one considered in [9], namely by choosing $b(t)=0,\ \varphi(y(l,t)) = y^2(l,t), a_y(t) = 0, a_u(t) = 0$.

We define the following discretization of problem (P).

$$\begin{aligned} \text{Minimize} \quad & f_h(y_h,u_h) = \frac{dx}{4}((y_{0,m}-y_T(x_0))^2 \\ & +2\sum_{j=1}^{n-1}(y_{j,m}-y_T(x_j))^2 + (y_{n,m}-y_T(x_n))^2) + \frac{\alpha dt}{4}\left(2\sum_{j=1}^{m-1}u_j^2 + u_m^2\right) \\ & +dt\left(\sum_{j=1}^{m-1}(a_y(t_j)y_{n,j}+a_u(t_j)u_j) + \frac{1}{2}a_y(T)y_{n,m} + a_u(T)u_m\right) \end{aligned}$$

subject to (P_h)

$$\begin{aligned} & \frac{y_{j,i+1}-y_{j,i}}{dt} = \frac{1}{2}(y_{j-1,i} - 2y_{j,i} + y_{j+1,i} \\ & +y_{j-1,i+1} - 2y_{j,i+1} + y_{j+1,i+1})/dx^2 \\ & \qquad i = 0,\dots,m-1, \quad j = 1,\dots,n-1 \\ & y_{j,0} = a(x_j), \quad j = 0,\dots,n \\ & y_{2,i} - 4y_{1,i} + 3y_{0,i} = 0, \quad i = 1,\dots,m \\ & (y_{n-2,i} - 4y_{n-1,i} + 3y_{n,i})/(2dx) + \beta y_{n,i} \\ & = u_i + b(t_i) - \varphi(y_{n,i}), \quad i = 1,\dots,m \\ & \alpha_1 \le u_i \le \alpha_2, \quad i = 1,\dots,m \\ & \alpha_3 \le y_{j,i} \le \alpha_4, \quad j = 1,\dots,n, i = 1,\dots,m. \end{aligned}$$

Here $x_j = jdx$, $dx = 1/n$, $t_j = jdt$, $dt = T/m$.

For the problem (P) above and specific data an analytic solution is given in [1] and this also permits the authors to verify the necessary and sufficient optimality conditions they had stated and proved.

Next, we consider the instationary Burgers equation

$$y_t = \nu y_{xx} - y y_x \quad \text{in } (0,l) \times (0,T)$$

which we discretize by

$$\begin{aligned}
&\frac{y_{j,i+1} - y_{j,i}}{dt} = \nu \frac{1}{2}(y_{j-1,i} - 2y_{j,i} + y_{j+1,i} \\
&+ y_{j-1,i+1} - 2y_{j,i+1} + y_{j+1,i+1})/dx^2 \\
&-\frac{1}{8}(y_{i+1,j} + y_{i,j})(y_{i,j+1} - y_{i,j-1} + \\
&y_{i+1,j+1} - y_{i+1,j-1})/dx \\
&\qquad i = 0, \ldots, m-1, \quad j = 1, \ldots, n-1
\end{aligned}$$

For the elliptic control problems, we consider the class defined in [18]. It includes boundary and distributed controls which are addressed separately in [16], respectively [17] as well as Dirichlet, Neumann, and mixed boundary conditions. In the case of boundary control the underlying continuous problem is

minimize

$$\begin{aligned}
F(y,u) &= \int_{\Omega_0} f(x, y(x))dx + \int_{\Gamma_1} g(x, y(x), u(x))ds \\
&\quad + \int_{\Gamma_2} k(x, u(x))ds
\end{aligned}$$

subject to (EB)

$$\begin{aligned}
&-\Delta y(x) + d(x, y(x)) = 0, \quad \text{for } x \in \Omega, \\
&\partial_\nu y(x) - b(x, y(x), u(x)) = 0, \quad \text{for } x \in \Gamma_1, \\
&y(x) = a(x, u(x)), \quad \text{for } x \in \Gamma_2,
\end{aligned}$$

and

$$\begin{aligned}
C(x, u(x)) \le 0, &\quad \text{for } x \in \Gamma, \\
S(x, y(x)) \le 0, &\quad \text{for } x \in \Omega.
\end{aligned}$$

Here Ω is a bounded, plane domain with piecewise smooth boundary Γ. $\Omega_0 \subset \Omega$ is equal to Ω unless noted otherwise. ∂_ν denotes the derivative in the direction of the outward unit normal ν on Γ and the boundary is partitioned as $\Gamma = \Gamma_1 \cup \Gamma_2$ with disjoint sets Γ_1, Γ_2 consisting of finitely many connected components. For the general formulation given above in [17, 18], necessary optimality conditions are stated, a discretization is described in full detail and the corresponding optimality conditions are related carefully to those for the continuous problem. With concrete applications in mind then, however, the following data of the problem are specialized, the objective function

$$F(y,u) = \frac{1}{2}\int_{\Omega_0}(y(x)-y_d(x))^2dx + \frac{\alpha}{2}\int_{\Gamma}(u(x)-u_d(x))^2ds \tag{2.1}$$

with $\alpha \geq 0$ and the inequality constraints

$$y(x) \leq \varphi(x), \quad x \text{ in } \Omega, \quad u_1(x) \leq u(x) \leq u_2(x) \quad \text{on } \Gamma. \tag{2.2}$$

3. Known SSC Results for the Continuous Problems

No attempt will be made to quote all relevant results on optimality conditions. In the case of problem (P) without state constraints the following set of SSC is stated in [1]. First, the Lagrange function

$$\begin{aligned} L(y,u,p) \quad = \quad & F(y,u) - \int_0^T\int_0^l (y_t - y_{xx})p(x,t)dxdt \\ & + \int_0^T y_x(0,t)p(0,t)dt \\ & - \int_0^T (y_x(l,t)+y(l,t))p(l,t)dt \\ & + \int_0^T (b(t)+u(t)-\varphi(y(l,t)))p(l,t)dt \end{aligned}$$

is defined with the Lagrange multiplier function p. Then the second derivative of L with respect to (y,u) is called L'' and is evaluated at a point $(\bar{y},\bar{u},\bar{p})$ satisfying the first order optimality conditions. The key requirement is the inequality

$$L''(\bar{y},\bar{u},\bar{p})(y,u)^2 \geq \|u\|^2_{L^2(0,T)} \tag{3.1}$$

which has to hold for all (y,u) which satisfy the linearized (at $(\bar{y},\bar{u})$) constraints. In the example considered in [1] and below, the inequality even holds for all (y,u). Locally, then, $(\bar{y},\bar{u})$ is a minimizer of problem (P).

SSC have not been stated exactly for the elliptic control problems (EB) in the previous section, but a series of papers address special cases. In [4], for example, a good overview of the literature is given and the boundary control problem covered in much technical detail is nearly identical to (EB) except for

$$d(x,y(x)) = y(x), \quad \Gamma_1 = \Gamma.$$

Finally, the state-constrained case is addressed in [5] for (EB) and in [21] for (P).

4. Second Order Sufficient Conditions for the Discretized Problems

The control problems (P) and (EB) defined in the previous sections lead after suitable discretization to nonlinear finite-dimensional optimization problems of the form

$$\min F^h(z) \quad \text{subject to} \quad G^h(z) = 0, \quad H(z) \leq 0 \tag{4.1}$$

where z comprises the discretized control and state variables.

A discretization (P_h) of (P) is given in section 2 while the elliptic problem is assumed to be discretized as described in detail in [16, 18]. $G^h(z)$ symbolizes the state equation and boundary conditions while $H(z)$ denotes both pointwise control and state constraints, the only constraints of inequality type prescribed above. Thus, alternatively, it can be written as

$$z_l \leq z \leq z_u \tag{4.2}$$

where components of $z_l(z_u)$ are taken as $-\infty(\infty)$ if no constraint is imposed on the component.

We state the well-known SSC for (4.1), assuming $z \in \mathbf{R}^n$, $G^h : \mathbf{R}^n \to \mathbf{R}^m$, $m < n$. Let z^* be an admissible point satisfying the first-order necessary optimality conditions with associated Lagrange multipliers μ^* and λ^*. Let further

$$N(z^*) = (\nabla G^h(z^*), \nabla H_a(z^*))$$

be a column-regular $n \times (m+p)$ matrix where $m + p < n$ and $\nabla H_a(z^*)$ denotes the gradients of the p active inequality constraints. Let finally $N = QR$ be a QR decomposition and $Q = (Q_1, Q_2)$ a splitting into the first $m+p$ and the remaining columns.

The point z^* is a strict local minimizer if a $\gamma > 0$ exists such that, see, for example, [26]

$$\lambda_{\min}(L_2(z^*)) = \gamma > 0. \tag{4.3}$$

Here $L_2(z^*)$ is the projected Hessian of the Lagrangian

$$L_2(z^*) = Q_2^T(\nabla^2 F^h(z^*) - \mu^{*T}\nabla^2 G^h(z^*))Q_2.$$

No Hessian of H appears on the right due to its linearity. Next, we will detail how condition (4.3) will be checked for the discrete versions of both the parabolic and the elliptic control problems. As was already done in [16, 18] the control problems are written in the form of AMPL [7] scripts. This way, a number of nonlinear optimization codes can be utilized for their solution. It had been an observation in our previous work that from the currently available codes only LOQO [27] is able to solve all the problems effectively and for sufficiently fine discretizations. The following is independent of the solver used.

After computing a solution an AMPL *stub* (or **.nl*) file is written as well as a file with the computed Lagrange multipliers. This allows to check the SSC (4.3) with the help of a Fortran, alternatively, a C or Matlab, program.

This program reads the files and verifies first the necessary first-order optimality conditions, the column regularity of $N(z^*)$ and the strict complementarity. For this, it utilizes routines provided by AMPL which permit evaluation of the objective and constraint gradients. Next, the the QR decomposition of $N(z^*)$ is computed by one of the methods exploiting sparsity. We have utilized the algorithm described in [22]. AMPL also provides a routine to multiply the Hessian of the Lagrangian times a vector. This is called with the columns of Q_2 and thus $L_2(z^*)$ can be formed. Its eigenvalues are computed with LAPACK routine DSYEV and the smallest eigenvalue $\gamma = \gamma_h$ is determined. The use of this eigenvalue routine is possible since the order of the matrices corresponding to the "free" control variables is moderate. In case of distributed control problems when this number may be on the order of the state variables, a sparse solver, preferably just for finding the minimal eigenvalue, will have to be used.

5. Verification of SSC Conditions

In this section numerical results will be reported for the application of the method outlined in the previous section to three parabolic control problems from [1, 9] and an exemplary elliptic control problems from [19]. The first parabolic problem is particularly interesting because for it in [1] an analytical exact solution is given and the continuous SSC conditions are verified. For the sake of completeness, the specification will be given for each problem. The discretizations used are those defined in section 2 for the parabolic and in [16]–[18] for the elliptic problem.

Problem (P) from section 2 is solved without state constraints and with the following data

$$\begin{aligned}
&l = \pi/4, \quad T = 1, \quad \alpha = \frac{\sqrt{2}}{2}(e^{2/3} - e^{1/3}) \\
&y_T(x) = (e + e^{-1}) \cos x, \quad \alpha_1 = 0, \quad \alpha_2 = 1 \\
&a(x) = \cos x, \quad a_y(t) = -e^{-2t}, \quad a_u(t) = \frac{\sqrt{2}}{2} e^{1/3} \\
&b(t) = \frac{1}{4} e^{-4t} - \min\left(1, \max\left(0, \frac{e^t - e^{1/3}}{e^{2/3} - e^{1/3}}\right)\right) \\
&\varphi(y) = y|y|^3, \quad \beta = 1.
\end{aligned} \tag{5.1}$$

As is shown in [1] a local optimum for this problem is the pair $(\bar{y}, \bar{u})$,

$$\begin{aligned}
\bar{y}(x,t) &= e^{-t} \cos x \\
\bar{u}(t) &= \min\left(1, \max\left(0, \frac{e^t - e^{1/3}}{e^{2/3} - e^{1/3}}\right)\right).
\end{aligned}$$

The discretization (P_h) of section 2 was coded in AMPL and solved with LOQO. Afterwards, the SSC were checked as described in the previous section resulting in the minimal eigenvalue γ_h, cf. (4.3).

$1/h$	u-error	y-error
100	7.444e-5	1.208e-5
200	1.884e-5	3.128e-6
300	9.279e-6	1.412e-6

TABLE 1. Solution errors for problem 5.1

$1/h$	γ_h	γ_h/h^2
100	1.774e-4	1.77
200	4.266e-5	1.71
300	1.923e-5	1.73

TABLE 2. Minimal eigenvalue for problem 5.1

In Table 1 the maximum errors over the grid points are listed for the computed control and state functions while in Table 2 the smallest eigenvalue is listed as well as the value scaled by h^{-2}. The errors exhibit quadratic convergence while the scaled eigenvalue stays nearly constant with a slightly increasing tendency. These facts justify the application of the technique to other control problems for which no exact solution is known and it also shows which scaling of γ_h is appropriate for problem (P_h). This scaling is also suggested by known quadratic convergence estimates for fixed control in L^∞ since the differential equation as well as all boundary conditions were discretized with second-order consistency. An argument could be made that the smallness of the eigenvalue may indicate that it is, in fact, negative and the relatively coarse discretization prevents it from exhibiting this fact. This seems highly unlikely in the light of the very uniform and even slightly growing behavior of the scaled value combined with the fact that all the eigenvalues were computed and the smallest one is not considerably smaller than the next ones but they grow only gradually.

As a second parabolic case one from [9] was chosen. The data for (P) are

$$
\begin{aligned}
& l = 1, \quad T = 1.58, \quad \alpha = .001 \\
& y_T(x) = .5(1 - x^2), \quad \alpha_1 = -1, \quad \alpha_2 = 1 \\
& a(x) = 0, \quad a_y(t) = 0, \quad a_u(t) = 0, \qquad (5.2) \\
\text{(I)} \quad & b(t) = 0, \quad \varphi(y) = 0, \quad \beta = 1 \\
\text{(II)} \quad & b(t) = 0, \quad \varphi(y) = y^2, \quad \beta = 0 \\
\text{(III) as (II) and} \quad & \alpha_3 = 0, \quad \alpha_4 = .675
\end{aligned}
$$

The case (I) leads to a linear-quadratic control problem and had already been considered in [23]. Here, both lower and upper control bounds are active in the solution while for (II) the lower control bounds are active only, and for (III) additionally the lower and upper state constraints. For all cases just the minimal eigenvalue is listed without and with the same scaling as in the previous example.

$1/h$	γ_h	γ_h/h^2
100	1.908e-6	1.91e-2
200	4.726e-6	1.89e-2

TABLE 3. Minimal eigenvalue for 5.2-I

$1/h$	γ_h	γ_h/h^2
100	9.431e-7	9.43e-3
200	2.447e-7	9.79e-3

TABLE 4. Minimal eigenvalue for 5.2-II

$1/h$	γ_h	γ_h/h^2
100	1.010e-6	1.01e-2
200	2.611e-7	1.04e-2

TABLE 5. Minimal eigenvalue for 5.2-III

We include a plot of the optimal control and state for this problem. Due to the consistent results obtained the computations for $1/h = 300$ were not done in these cases.
Finally, we solve a boundary control problem for the instationary Burgers equation, see also [30]. For simplicity we used the same data as for case (II) above with $\nu = .01$ and the control bounds $\alpha_1 = 0.1, \alpha_2 = 0.6$ both of which become active in the solution. This case is denoted by 5.2-IV. It clearly shows that for this problem the suitable normalization of the eigenvalue is h^{-1}. This will have to be confirmed by a careful comparison of the continuous and discrete SSC as had been done for the first order necessary conditions in [16, 17]. A similar situation arose in the elliptic boundary control problem in [18] which is addressed next below. For it in [19] the scaling by h^{-2} seemed appropriate in contrast to the other such problems considered.

Next, we present the data for the elliptic boundary control problem from [18] and the eigenvalues obtained. The domain is the unit square. The problem (EB) together with (2.1) and (2.2) is considered and $u_d = 0$.

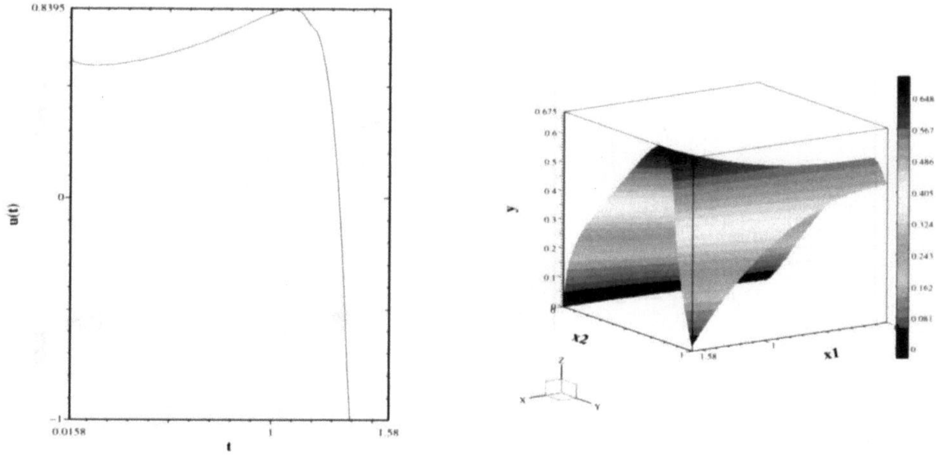

FIGURE 1. Example 5.2-III, Optimal control and state

$1/h$	γ_h	γ_h/h
100	1.211e-5	1.21e-3
200	5.939e-6	1.19e-3

TABLE 6. Minimal eigenvalue for 5.2-IV

$$
\begin{aligned}
&d(x,y) \equiv 0; \quad g(x,y,u) \equiv 0; \quad b(x,y,u) = 0, \quad \text{on} \quad x_2 = 0,\\
&b(x,y,u) = y - 5 \quad \text{on} \quad x_1 = 0, \quad x_1 = 1,\\
&a(x,u) = u \quad \text{on} \quad \{x_2 = 1\} = \Gamma_2,\\
&\psi(x) = 3.15 \quad \text{in} \quad \Omega_0, \quad \psi(x) = 10 \quad \text{in} \quad \Omega\backslash\Omega_0,\\
&u_1 = 0, \quad u_2 = 10 \quad \text{on} \quad \Gamma_2\\
&y_d \equiv 1, \quad \Omega_0 = [.25,.75]^2, \quad \alpha = .005
\end{aligned}
$$

$1/h$	γ_h	γ_h/h^2
120	1.472e-5	2.12e-1
180	6.525e-6	2.11e-1

TABLE 7. Minimal eigenvalue for (EB)

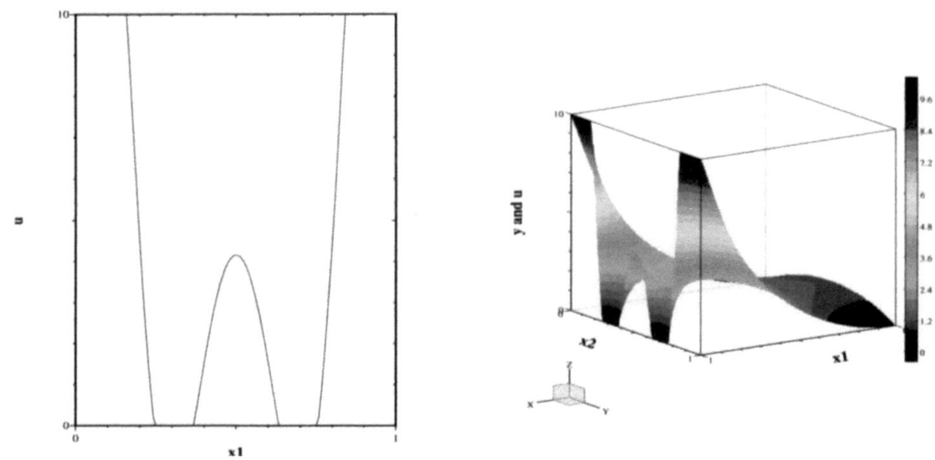

FIGURE 2. Example 5.3, Optimal control and state

We also include a plot of the optimal control and state for this problem. For plots of the adjoint variables and further details, we refer to [18].

6. Conclusion

In this work the second-order sufficient optimality conditions (SSC) were verified numerically for several optimal control problems taken from recent papers. For a parabolic problem with analytically known solution a second-order finite-difference discretization was shown to have good accuracy already for moderate discretizations. Computational results on coarse grids from [19] were confirmed. They show a very consistent trend: the smallest eigenvalue of the projected Hessian of the Lagrangian suitably scaled by the discretization parameter behaves nearly constant indicating that the computed stationary solution appears to be a strict local minimizer. Subsequently, the procedure is applied to three other parabolic problems including one with pointwise state constraints and to one elliptic control problem, confirming the SSC in each case.

Open questions that should be addressed in future work include the following: can a formal proof be given of the satisfaction of the SSC which were numerically verified above? Can the results be generalized to the singular and bang-bang controls observed in [16]–[18]? Which numerical results can be obtained for state-constrained parabolic control problems? It is to be expected that such PDE-constrained optimization problems as considered in this paper will be subject of intense research efforts in the near future. We refer again to [24]; see also the short survey [10].

References

[1] N. Arada, J.-P. Raymond, and F. Tröltzsch, *On an augmented Lagrangian SQP method for a class of optimal control problems in Banach spaces*, to appear.

[2] J. F. Bonnans, *Second-order analysis for control constrained optimal control problems of semilinear elliptic systems,* Appl. Math. Optim., **38** (1998), pp. 303–325.

[3] Ch. Büskens and H. Maurer, *SQP-methods for solving optimal control problems with control and state constraints; adjoint variables, sensitivity analysis, and real-time control,* J. Comp. Appl. Math., **120** (2000), pp. 85–108.

[4] E. Casas, F. Tröltzsch, and A. Unger, *Second order sufficient optimality conditions for a nonlinear elliptic control problem,* J. Anal. Appl., **15** (1996), pp. 687–707.

[5] E. Casas, F. Tröltzsch, and A. Unger, *Second order sufficient optimality conditions for some state-constrained control problems of semilinear elliptic equations,* to appear in SIAM J. Control Optim.

[6] A. L. Dontchev, W. W. Hager, A. B. Poore, and B. Yang, *Optimality, stability, and convergence in optimal control,* Appl. Math. Optim., **31** (1995), pp. 297–326.

[7] R. Fourer, D. M. Gay, and B. W. Kernighan, *AMPL: A modeling language for mathematical programming,* Duxbury Press, Brooks/Cole Publishing Company, Pacific Grove, CA, 1993.

[8] H. Goldberg and F. Tröltzsch, *Second order sufficient optimality conditions for a class of nonlinear parabolic boundary control problems,* SIAM J. Control Optim., **31** (1993), pp. 1007–1025.

[9] H. Goldberg and F. Tröltzsch, *On a Lagrange-Newton method for a nonlinear parabolic boundary control problem,* Optim. Meth. Software, **8** (1998), pp. 225–247.

[10] M. Heinkenschloss, *SQP interior-point methods for distributed optimal control problems,* to appear in Encyclopedia of Optimization, P. Pardalos and C. Floudas, eds., Kluwer Academic Publishers.

[11] A. D. Ioffe, *Necessary and sufficient conditions for a local minimum, part 3: Second order conditions and augmented duality,* SIAM J. Control Optim., **17** (1979), pp. 266–288.

[12] K. Ito and K. Kunisch, *Augmented Lagrangian-SQP methods for nonlinear optimal control problems of tracking type,* SIAM J. Control Optim., **34** (1996), pp. 874–891.

[13] K. Ito and K. Kunisch, *The Newton algorithm for a class of weakly singular optimal control problems,* to appear in SIAM J. Optim.

[14] K. Malanowski, *Sufficient optimality conditions for optimal control problems subject to state constraints,* SIAM J. Control Optim., **35** (1997), pp. 205–227. 1994.

[15] H. Maurer, *First and second order sufficient optimality conditions in mathematical programming and optimal control,* Math. Programming Study, **14** (1981), pp. 163–177.

[16] H. Maurer and H. D. Mittelmann, *Optimization techniques for solving elliptic control problems with control and state constraints. Part I: Boundary control,* Comput. Optim. Appl., **16** (2000), pp. 29–55.

[17] H. Maurer and H. D. Mittelmann, *Optimization techniques for solving elliptic control problems with control and state constraints. Part II: Distributed control,* to appear in Comput. Optim. Appl.

[18] H. D. Mittelmann and H. Maurer, *Solving elliptic control problems with interior and SQP methods: control and state constraints,* J. Comp. Appl. Math., **120** (2000), pp. 175–195.

[19] H. D. Mittelmann, *Verification of Second-Order Sufficient Optimality Conditions for Semilinear Elliptic and Parabolic Control Problems,* to appear in Comput. Optim. Appl.

[20] H. Maurer and S. Pickenhain, *Second-order sufficient conditions for control problems with mixed control-state constraints,* J. Optim. Theory Appl., **86** (1995), pp. 649–667.

[21] J.-P. Raymond and F. Tröltzsch, *Second order sufficient optimality conditions for nonlinear parabolic control problems with state constraints,* to appear.

[22] Th. H. Robey and D. L. Sulsky, *Row ordering for Sparse QR Decomposition,* SIAM J. Matrix Anal. Applic., **15** (1994), pp. 1208–1225.

[23] K. Schittkowski, *Numerical solution of a time-optimal parabolic boundary-value control problem,* J. Optim. Theory Appl., **27** (1979), pp. 271–290.

[24] V. H. Schulz (ed.), *SQP-based direct discretization methods for practical optimal control problems,* J. Comput. Appl. Math. **120,** (2000) (special issue).

[25] A. R. Shenoy, M. Heinkenschloss, and E. M. Cliff, *Airfoil design by an all-at-once method,* Intern. J. Comp. Fluid Dynam., **11** (1998), pp. 3–25.

[26] P. Spellucci, *Numerische Verfahren der nichtlinearen Optimierung,* Birkhäuser-Verlag, Basel, 1993.

[27] R. J. Vanderbei and D. F. Shanno, *An interior-point algorithm for nonconvex nonlinear programming,* Comput. Optim. Appl., **13** (2000), pp. 231–252.

[28] S. Volkwein, *Distributed control problems for the Burgers equation,* to appear in Comput. Optim. Appl.

[29] S. Volkwein, *Application of the augmented Lagrangian-SQP method to optimal control problems for the stationary Burgers equation*, Comput. Optim. Appl., **16**, No. 1 (2000), pp. 57–81.

[30] S. Volkwein, *Mesh-independence for an augmented Lagrangian-SQP method in Hilbert spaces*, SIAM J. Control Optim., **38**, No. 3 (2000), pp. 767–785.

[31] V. Zeidan, *The Riccati equation for optimal control problems with mixed state-control constraints: Necessity and Sufficiency,* SIAM J. Control Optim., **32** (1994), pp. 1297–1321.

Global Quadratic Dynamic Minimization Approaches for Shape Design with Incomplete Sensitivities on CAD-Free Parameterization of Multi-Disciplinary Configurations

Bijan Mohammadi

Abstract. We would like to show how to perform optimization for realistic configurations at a cost comparable to the computational cost of the direct problem. To this end, we impose to use only a set of informations on the cost function and state and also incomplete sensitivities not requiring the solution of any linearized problem.

1. Introduction

Control of distributed systems has various possible industrial applications as we are usually interested in keeping complex systems in some given states. In this paper, we consider the control of the behavior of a complex structure (for instance an aircraft) under external aerodynamical forces. The problem is therefore multi-disciplinary as it involves several coupled physical models.

Control space definition or parameterization is the first main issue we face when targeting to formulate a control problem. Usually, one wishes to keep the parameterization space dimension as low as possible and hence limit the complexity of the problem. In addition, for any control approach to be efficient, it has to be realizable during the time the system is still controllable. Computational cost is therefore another essential issue. Our aim through the paper is to discuss alternatives to these two difficulties.

The first point we would like to discuss is the behavior of our platform on two complementary situations: where the number of control is small (basically one) and where the number of control is large (assimilated to all the points in the CAD-free parameterization for instance). Our aim is to present our sub-optimal control technique using accurate gradient evaluation for the first class of problem and for the second class of applications to show that the sub-optimal control is also efficient using incomplete evaluation of these gradient, but this only for some class of cost functions.

The second point of interest is to formulate minimization algorithms as dynamic or pseudo-unsteady systems. This formulation makes easier to understand the coupling between the different ingredients. We try to reformulate our optimization problem as the fixed point (or steady solution) of a second order dynamic system. In addition, we try for the system to have global search features using the natural instability of second order hyperbolic systems [2, 3].

The problem being multi-disciplinary, we need to couple different state equations during control. In that context, the dynamic system above is seen as a state equation for the parameterization.

For a control algorithm based on some gradient method to be efficient, we need a cheap and easy gradient evaluation. But, as we said, we would like the design to have the complexity of the direct problem. This implies to somehow avoid the adjoint equation solution.

The third key idea is therefore our approximate gradient evaluation where the sensitivity with respect to the state is neglected in the gradient for cost functions lying on the shape. This being often the case in applications, as cost functions are often based on boundary integrals, a cheap evaluation of the gradient avoiding the state equations solvers differentiation is therefore possible.

2. Dynamic shape optimization and state control

Consider the following optimization or control problem:

$$\min_{x(t)} J(x(t), q(x), U(q), \nabla U(q)), \quad E(x(t), q(x), U(q), \nabla U(q)) = 0, \tag{1}$$

$$g_1(x(t)) \leq 0, g_2(q) \leq 0, g_3(q, U(q)) \leq 0,$$

where $x \in R^n$ describes our parameterization. This is for optimization problems a geometrical CAD-free model [7] and for control problems, the amount of the injection/suction velocity or a flap deflection [9, 10, 11, 1]. In the first and second cases the size n of the control space is large and for the last case it is small. When using injection/suction devices, the locations of the injection devices are optimized following the local norm of the sensitivities for the unperturbated flow. Once the location is decided, it is kept fixed during control. q describes geometrical entities (normals, surfaces, volumes,...). $U \in R^N$ denotes the state variables. $E \in R^N$ is the time dependent state equation. In this platform, we consider for the fluid side the Euler system in conservation variables or the Navier-Stokes and the $k-\varepsilon$ system and on the structure side a rigid body movement model or a CAD-free elastic model [9, 11]. However, other simpler equations are also available which are useful for sensitivity evaluation in the sense that even if the state variables are computed using complex state equations, reduced physics can be accounted using simplified models [11, 1]. g_1 defines the constraints on the parameterization, g_2 those on geometrical quantities and g_3 state constraints. Details of these are available in [1].

2.1. Closure equation for $x(t)$

In our approach minimization algorithms are seen as closure equation for the parameterization. In other words, we introduce a new time dependent problem for $x(t)$. This can also be seen as an equation for the structure. We can show that most linear or quadratic gradient based minimization algorithms can be put under the following form:

$$-\dot{x} + \epsilon\, \ddot{x} = F(\Pi, M^{-1}, (\nabla_{xx}J)^{-1}, \nabla_x J), \tag{2}$$

where F is a function of the exact or incomplete gradient and of the inverse of the Hessian of the cost function. It also takes into account the projection over the admissible space Π and the smoothing operator we use when using the CAD-free parameterization [7, 1]. Usually, Π does not depend on p except when using mesh adaptation [6, 4, 5, 7].

Let us consider the particular case of $\epsilon > 0$, where we recover the so called heavy ball method [2, 3, 1, 12]. The aim in this approach is to access different minima of the problem and not only the nearest local minimum by helping to escape from local minima after introduction of second order perturbation term. The difference with the original heavy ball method is based on the fact that here the method is seen as a perturbation of the first order derivative while in the original heavy ball method the steepest descent is seen as a perturbation of the hyperbolic second order system. This reformulation is suitable for numerical experiences as it enables to tune the perturbation to be the weakest possible. Indeed, otherwise especially for complex applications, the optimization process becomes hard to control.

Another interesting feature of dynamic minimization algorithm is a possible coupling between several balls (point in the admissible control space) to improve the global search ability of the method by communicating informations between balls on their respective state. The idea is therefore to solve the pseudo-unsteady system (2) from different ball positions and to couple the paths using cross informations involving a global gradient [12]. Consider, q balls $x_j, j = 1, .., q$, following the motion prescribed by q pseudo-unsteady systems:

$$-\dot{x_j} + \epsilon\, \ddot{x_j} = -(F_j + G_j), \tag{3}$$

where F_j is as in (2) and G_j a global gradient representing the interaction between balls (recall that each ball is a design configuration). This has also shown a stability improvement. To be able to reach the global minima, the number of balls has to be enough large. A good estimation for this number is given by the dimension of the design space (n). Even with this number the complexity is negligible compared to those of evolutionary algorithms. Our experience shows that the following choice of G_j gives satisfaction (see example below):

$$(G_j)_i = \sum_{k=1, k\neq j}^{q} \frac{J_j - J_k}{||x_j - x_k||^2}(x_{j_i} - x_{k_i}), \quad \text{for} \quad j = 1, .., q, \quad i = 1, .., n.$$

However, in CAD-free parameterization, n can be quite large and, due to the required computational effort for one simulation, we cannot afford for more than a few (say 3 or 4) shape evolutions at the same time. This approach can be seen therefore as an improvement of the search capacity of the original algorithm. In addition, the process is suitable for a distributed treatment as in evolutionary type minimizations.

We show below the behavior of the pseudo-unsteady systems (3) with two balls and constant λ and ϵ, is shown in Figure (1-left) for the minimization of the following function. This function has several local minima (the global minimum is reached at (0,0)). For $(x, y) \in]-10, 10[\times]-10, 10[$ consider J defined by:

$$J(x,y) = 1 - \cos(x)\cos(\frac{y}{\sqrt{2}}) + \frac{1}{50}((x - 0.5 * y)^2 + 1.75y^2).$$

The aim is to show that heavy ball method improves global minimum search by helping to escape from local minima (Figure 2-left). Finding the global minima requires however several trials. But coupling several heavy balls can help finding the global minima (Figure 2) using balls not converging to the global minima individually (Figure 1-right).

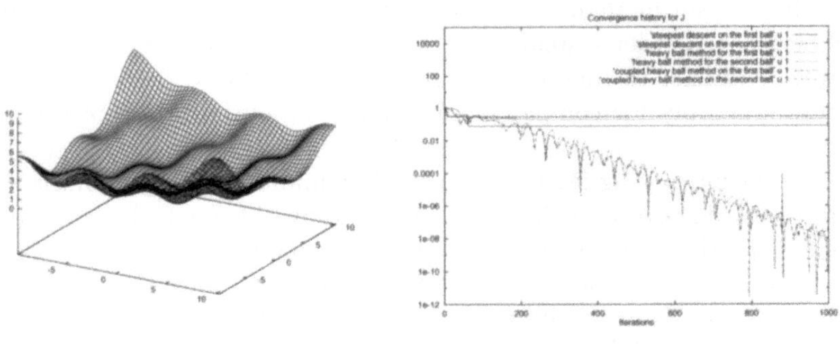

FIGURE 1. *Convergence histories for the steepest descent and heavy ball methods starting from two different points, all captured by local minima.*

3. Sensitivities and incomplete sensitivities

Consider the general simulation loop, involved in (1), leading from shape parameterization or control variables to the cost function:

$$J(x) : x \to q(x) \to U(q(x)) \to J(x, q(x), U(q(x))).$$

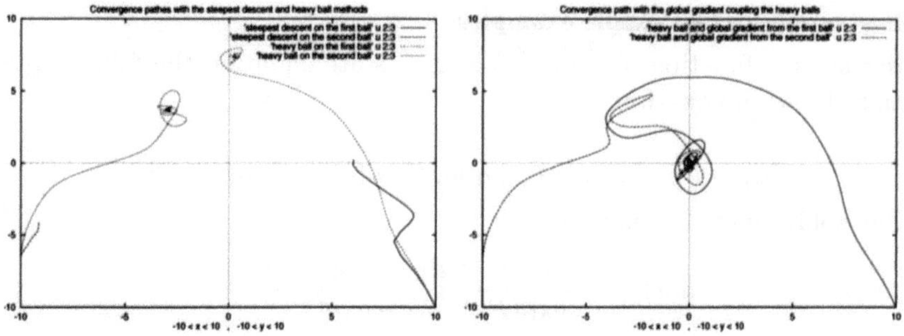

FIGURE 2. *Paths for the steepest descent and heavy ball methods starting from two different points (left). When coupling the two balls in heavy ball method using the global gradient, the global minimum is reached (right).*

The gradient of J with respect to x is:

$$\frac{dJ}{dx} = \frac{\partial J}{\partial x} + \frac{\partial J}{\partial q}\frac{\partial q}{\partial x} + \frac{\partial J}{\partial U}\frac{\partial U}{\partial q}\frac{\partial q}{\partial x}. \tag{4}$$

If the following requirements hold, we can introduce incomplete evaluation of this gradient, reducing its evaluation cost:

- If both the cost function and control space are defined on the shape (or a same part of it),
- If J is of the form

$$J(x) = \int_{\text{shape or part of the shape}} f(x)g(u)d\gamma,$$

- If the local curvature of the shape is not too large.

The incomplete sensitivity approach means that we can drop the last term in (4). This idea has been discovered using automatic differentiation [14, 15, 13] which enabled for the analysis of the impact of the different operators involved in our simulation loop on the cost function. We noticed that at convergence of the optimization procedure we would like to have $\nabla J_h^\infty = 0$. Therefore, for the optimization to converge, it is sufficient for (∇J_h^n) to be strictly decreasing and experiences showed that this was achieved for steady configurations with only one backward iteration in the adjoint variable calculation. We noticed that the previous remark was valid for cost functions based on informations on the shape (for instance through boundary integrals). This does not mean that a precise evaluation of the state is not necessary, but that for a small change in the shape the state will remain almost unchanged, while geometrical quantities have variations of the same order.

3.1. Illustration on two simple examples

Consider as cost function $J = au_x(a)$ and as state equation the following steady advection-diffusion equation:

$$u_x - Pe^{-1}\ u_{xx} = 0,\ \text{on }]a,1[,\ u(a) = 0,\ u(1) = 1.$$

The solution of this equation is

$$u(x) = \frac{\exp(Pe^{-1}\ a) - \exp(Pe^{-1}\ x)}{\exp(Pe^{-1}\ a) - \exp(Pe^{-1})}.$$

We are looking for $J_a(a) = u_x(a) + a(u_x)_a(a)$. We are in the domain of application of the incomplete sensitivities, where the cost function involves products of state and geometrical quantities and is defined at the boundary.

$$J_a(a) = u_x\ (a)\left(1 + a\frac{Pe^{-1}\ \exp(Pe^{-1}\ a)}{\exp(Pe^{-1}\ a) - \exp(Pe^{-1})}\right).$$

The second term in the parenthesis is the state linearization contribution which is negligible for large Peclet number. In all case, the sign of the sensitivity is always correct.

Another example concerns the sensitivity analysis for the debit of a Poiseuille flow in a channel driven by a constant pressure gradient (p_x) with respect to the channel width. The walls are at $y = +/-a$. The flow velocity satisfies:

$$u_{yy} = \frac{p_x}{\nu}, \quad u(-a) = u(a) = 0. \tag{5}$$

The analytical solution satisfying the boundary conditions is: $u(a,y) = \frac{p_x}{2\nu}(y^2 - a^2)$.

The flow debit is given by $J(a) = \int_{-a}^{a} u(a,y)dy \quad (= \frac{-2p_x a^3}{3\nu})$. The gradient is given by (using the boundary conditions in (5)):

$$\frac{dJ}{da} = \int_{-a}^{a} \partial_a U(a,y)dy = \frac{-2a^2 p_x}{\nu},$$

while the incomplete sensitivity vanishes. Indeed, in this example we are not in the validity [1] domain of sensitivity analysis requiring for cost functions to involve product of state and geometrical entities.

Now consider the following cost function obtained multiplying the flow rate by a: $\tilde{J}(a,u) = \int_{-a}^{a} au(a,y)dy$, which has sensitivity given by:

$$\frac{d\tilde{J}}{da} = J + a\frac{dJ}{da} = \frac{-p_x a^3}{\nu}(\frac{2}{3} + 2),$$

and where the incomplete sensitivity is $\frac{-2p_x a^3}{3\nu}$ which has always the right sign. This is a good example of how to reduce the cost of sensitivity evaluation by an appropriate redefinition of cost functions.

The previous arguments show for instance the importance of including viscous effects in state evaluations for optimization [8] but not necessarily to take them into account in the sensitivities [1]. These remarks are essential, as usually users are penalized by the cost of sensitivity evaluations which forces them to use coarser meshes in optimization than the meshes they can afford for simulation.

4. Some applications

As we said, we would like to show two multi-disciplinary control problems. In the first case, only one control parameter is involved and we compute the cost and its exact sensitivity in time using the complex variable method [16, 17, 1]. The aim in this case is to show that sub-optimal control based on instantaneous gradient is efficient. In this case, the cost function is given by:

$$J(\alpha(t)) = \int_{\partial\Omega} (X(t) - X_{target}(t))^2 d\gamma + \beta \int_0^T \alpha(t)^2 dt, \tag{6}$$

where $-2 < \alpha(t) < 2$ in degree, is a flap incidence changing the wing incidence. X is the CAD-free parameterization of the wing. Our aim is to reduce its deformation. By instantaneous control, we mean $\beta = 0$ which produces a sub-optimal control. The admissible control update frequency is 4 Hz.

In the second case, the number of control parameters is large. We therefore combine the instantaneous control law above with an incomplete evaluation of sensitivities. Control is performed by mass injection over the surface of the wing. The maximum amount on instantaneous injection is less than one percent of the inflow Mach number (i.e. about 2 m/s). This injection velocity is prescribed from the deformation defined by the time dependent minimization using the Hadamard's equivalent transpiration boundary condition technique [1].

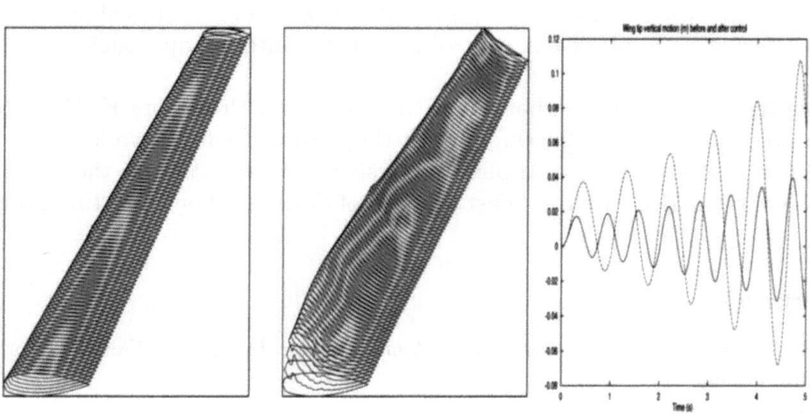

FIGURE 3. Aeroelastic simulation and control for an M6 wing. Initial wing position and a deformed wing position without control. On the right, wing tip vertical evolutions with and without the control.

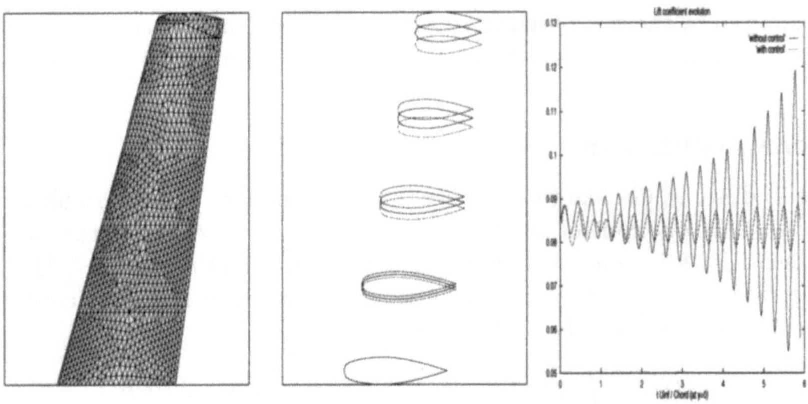

FIGURE 4. Aeroelastic flow over an M6 wing at transonic regime. Control by mass injection over the wing prescribed using instantaneous incomplete gradients. CAD-free parameterization (left), snapshots of the shape (middle), lift coefficient histories with and without control.

5. Concluding Remarks

The application of our control and optimization platform combining instantaneous and incomplete sensitivities with dynamic minimization algorithms has been presented. Our aim was to make the cost of control and simulation problems equivalent. This enables therefore to propose real time control tools based on sensitivities as an alternative to black-box control tools using filtering approach.

Acknowledgments. The author would like to thank Professors K.-H. Hoffmann, R. Hoppe and V. Schulz for the interest they brought to this work. Many thanks to Weierstrass Institute for Applied Analysis and Stochastics for the organization of the workshop dedicated to "Fast solution of discretized optimization problems".

References

[1] B. Mohammadi and O. Pironneau, *Applied Shape Design for Fluids*, Oxford Univ. Press, 2000.

[2] H. Attouch, P. Redont, and X. Goudou, *The Heavy Ball with Friction Method. I. The Continuous Dynamical System: Global Exploration of the Local Minima of a Real-valued Function by Asymptotic Analysis of a Dissipative Dynamical System*, Commun. Contemp. Math., **2**, No. 1 (2000), pp. 1–34.

[3] H. Attouch and R. Cominetti, *A Dynamical Approach to Convex Minimization Coupling Approximation with the Steepest Descent Method,* J. Differential Equations, **128** (2) (1996), pp. 519–540.

[4] H. Borouchaki, P. L. George, and B. Mohammadi, *Delaunay Mesh Generation Governed by Metric Specifications. Part II: Applications*, Finite Element in Analysis and Design, special issue on mesh adaption, 1996.

[5] M. Castro-Diaz, F. Hecht, and B. Mohammadi, *Anisotropic Grid Adaptation for Inviscid and Viscous Flows Simulations*, Internat. J. Numer. Methods Fluids, **25** (1995), pp. 475-491.

[6] F. Hecht and B. Mohammadi, *Mesh Adaption by Metric Control for Multi-scale Phenomena and Turbulence*, AIAA paper 97-08, 1997.

[7] B. Mohammadi, *Practical Applications to Fluid Flows of Automatic Differentiation for Design Problems*, VKI lecture series, 1997-05.

[8] B. Mohammadi, *A New Optimal Shape Design Procedure for Inviscid and Viscous Turbulent Flows*, Internat. J. Numer. Methods Fluids, **25** (1997), pp. 183–203.

[9] B. Mohammadi, *Flow Control and Shape Optimization in Aeroelastic Configurations*, AIAA, 99-0182, 1999.

[10] B. Mohammadi, *Dynamical Approaches and Incomplete Gradients for Shape Optimization and Flow Control*, AIAA, 99-3374, 1999.

[11] B. Mohammadi, *Sensitivity Analysis for Design and Control in an Elastic CAD-free Framework for Multi-Model Configurations*, Journal Europèene des Elements Finis, special issue on fluid/structure interaction, 1999.

[12] H. Attouch, M. Cabot, M. Masmoudi, B. Mohammadi, and P. Redont, *Coupling Dynamic Approaches for Global Minimization*, to appear in Journal of Convex Analysis, 2001.

[13] C. Faure, *Splitting of Algebraic Expressions for Automatic Differentiation*, in: Proc. of the Second SIAM Internat. Workshop on Computational Differentiation, Santa Fe, 1996.

[14] J. C. Gilbert, G. Le Vey, and J. Masse, *La différentiation automatique de fonctions représentées par des programmes*, Rapport de Recherche INRIA 1557, 1991.

[15] N. Rostaing-Schmidt, *Différentiation automatique: Application à un problème d'optimisation en météorologie*, Ph.D. Thesis, University of Nice, 1993.

[16] W. Squire and G. Trapp, *Using Complex Variables to Estimate Derivatives of Real Functions*, siam review, **10**, No. 1 (1998), pp. 110–112.

[17] K. Anderson, J. Newman, D. Whitfield, and E. Nielsen, *Sensitivity Analysis for the Navier-Stokes Equations on Unstructured Grids using Complex Variables*, AIAA paper 99-329, 1999.

Convergence of Runge-Kutta Discretization for Optimal Control Problems

Appolinaire Nzali

Abstract. We present a discretization scheme for optimal control with an ordinary differential equation (ODE). It is based on Runge-Kutta methods for ODE and on a piecewise linear parameterization of the control. The order of convergence of the discrete solutions is obtained under regularity conditions and second order optimality conditions. Furthermore, an approach for the calculation of the gradients via the adjoint system is presented. Finally some numerical results are given and relations to the theoretical convergence results are discussed.

1. Introduction

Results of convergence for Runge-Kutta discretization of optimal control problems occur in a relatively small number of papers in the literature compared to Euler discretization. Schwartz and Polak [22] analyze the consistency of the explicit Runge-Kutta approximation. In their article [22] a convergence for the global solutions of the discrete problem to the global solution of the continuous problem is proven. The analysis is based on a Runge-Kutta scheme, in which certain coefficients must be positive. Velio [26] presents the convergence of the optimum value. He obtains an approximation error of order $O(h^2)$ for control problems with constraints and linear differential equations, without assuming a condition on the regularity of the optimal control. This result is extended later in [27] to systems, which are nonlinear with respect to the state variable. In [12, 11] Hager imposes conditions on the coefficients of a Runge-Kutta integration scheme, which guarantee a given order of accuracy in the optimal control for orders up to four.

The most important point of this paper is the investigation of a combination of a two grid discretization with a Runge-Kutta method for ODE. Contrary to Hager no restrictive assumptions on the coefficients of the integration scheme are made here. The adjoint discrete system considered here is deduced from the discrete problem. Assumptions on the convergence are made for the discrete adjoint system.

2. Problem statement

We will denote by $\emptyset$ the empty set, $\mathbb{N} := \{0, 1, 2, \cdots\}$, $\mathbb{R}^n$ the n-dimensional Euclidian space, $\mathbb{R}_+$ the set $[0, +\infty]$, $\mathcal{L}^\theta([0,1], \mathbb{R}^n)$ the Lebesgue space, $1 \leq \theta \leq \infty$, $\mathcal{W}^{k,\theta}([0,1], \mathbb{R}^r)$ the Sobolev space, $C^k([0,1]; \mathbb{R}^n)$ the space of k-times continiuously differentiable functions on the segment $[0,1]$, $\mathbb{P}_\mu$ the set of polynomials of degree strictly smaller than μ, $\mathbb{P}_{\mu,\mathcal{I}_h,\eta}$ the set of piecewise polynomials of order μ with points of discontinuity τ_i and order of smoothness η_i at τ_i, $\tau_i < \tau_{i+1}$, $i = 1, \cdots, M-1$, $\mathcal{I}_h := \{\tau_0, \tau_1, \cdots, \tau_{M-1}, \tau_M\}$, $M \in \mathbb{N}\backslash\{0\}$ (i.e. the set all functions, which coincide on the subinterval $(\tau_i\ ,\ \tau_{i+1})$ with an element of $\mathbb{P}_\mu$ and for which the jth derivatives for $j < \eta_i$ are continuous at the point τ_i), ∇ the differential operator, defined by $\nabla f(x) = (\frac{\partial f}{\partial x_1}, \cdots, \frac{\partial f}{\partial x_n})^\top$, $f : \mathbb{R}^n \to \mathbb{R}$ (in case it is not clear, with respect to which argument one differentiates, then this argument appears as an index of the operator, e.g., ∇_x), U_a a nonempty convex and closed set of feasible controls, V_a a nonempty convex and closed set in $\mathbb{R}^{d_o}$, $d_o \in \mathbb{N}\backslash\{0\}$, $\|\cdot\|$ the norm in an appropriate space, $\langle\cdot,\cdot\rangle$ the inner product in the appropriate space, $d(A, B) := \inf\limits_{a\in A,\ b\in B} \|a - b\|$.

Let the functions $f : \mathbb{R}^{n+r+d_o} \to \mathbb{R}^n$ and $g_0 : \mathbb{R}^{n+d_o} \to \mathbb{R}$ be given. We consider the optimal control problem $(\mathbf{P}_*)$ governed by an ordinary differential equation and with control and parameter constraints given by:

$$\min_{u,v} g_0(x(1), v) \quad \text{s.t.} \tag{1}$$

$$\dot{x}(t) = f(x(t), u(t), v),\ 0 \leq t \leq 1,\ x(0) = x_0(v), \tag{2}$$

$$u(t) \in U_a, \qquad \text{a.e. } t \in [0,1], \quad v \in V_a, \tag{3}$$

where the state $x \in \mathcal{X} := \mathcal{W}^{1,\infty}([0,1], \mathbb{R}^n)$, the space of Lipschitz continuous functions (the space of essentially bounded functions with essentially bounded derivatives), the control $u \in \mathcal{U} := \mathcal{L}^\infty([0,1], \mathbb{R}^r)$, the space of essentially bounded functions, $\dot{x} \equiv \frac{d}{dt}x$ and $x_0(v)$ the initial state depending on the parameter v. We assume:

A 2.1. *There exists a solution (u^*, v^*) to the problem $(\mathbf{P}_*)$.*

3. Parameterization of the control problem

The choice of the control function u from the set $\mathcal{L}^\infty([0,1], \mathbb{R}^r)$ makes the numerical investigation of the control problem more difficult. One of the main difficulties is the handling of the possibly occurring points of discontinuity. To overcome this problem, we first approximate the solution of the original problem by means of a parameterizing method and reformulate the control problem in the parameterized space. The method of parameterizing the control is based on replacing the set of the admissible controls (see [1])

$$\{u \in \mathcal{U}\ :\ u(t) \in U_a \quad \text{a.e.} \quad t \in [0,1]\} \tag{4}$$

by the set given by:

$$\{u(\cdot,w) \in \mathcal{U} \ : \ u(t,w) \in U_a, \ t \in [0,1], \ w \in I\!R^L, \ L \in I\!N \backslash \{0\}\}. \tag{5}$$

Let W be the subset of $I\!R^L$ determined by all parameters w, which describe the set in (5). That means

$$W := \{w \in I\!R^L \ : \ u(t,w) \in U_a \quad \text{a.e.} \quad t \in [0,1]\}. \tag{6}$$

By parameterizing the control space each element $u(\cdot,w)$ is uniquely determined by the finite-dimensional parameter $w \in W$.
Let the number M, the interval I, and the control grid be fixed and an arbitrary partition of the interval $I = [0,1]$ be given by:

$$\mathcal{I}_h := \{\tau_0, \tau_1, \tau_2, \cdots, \tau_{M-1}, \tau_M\}, \quad \tau_0 := 0 < \tau_1 < \tau_2 < \cdots < \tau_{M-1} < \tau_M := 1. \tag{7}$$

where the index h is defined as follows:

$$h := \max_{1 \le i \le M} (\tau_i - \tau_{i-1}). \tag{8}$$

For our investigation we consider a set W given by:

$$W := \mathop{\times}_{j=1}^{M} (U_a \times U_a),$$

and a parameter w defined by:

$$w := (u_1^+, u_1^-, u_2^+, u_2^-, \cdots, u_M^+, u_M^-) \in W$$

where u_j^+ resp. u_j^- denotes the right limit value at the point τ_{j-1} resp. the left limit value at the point τ_j $(j = 1, \cdots, M)$ of the function $u(\cdot,w)$ as well as $u(\tau_j, w) = u_j^+$, $j = 0, \cdots, M-1$ and $u(\tau_M, w) = u_M^-$. For the approximation $u(\cdot,w) \in \mathcal{U}$, we consider a piecewise linear function of the form:

$$u(t,w) := \sum_{j=1}^{M} \frac{1}{\tau_j - \tau_{j-1}} (u_j^-(t - \tau_{j-1}) + u_j^+(\tau_j - t))\chi_j(t), \ t \in [0,1]. \tag{9}$$

where $\chi_j(t)$, $j = 1, \cdots, M$ denote the characteristic functions of the subintervals $[\tau_{j-1}, \tau_j)$, $j = 1, \cdots, M-1$ and $[\tau_{M-1}, \tau_M]$.

The special feature in the approximation of the control by the formula (9) is the flexibility of the piecewise linear approximation. It enables the treatment of a possibly occurring point of discontinuity of the expected solution as well as the reproduction of continuous parts of the function. For example, if it is known that the expected function has a simple shape on some subintervals and on other subintervals a complicated shape, then the function can be better approximated on the subintervals separately with functions $u(\cdot,w) \in I\!P_{\mu,\mathcal{I}_h,\eta}$.

Let $\mathcal{U}_w \subset \mathcal{L}^\infty([0,1], I\!R^r)$ denote the set of all elements, which can be represented according to the model in (5) with $w \in I\!R^L$. The space $\mathcal{U}_w$ is equipped with the norm of the space $\mathcal{U}$.

Considering the control in $\mathcal{U}_w$, the reformulation of the problem $(\mathbf{P}_*)$ in the new spaces leads to the following problem $(\mathbf{P}_w)$:

$$\min_{w,v} g_0(x(1), v) \quad \text{s.t.} \tag{10}$$

$$\dot{x}(t) = f(x(t), u(t,w), v), \; 0 \le t \le 1, \; x(0) = x_0(v), \tag{11}$$

$$u(t,w) \in U_a, \quad t \in [0,1], \quad v \in V_a. \tag{12}$$

For our investigation, it is of substantial relevance to know whether the solution in question can be well approximated by an element from the new space $\mathcal{U}_w$.

In order that the formulation of the original problem $(\mathbf{P}_*)$ in the new space $\mathcal{U}_w$ makes sense, we assume the following condition:

A 3.1. *There exists a function* $\epsilon : [0,1] \to \mathbb{R}_+$ *with* $\lim_{h\to 0} \epsilon(h) = 0$ *such that*

$$d(u^*, \mathcal{U}_w)_{\mathcal{L}^1} \le \epsilon(h), \tag{13}$$

where $d(A,B)_{\mathcal{L}^1} := \inf_{a\in A,\; b\in B} \|a-b\|_{\mathcal{L}^1}$.

The inequality (13) presupposes that there is an element $w^o \in W$ such that $u(\cdot, w^o) \in \mathcal{U}_w$ is sufficiently close to the expected optimal solution u^*. The set V_a is independent of the parameterization and remains unchanged during the whole parameterizing process. Thus v^o, the appropriate parameter to w^o, can be directly set equal to v^*. Now the question of the realization of (13) arises. The feasibility of (13) consists in the investigation for which functions condition (13) can be fulfilled. The answer to the question will be expressed in terms of the averaged modulus of smoothness.

Let the *kth* difference of the function u with the step size h at the point t be defined by:

$$\Delta_h^k u(t) := \sum_{i=0}^{k} (-1)^{i+k} \binom{k}{i} u(t+ih), \text{ with } \binom{k}{i} = \frac{k!}{i!(k-i)!}, \quad i = 0, \cdots, k$$

where $\binom{k}{i}$ is the binomial coefficients. Let $\omega_k(u,t;\gamma)$ denote the local modulus of continuity of the function u defined by:

$$\omega_k(u,t;\gamma) := \sup\left\{ |\Delta_h^k u(s)| : s, s+kh \in [t - \frac{k\gamma}{2}, t + \frac{k\gamma}{2}] \bigcap [0,1] \right\} \text{ for } \gamma \in [0, \frac{1}{k}],$$

and $\tau_k(u;\gamma)$ denotes the averaged modulus of smoothness of the function u of the order k over the interval $[0, \frac{1}{k}]$ given by:

$$\tau_k(u;\gamma) := \|\omega_k(u,\cdot;\gamma)\|_{\mathcal{L}^e} = \left[\int_1^0 (\omega_k(u,t;\gamma))^e dt \right]^{\frac{1}{e}}, \; \gamma \in [0, \frac{1}{k}].$$

For functions u^* from the set of bounded and measurable functions on the interval $[0,1]$ an estimation of the form

$$d(u^*, \mathcal{U}_w)_{\mathcal{L}^1} = \tau_2(u^*; h)$$

is attainable by spline functions of first degree (see Theorem 4.4 [23]). Let $\overset{1}{\underset{0}{V}}(u^*)$ be the variation [1] of the u^* on the interval $[0,1]$. If in addition the derivative ∇u^* exists and is a function of bounded variation, then

$$\epsilon(h) = \tau_2(u^*;h) \leq h\tau_1(\nabla u^*;h) \leq h^2 \overset{1}{\underset{0}{V}}(\nabla u^*)$$

follows from the characteristization (4); page 8 and (7); page 10 in [23]. That means, the order of approximation in (13) can be improved with the aid of splines of higher order, under the assumption that the kth derivative $\nabla^{(k)}u^*$ of the function u^* is of bounded variation.

Next we assume some differentiability conditions. These conditions are needed for the formulation of regularity conditions, which ensure the existence of the Lagrange-multipliers.

A 3.2. *There exist a positive number δ_* and a closed set $\mathcal{C} = W \times V_a$ such that*

- **(a)** $(w^o, v^o) \in \mathcal{C}$ *and is the solution of* $(\mathbf{P}_w)$,
- **(b)** *the function $f(x,\cdot,\cdot)$ resp. $f(\cdot,\cdot,v)$ is twice continuously Fréchet differentiable for all $x \in I\!R^n$ resp. for all $v \in V_a$. The function g_0 is continuously Fréchet differentiable with respect to x and twice continuously Fréchet differentiable with respect to v. The functions $\nabla_x f$, $\nabla_u f$, $\nabla_v f$ are Lipschitz continuous in all variables,*
- **(c)** $d((w^o,v^o), \partial\mathcal{C}) \geq \delta_*$,

where $\partial\mathcal{C}$ denotes the boundary of $\mathcal{C}$.

The regularity condition occurs in different form in the literature. We mention for example a general formulation given by Robinson in [20], Zowe and Kurcyusz [28] who give sufficient conditions for the regularity condition of one point using the equality of sets. Tröltzsch presents in [25] necessary conditions which are relatively easy to verify, if inner points are present in the considered sets. Let $(\bar{w},\bar{v})$ be a feasible point of the problem $(\mathbf{P}_w)$. We define the normal cone function $\mathcal{N}_c[W \times V_a]$ of the set $W \times V_a$ into the set $I\!R^L \times I\!R^{d_0}$ by:

$$\mathcal{N}_c[W\times V_a](\bar{w},\bar{v}) := \begin{cases} \left\{ \begin{array}{l} y = (y^w, y^v) \in I\!R^L \times I\!R^{d_0} : \\ (w-\bar{w})^\top \cdot y^w + (v-\bar{v})^\top \cdot y^v \geq 0, \\ \forall\, (w,v) \in W \times V_a \end{array} \right\} & : (\bar{w},\bar{v}) \in W \times V_a, \\ \emptyset & : (\bar{w},\bar{v}) \notin W \times V_a. \end{cases}$$

Let the set $\mathcal{Y}$ resp. $\mathcal{Z}$ be given by $\mathcal{Y} := \mathcal{L}^\infty([0,1], I\!R^n) \times I\!R^n$ resp. $\mathcal{Z} := I\!R^L \times I\!R^{d_o}$, and the function G be defined from $\mathcal{Z}$ to $\mathcal{Y}$ by

$$G(z) := (\dot{x}(\cdot) - f(x(\cdot), u(\cdot,w), v), x(0) - x_o(v)), \quad z = (w,v) \in \mathcal{Z}.$$

In the context of this paper we consider the following definition.

[1] The variation (or the total variation) of the function u on the interval $[0,1]$ is the upper bound of the sum $\sum_{k=0}^m |u(\tau_{k+1}) - u(\tau_k)|$ for any sequence of numbers $\{\tau_k\}$ with $\tau_0 := 0 < \tau_1 < \tau_2 < \cdots < \tau_{m-1} < \tau_m := 1$.

Definition 3.3. *A point (w^o, v^o) is regular, if for all $a \in \mathcal{Y}^*$, the only element a, which satisfies*

$$-a^\top \nabla_{(w,v)} G(w^o, v^o) \in \mathcal{N}_c[W \times V_a](w^o, v^o) \tag{14}$$

is $a = 0$.

A similar formulation of the regularity condition is found in corollary 6.15, [21]. Generally the restriction (12) is described by finitely many equalities and inequalities for optimization problems. If this holds for the problem $(\mathbf{P}_w)$ and if the functions, which describe the finitely many equalities and inequalities, are continuously Fréchet differentiable, then equation (14) is a form of the Mangasarian-Fromowitz Constraint Qualification. From Lemma 2.3 [18] the equation (14) is equivalent to the following equation

$$\nabla_{(w,v)} G(w^o, v^o)\left((W - w^o) \times (V_a - v^o)\right) = \mathcal{Y}. \tag{15}$$

Equation (15) expresses the surjectivity of the operator $\nabla_{(w,v)} G(w^o, v^o)$. For the problem $(\mathbf{P}_w)$ each admissible point is regular (see Example 9,2 in [8]), and thus the surjectivity condition (15) is fulfilled.

The next preparations are made for the formulation of both the necessary optimality conditions and the coercivity condition. Let the Lagrangian function $\mathcal{L}^w$ for the problem $(\mathbf{P}_w)$ be defined by:

$$\mathcal{L}^w(w, v, \lambda) = g_0(x(1), v) - \lambda^\top(\dot{x}(\cdot) - f(x(\cdot), u(\cdot, w), v)). \tag{16}$$

We consider the matrices $\breve{\mathbf{a}}$, $\breve{\mathbf{p}}$, $\breve{\mathbf{q}}$, $\breve{\mathbf{w}}$, $\breve{\mathbf{r}}$, $\breve{\mathbf{s}}$, $\breve{\mathbf{v}}$, $\mathbf{H}$ defined by:

$$\begin{aligned}
\breve{\mathbf{a}}[t] &:= \nabla_x f(x, u(\cdot, w), v)(t)|_{(w^o, v^o, x^o)},\\
\breve{\mathbf{p}}[t] &:= \nabla_u f(x, u(\cdot, w), v) \nabla_w u(\cdot, w)(t)|_{(w^o, v^o, x^o)},\\
\breve{\mathbf{q}}[t] &:= \nabla_v f(x, u(\cdot, w), v)(t)|_{(w^o, v^o, x^o)},\\
\breve{\mathbf{w}}[t] &:= [\langle \nabla_w u(\cdot, w), \lambda^\top \nabla^2_{uu} f(x, u(\cdot, w), v) \nabla_w u(\cdot, w)\rangle\\
&\quad + \langle \nabla_u f(x, u(\cdot, w), v), \nabla^2_{w,w} u(\cdot, w)\rangle](t)|_{(w^o, v^o, x^o, \lambda^o)},\\
\breve{\mathbf{r}}[t] &:= \lambda^\top \nabla^2_{u,v} f(x, u(\cdot, w), v) \nabla_w u(\cdot, w)(t)|_{(w^o, v^o, x^o, \lambda^o)},\\
\breve{\mathbf{s}}[t] &:= \lambda^\top \nabla^2_{v,u} f(x, u(\cdot, w), v) \nabla_w u(\cdot, w)(t)|_{(w^o, v^o, x^o, \lambda^o)},\\
\breve{\mathbf{v}}[t] &:= \lambda^\top \nabla^2_{v,v} f(x, u(\cdot, w), v)(t)|_{(w^o, v^o, x^o, \lambda^o)},\\
\mathbf{H}[t] &:= \nabla^2_{(w,v),(w,v)} \mathcal{L}^w(w, v, \lambda)|_{(w^o, v^o, \lambda^o)},
\end{aligned} \tag{17}$$

where the appropriate state x^o resp. the vector λ^o satisfies the following equations:

$$\dot{x}^o(t) = f(x^o(t), u(t, w^o), v^o), \quad x^o(0) = x_0(v^o), \tag{18}$$

resp.

$$\dot{\lambda}^o(t) = -(\lambda^o(t))^\top \nabla_x f(x^o(t), u(t, w^o), v^o), \quad \lambda^o(1) = [\nabla_x g_0(x^o(1), v^o)]^\top. \tag{19}$$

The calculation of the operators $\nabla_{(w,v)}\mathcal{L}^w|_{(w^o,v^o,x^o,\lambda^o)}$ and $\mathbf{H}[t]$ results in the equations:

$$\nabla_{(w,v)}\mathcal{L}^w|_{(w^o,v^o,x^o,\lambda^o)} = \begin{pmatrix} (\lambda^o)^\top(t)\breve{\mathbf{p}}[t] \\ \nabla_v g_0(x^o(1),v^o) + (\lambda^o)^\top(t)\breve{\mathbf{q}}[t] \end{pmatrix}^\top,$$
$$\mathbf{H}[t] = \begin{pmatrix} \breve{\mathbf{w}}[t] & \breve{\mathbf{r}}[t] \\ \breve{\mathbf{s}}[t] & \nabla^2_{v,v} g_0(x^o(1),v^o) + \breve{\mathbf{v}}[t] \end{pmatrix}.$$

The constraint for the problem $(\mathbf{P}_w)$ describes a cone $\mathcal{K}$. Let $\mathcal{K}^+$ be a subset of the dual space $\mathcal{Y}^*$ given by:

$$\mathcal{K}^+ := \{\lambda \in \mathcal{Y}^* \ : \ \langle \lambda, y\rangle \geq 0, \ \forall y \in \mathcal{K}\} = \mathcal{Y}^*.$$

Theorem 3.4 (First order necessary optimality conditions). *Let (w^o, v^o) be a regular point and assume that* **A3.2** *is valid. Then there exists an element $\lambda^o \in \mathcal{K}^+$, so that the following holds:*

$$-\begin{pmatrix} (\lambda^o)^\top(t)\breve{\mathbf{p}}[t] \\ \nabla_v g_0(x^o(1),v^o) + (\lambda^o)^\top(t)\breve{\mathbf{q}}[t] \end{pmatrix} \in \mathcal{N}_c[W \times V_a](w^o, v^o). \tag{20}$$

For the proof of Theorem 3.4 we refer to § 2.4.2 in [14].

We assume the following coercivity condition:

A 3.5. *For all $t \in [0,1]$, $w = w_1 - w_2$, with w_1, $w_2 \in W$, $v = v_1 - v_2$, with v_1, $v_2 \in V_a$, there exists a positive constant α such that*

$$\begin{pmatrix} w \\ v \end{pmatrix}^\top \mathbf{H}[t] \begin{pmatrix} w \\ v \end{pmatrix} \geq \alpha(||w||^2_{\mathbb{R}^L} + ||v||^2_{\mathbb{R}^{d_o}}), \tag{21}$$

where the appropriate control $u(\cdot, w)$ belongs to $\mathcal{L}^2([0,1], \mathbb{R}^r)$ and the state $x \in \mathcal{W}^{1,2}([0,1], \mathbb{R}^r)$ meet the following equations:

$$\dot{x}(t) = \breve{\mathbf{a}}[t]x(t) + \breve{\mathbf{p}}[t]w + \breve{\mathbf{q}}[t]v, \quad x(0) = 0. \tag{22}$$

4. Runge-Kutta discretization

Let $N \in \mathbb{N}\backslash\{0\}$. We consider a fixed state grid $\mathcal{G}_h$ on the interval $I = [0,1]$ and the local step sizes h_l given by:

$$\mathcal{G}_h := \{t_0, t_1, \cdots, t_N\}, \ t_0 := 0 < t_1 < \cdots < t_N := 1, \tag{23}$$
$$h_l := t_l - t_{l-1}, \ l = 1, \cdots, N. \tag{24}$$

The following technical properties are necessary for the selection of the state grid $\mathcal{G}_h$:

A 4.1. *The state grid $\mathcal{G}_h$ is to be selected as a refinement of the control grid $\mathcal{I}_h$ (i.e. $\tau_j \in \{t_0, t_1, \cdots, t_N\}$, $j = 1, \cdots, M$) and the local step sizes fulfill the following conditions:*

$$h_{min} := \frac{c_1}{N} \leq h_l \leq \frac{c_2}{N} =: h_{max}, \quad l = 1, \cdots, N, \tag{25}$$

where c_1 and c_2 are suitable positive constants.

The condition (25) introduced to avoid the numerical singularity. Let us consider a discrete approximation of the ordinary differential equation using the Runge-Kutta scheme (see [13, 5, 10]) given by:

$$x_l = x_{l-1} + h_l \varphi_l(x_l, x_{l-1}, w, v), \quad x_0 = x_0(v), \ l = 1, \cdots, N, \tag{26}$$

$$\varphi_l(x_l, x_{l-1}, w, v) := \sum_{i=1}^{q} b_i K_{li}, \ l = 1, \cdots, N, \tag{27}$$

$$K_{li} = f\left(x_{l-1} + h_l \sum_{j=1}^{q} \alpha_{ij} K_{lj}, u(t_{l-1} + c_i h_l, w), v\right), \ i = 1, \cdots, q, \tag{28}$$

with $\sum_{i=1}^{q} b_i = 1$, $\sum_{j=1}^{q} \alpha_{ij} = c_i$, $i = 1, \cdots, q$, where $(\alpha_{i,j})$, $b = (b_1, \cdots, b_q)$, $c = (c_1, \cdots, c_q)$ denote the parameters of the Runge-Kutta scheme. We formulate the discrete problem $(\mathbf{P}_{wd})$

$$\min_{w,v} g_0(x_N, v) \quad \text{s.t.} \tag{29}$$

$$x_l = x_{l-1} + h_l \varphi_l(x_l, x_{l-1}, w, v), \ l = 1, \cdots, N, \ x_0 = x_0(v), \tag{30}$$

$$w \in W, \ v \in V_a. \tag{31}$$

In a similar way as for the continuous case, we formulate the regularity condition for a point and the first order necessary optimality conditions for the discrete problem.

Let the grids $\mathcal{I}_h$ and $\mathcal{G}_h$ be given and (w^h, v^h) be an admissible point of the problem $(\mathbf{P}_{wd})$. Let $G_{h,l} : \mathbb{R}^L \times \mathbb{R}^{d_o} \to \mathbb{R}^n$, $l = 0, \cdots, N$, be a sequence of functions defined by

$$G_{h,0}(w, v) := x_0 - x_0(v),$$
$$G_{h,l}(w, v) := x_l - x_{l-1} - h_l \varphi_l(x_l, x_{l-1}, w, v), \ l = 1, \cdots, N.$$

Definition 4.2. *A point (w^h, v^h) is said to be regular, if for all $a = (a_0, \cdots, a_N) \in \mathbb{R}^{(N+1)(L+d_o)}$, with $a_l \geq 0$, $a_l \in \mathbb{R}^{L+d_o}$, $l = 0, \cdots, N$, the only element a, which satisfies*

$$-\sum_{l=0}^{N} a_l^\top \nabla_{(w,v)} G_{h,l}(w^h, v^h) \in \mathcal{N}_c[W \times V_a](w^h, v^h) \tag{32}$$

is $a = 0$.

Let $\lambda = \{\lambda_0, \cdots, \lambda_N\}$ be an element from $\mathbb{R}^{(N+1)n}$ and the discrete Lagrangian function $\mathcal{L}^d$ for the problem $(\mathbf{P}_{wd})$ be given by:

$$\mathcal{L}^d(w, v, \lambda) = g_0(x_N, v) - \sum_{l=1}^{N} \lambda_l^\top (x_l - x_{l-1} - h_l \varphi_l(x_l, x_{l-1}, w, v)). \tag{33}$$

According to the definition of the discretization scheme in (27) - (28) the functions φ_l, $l = 1, \cdots, N$, are twice continuously Fréchet differentiable with respect to w, v and continuously differentiable with respect to x_{l-1} and x_l, $l = 1, \cdots, N$.

Let us consider the sets $\mathcal{K}_h = \{0\}$ a cone describing the constraint of the problem $(\mathbf{P}_{wd})$ and $\mathcal{K}_h^+$ given by $\mathcal{K}_h^+ = \{\lambda \in I\!R^{(N+1)n} \;:\; \langle \lambda, k\rangle \geq 0,\ \forall\, k \in \mathcal{K}_h\} = I\!R^{(N+1)n}$. Next we formulate the discrete first order necessary optimality condition.

Theorem 4.3. *Let $(w^h, v^h) \in \mathcal{C}$ be a regular point and a local optimal solution of the problem $(\mathbf{P}_{wd})$, and assume that* **A3.2** – **A4.1** *are valid. Then there exists a vector $\lambda^h = (\lambda_1^h, \cdots, \lambda_n^h) \in \mathcal{K}_h^+$, such that the following system is fulfilled:*

$$-\nabla_{(w,v)}\left(g_0(x_N, v) + \sum_{l=1}^{N} \lambda_l^\top (h_l \varphi_l(x_l, x_{l-1}, w, v))\right)\Bigg|_{z^h} \in \mathcal{N}_c[W \times V_a](w^h, v^h), \tag{34}$$

with $z^h = (w^h, v^h, x^h, \lambda^h)$, where the appropriate state x^h and the appropriate Lagrange multiplier λ^h satisfy the following equations:

$$\begin{array}{c} x_l - x_{l-1} - h_l \varphi_l(x_l, x_{l-1}, w, v) = 0, \quad x_0 = x_0(v) \\ l = 1, \cdots, N, \end{array} \tag{35}$$

$$\left.\begin{array}{c} \lambda_l - \lambda_{l+1} - h_l[\nabla_{x_l}\varphi_l]^\top \lambda_l - h_{l+1}[\nabla_{x_l}\varphi_{l+1}(x_{l+1}, x_l, w, v)]^\top \lambda_{l+1} = 0, \\ l = N-1, \cdots, 0, \\ \lambda_N - \nabla_x g_0(x_N, v)^\top - h_N \left(\nabla_{x_N}\varphi_N(x_N, x_{N-1}, w, v)\right)^\top \lambda_N = 0. \end{array}\right\} \tag{36}$$

Theorem 4.3 is an analogue of (1.1) in [28] or theorem 2,1 in [17] in finite dimensional spaces, because the problem $(\mathbf{P}_{wd})$ resp. the appropriate sets $I\!R^L \times I\!R^{d_o}$, $I\!R^{(N+1)n}$, $W \times V_a$ have a similar shape as the problem $(\mathbf{P})$ resp. the appropriate sets in [28, 17].

If one follows the lines of the proof for the determination of the order of accuracy of an integration procedure, then one can observe the following. The order of accuracy of the solution of the integration procedure (35) is not always equal to the order of accuracy of an approximate solution of the discrete dual equations (36). This justifies the introduction of the following condition.

A 4.4. *We assume: The solution λ^o of (19) as well as λ^h the solution of the discrete dual system (36) fulfill the following discrete approximation property:*

$$\max_{l \in \{1, \cdots, N\}} ||\lambda^o(t_l) - \lambda_l^h|| = O(h). \tag{37}$$

In order to obtain the third or fourth order accuracy, the order-discrepancy between the primary and the dual discretization leads in [12, 11] to additional conditions on the coefficients of the Runge-Kutta discretization. The transformation of the adjoint system in [12, 11] leads to a Runge-Kutta discretization for the adjoint equation of the optimal control problem, which is frequently different from the original Runge-Kutta discretization of the state equation. In particular the paper [12] concentrates on Runge-Kutta discretizations, whose coefficients are all positive in the last calculation phase. We also note that the error estimate of the discrete control in (14) [12] does not reflect the error estimate of the control solution of the discrete problem (see remark 2.2 [12]).

Theorem 4.5. *Let the function f be Lipschitz-continuous on $\mathbb{R}^{n+r+d_o}$, $x^* \in \mathcal{W}^{1,\infty}$ resp. $\lambda^* \in \mathcal{W}^{1,\infty}$ be the appropriate state resp. Lagrange multiplier to the optimal solution (u^*, v^*) and a function ϵ be given as described in the assumption* **A3.1**. *Let the assumptions* **A2.1** - **A4.4** *and the regularity condition (32) be satisfied. Then for each N sufficiently large, there exists a local minimum (w^h, v^h) to the problem $(\mathbf{P}_{wd})$ as well as x^h resp. λ^h i.e. the discrete appropriate state resp. Lagrange multiplier, and the following error estimate holds:*

$$||u^* - u(\cdot, w^h)||_{\mathcal{L}^\infty} + |v^* - v^h| + ||x^* - x^h||_{\mathcal{L}^\infty} + ||\lambda^* - \lambda^h||_{\mathcal{L}^\infty} = O(h) + \epsilon(h). \quad (38)$$

Theorem 4.5 expresses the global error as a sum of the discretization and approximation errors. To prove theorem 4.5 the first order necessary optimality conditions of both the parameterized and the discrete problem are reformulated in form of generalized equations. Afterwards results of the theory of approximation for the investigation of optimization problems in Banach spaces from Dontchev and Hager [7, 6] are applied. In our case we use Corollary 1 in [6]. Corollary 1 in [6] gives conditions for the existence of a solution of problem $(\mathbf{P}_w)$ for different values of h and at the same time, a possibility for the approximation of a solution of the problem $(\mathbf{P}_*)$. A complete proof of the main theorem 4.5 is given in [19]. The next paragraph presents a procedure for the exact calculation of gradients of the objective function. The system of equations (36) plays a special role in the calculation.

5. The method of internal numerical differentiation

The main point in the performance of an optimization method is the calculation of the derivatives. For the calculation of derivatives a huge variety of methods exists in the literature. We mention for example the numerical calculation of the gradients by finite differences [15], the numerical approximation of continuous gradients by integration of the adjoint equation [9, 24] and the internal numerical differentiation [1, 2, 3] (**IND**). In [1] the methods specified above are discussed in detail. A numerical comparison of the internal numerical differentiation and the other methods is given in Figure 8.3 [2]. According to Bock [2, 3] one of the most important advantages of the **IND** is "the drastic reduction of the computing time (60-80 p.c. especially for low tolerances) compared to external numerical differentiation because of substantial overhead savings ..., and much lower accuracy requirements for the basic discretization scheme". Moreover, "the method is stable in the sense that the exact derivative of an adaptively chosen discretization scheme is generated". We present the extended approach of the internal numerical differentiation developed in [1] for the special case of Runge-Kutta discretization (see (26) – (28)).

Let the objective function $g_0(x_N, v)$ be viewed as an explicit function of the parameter $w \in \mathbb{R}^L$ as well as the parameter $v \in \mathbb{R}^{d_o}$ and the discrete state variables x_l $l = 1, \cdots, N$. We consider the function F given by:

$$F(z) = g_0(x_N, v), \quad \text{with} \quad z = (z_1, \cdots, z_{L+d_o}) := (w, v).$$

Lemma 5.1. *Let g_0 be continuously differentiable in all arguments, $l(k) \in \{0, \cdots, N\}$ the index of state grid points with the $t_{l(k)} = \tau_k$, $k = 0, \cdots, M$. If the partial derivatives $\nabla_{z_k} x_l$, $l = 1, \cdots, N$, exist for some $k \in \{1, \cdots, L + d_o\}$, then*

$$\nabla_{w_j} F = \sum_{l=l(j-1)+1}^{l(j)} [h_l \lambda_l^\top \nabla_{w_j} \varphi_l] + \nabla_{w_j} g_0(x_N, v), \; j = 1, \cdots, L, \tag{39}$$

$$\nabla_{v_j} F = \sum_{l=1}^{N} [h_l \lambda_l^\top \nabla_{v_j} \varphi_l] + \nabla_{v_j} g_0(x_N, v), \quad 1 \le j \le d_o, \tag{40}$$

where the discrete adjoint variables λ_l, $l = N-1, \cdots, 0$, are computed according to (36).

The statements of Lemma 5.1 follow from theorem 3.1 in [1] with the choice $\alpha_{l0} = 1$, $\alpha_{l1} = -1$, $k_l = 1$ for all $l = 1, \cdots, N-1$ and $\alpha_{l,j} = 0$ for all other coefficients.

In the case of a Runge-Kutta scheme the following steps are necessary for the computation of the gradients:

INPUT DATA

$$L = 2Mr, \quad (u_1^+, u_1^-, \cdots, u_j^+, u_j^-, \cdots, u_M^+, u_M^-, v_1, \cdots, v_{d_o})^\top \in I\!R^{2Mr+d_o}.$$

FIRST STEP Computation of the solution of the discrete primary system (30)

SECOND STEP Computation of the solution of the discrete adjoint system according to the model in (41) – (42). The dual system is described by:

$$\lambda_N = [\nabla_x g_0(x_N, v)]^\top + h_N [\nabla_{x_N} \varphi_N]^\top \lambda_N, \tag{41}$$

$$\lambda_l = \lambda_{l+1} + \sum_{i=1}^{q} b_i \left(h_l (Y_{l,i}^a)^\top \lambda_l + h_{l+1} (Y_{l+1,i}^b)^\top \lambda_{l+1} \right), \tag{42}$$

$$l = N-1, \cdots, 0,$$

with

$$Y_{l,i}^a = \nabla_{x_l} K_{l,i} = [\nabla_x f|_{l,i}] \cdot \left[h_l \sum_{j=1}^{q} \alpha_{ij} Y_{l,j}^a \right], \; i = 1, \cdots, q, \tag{43}$$

$$Y_{l+1,i}^b = \nabla_{x_l} K_{l+1,i} = [\nabla_x f|_{l+1,i}] \cdot \left[I + h_{l+1} \sum_{j=1}^{q} \alpha_{ij} Y_{l+1,j}^b \right], \; i = 1, \cdots, q, \tag{44}$$

where for $l = 1, \cdots, N$; $i = 1, \cdots, q$ it holds that:

$$[\nabla_x f|_{l,i}] = \nabla_x f \left(x_{l-1} + h_l \sum_{j=1}^{q} \alpha_{ij} K_{lj}, u(t_{l-1} + c_i h_l, w), v \right),$$

$$[\nabla_u f|_{l,i}] = \nabla_u f \left(x_{l-1} + h_l \sum_{j=1}^{q} \alpha_{ij} K_{lj}, u(t_{l-1} + c_i h_l, w), v \right).$$

THIRD STEP Computation of the derivative of K_{li}, $l = 0, \cdots, N-1$, $i = 1, \cdots, q$, with respect to z_k, $k = 1, \cdots, 2Mr + d_o$.

case (*control-dependent parameters*) $k \leq 2Mr$, $t = t_{l-1} + c_i h_l$:

$$\begin{aligned}
\nabla_{z_k} K_{li} &= \nabla_{z_k} f\left(x_{l-1} + h_l \sum_{j=1}^{q} \alpha_{ij} K_{lj}, u(t,w), v\right) \\
&= [\nabla_x f|_{l,i}] \cdot \left[h_l \sum_{j=1}^{q} \alpha_{ij} \nabla_{z_k} K_{lj}\right] + [\nabla_u f|_{l,i}] \nabla_{z_k} u(t,w).
\end{aligned}$$

It holds for $t \in [\tau_{j-1}, \tau_j)$, $z_k \in \{u_j^+, u_j^-\}$, $j = 1, \cdots, M-1$, that

$$\begin{aligned}
u(t,w) &= \frac{1}{\tau_j - \tau_{j-1}}[u_j^-(t - \tau_{j-1}) + u_j^+(\tau_j - t)], \\
\nabla_{z_k} u(t,w) &= \nabla_t u(t,w) \cdot \nabla_{z_k} t + \nabla_w u(t,w) \cdot \nabla_{z_k} w, \\
\nabla_{z_k} t &= 0, \quad z_k \in \{u_j^+, u_j^-\},\ j = 1, \cdots, M, \\
\nabla_z u(t,w) &= \left(0, \cdots, 0, \frac{\tau_j - t}{\tau_j - \tau_{j-1}}, \frac{t - \tau_{j-1}}{\tau_j - \tau_{j-1}}, 0, \cdots, 0\right), \\
\nabla_{z_k} w &= 1_k,
\end{aligned}$$

where 1_k is the kth unit vector in $\mathbb{R}^{2mr+d_o}$ and for $j = M$ we consider the interval of $[\tau_{M-1}, \tau_M]$.

case (*control-independent parameters*) $k > 2Mr$, $t = t_{l-1} + c_i h_l$:

$$\begin{aligned}
\nabla_{v_k} K_{li} &= \nabla_{v_k} f\left(x_{l-1} + h_l \sum_{j=1}^{q} \alpha_{ij} K_{lj}, u(t,w), v\right) \qquad (45) \\
&= [\nabla_x f|_{li}] \left[h_l \sum_{j=1}^{q} \alpha_{ij} \nabla_{v_k} K_{lj}\right] + [\nabla_v f|_{li}] \nabla_{v_k} v.
\end{aligned}$$

In the implicit case the equation (45) forms a nonlinear system of equations. In the explicit case the vectors $\nabla_{z_k} K_{li}$, $i = 1, \cdots, q$, are calculated recursively.

FOURTH STEP Computation of (39), (40).

6. Some numerical results

We consider the tunnel diode oscillator **case (b)**, [16, 4] to illustrate our convergence result. The tunnel diode oscillator represents an electrical circuit, whose structure is given in Figure 1 where t denotes the time, L the inductance of a coil, C the capacity of a condenser, R the resistance, I the current in the system, U the voltage, U_o the voltage at the generator, D the tunnel diode. The target of the

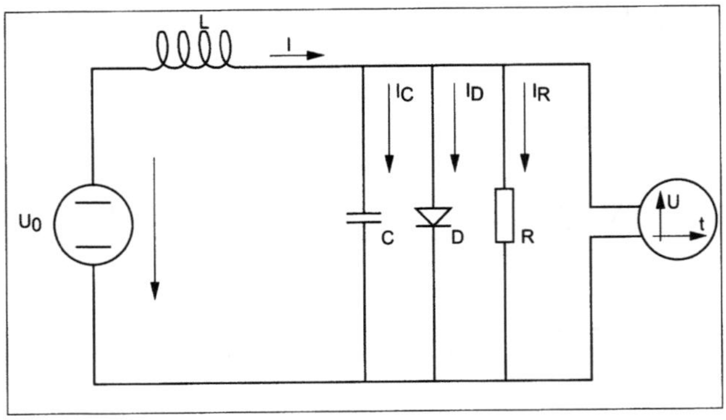

FIGURE 1. Tunnel diode oscillator

problem is to minimize both the total voltage at the generator and the total value of the current. The rate of convergence ρ_h is computed by an approximation of the estimate

$$||u^h - u^*||_{L^\infty} \leq c_h \; (h)^{\rho_h}. \tag{46}$$

The numerical results given in table 1 exceed our theoretical results ($\rho_h = 1$). In accordance with our observation, this is due to the flexible control (see definition (9)) and to the flexible state grid.

These calculations are executed on a Compute server HP 9000 (780/J280) with 180 MHz frequency and 768 MByte RAM under HP-UX 10.20 using the code RINDOPT implemented in [19].

TABLE 1. Rate of convergence

$h = \frac{t_e=2.5}{2^i}$	$\|\|u^h - u^*\|\|_{L^\infty}$	ρ_h	c_h
i=1	.26937	-	-
i=2	.31513	2.27662	173.70893
i=3	8.3606E-02	1.91428	63.60964
i=4	7.0345E-02	.24916	.19827
i=5	9.3878E-03	2.90559	12452.79352
i=6	7.0448E-03	.41422	7.00530E-02
i=7	6.9765E-08	16.62368	-

Conclusion

A combination of a two grid discretization with a Runge-Kutta method for ODE is presented here. In addition the control is parameterized. The presented discretization is particularly suitable for control functions with jumps. For the example presented above, the control grid is selected following our observation. It still remains open, finding a systematic way for the selection of the control grid.

Acknowledgement

I would like to thank *Prof. Werner Römisch*[2] cordially for the valuable hints during the completion of this paper.

References

[1] M. L. Baguer and W. Römisch, *Computing gradients in parametrization-discretization schemes for constrained optimal control problems*, in: Florenzano et al., Approximation and optimization in the aribbean II; Verlag Peter Lang, Frankfurt am Main, 1995, pp. 14–34.

[2] H. G. Bock, *Numerical treatement of the inverse problems in chemical reaction kinetics*, in: Modelling of chemical reaction systems, K. H. Ebert, P. Deuflhard, and W. Jäger, eds., Springer-Verlag, Berlin, 1981, pp. 102–125.

[3] H. G. Bock, *Recent advances in parameteridentification techniques for o.d.e.*, in: Numerical treatement of the inverse problems in differential and integral equations, P. Deuflhard, E. Hairer, eds., Birkhäuser, Boston, 1983, pp. 95–121.

[4] C. Büskens, *Optimierungsmethoden und Sensitivitätsanalyse für optimale Steuerprozesse mit Steuer- und Zustands-Beschränkung*, PhD thesis, Westfälische Wilhelms-Universität Münster, Fachbereich Mathematik und Informatik, 1998.

[5] J. C. Butcher, *The numerical analysis of ordinary differential equation: Runge-Kutta and general linear methods*, John Wiley & Sons Ltd, 1987.

[6] A. L. Dontchev and W. W. Hager, *Lipschitzian stability in nonlinear control and optimization*, SIAM J. Control Optim., **31** (9) (1993), pp. 569–603.

[7] A. L. Dontchev and W. W. Hager, *Lipschitzian stability for state constrained nonlinear optimal control*, SIAM J. Control Optim., **36** (2) (1998), pp. 696–718.

[8] I. V. Girsanov, *Lecture on mathematical theory of extreme problems*, Lecture Notes in Economic and Mathematical Systems, **67**, Springer, Berlin, 1972.

[9] C. J. Goh and K. L. Teo, *MISER, a FORTRAN program for solving optimal control problems*, Adv. Engineering Software, **10** (1988), pp. 90–99.

[10] R. D. Grigorieff, *Numerik für gewöhnliche Differentialgleichungen 1*, B. G. Teubner, Stuttgart, 1972.

[11] W. W. Hager, *Runge-Kutta diskretisation of optimal control problem*, Department of Mathematics, University of Florida, Gaineville, FL 32611, May 1999, www.math.ufl.edu/ hager.

[12] W. W. Hager, *Runge-Kutta methods in optimal control and the transformed adjoint system*, Numer. Math., **87**, No. 2 (2000), pp. 247–282, www.math.ufl.edu/ hager.

[13] E. Hairer, S. P. Norsett, and G. Wanner, *Solving Ordinary Differential Equations I*, Springer-Verlag, Berlin, Heidelberg, 1993.

[14] A. D. Ioffe and V. M. Tichomirov, *Theorie der Extremalaufgaben*, VEB Deutscher Verlag der Wissenschaften, Berlin, 1979.

[15] D. Kraft, *Finite difference gradients versus error-qradrature gradients in the solution of parametrized optimal control problem*, Optimal Control Applications & Methods, 1981, pp. 191–199.

[2] Institute of Mathematics, Humboldt University Berlin, Germany

[16] K. Malanowski, C. Büskens, and H. Maurer, *Convergence of approximations to nonlinear optimal control problems*, Mathematical Programming with Data Perturbations, A. V. Fiacco, ed., Lecture Notes in Pure and Applied Mathematics, Marcel Dekker, Inc., New York, **195**, 1997, pp. 253–284.

[17] H. Maurer, *First and second order sufficient optimality conditions in mathematical programming and optimal control*, Mathematical programming study, **14** (1981), pp. 163–177.

[18] H. Maurer and J. Zowe, *First and second-order necessary and sufficient optimality conditions for infinite-dimensional programming problems*, Mathematical Programming, **16** (1979), pp. 98–110.

[19] A. Nzali, *Zur Lösung optimaler Steuerungsprobleme: Diskretisierung, Konvergenz, Anwendung*, PhD thesis, Institute of Mathematics, Humboldt University Berlin, to appear.

[20] S. M. Robinson, *Stability theory for systems of inequalities, part II, differentiable nonlinear systems*, SIAM J. Numer. Anal., **13** (1976), pp. 497–513.

[21] R. T. Rockafellar and R. J. B. Wets, *Variational analysis*, Springer-Verlag, Berlin, Heidelberg, 1998.

[22] A. Schwartz and E. Polak, *Consistent approximations for optimal control problems based on Runge-Kutta intagration*, SIAM J. Control Optim., **34** (1996), pp. 1235–1269.

[23] B. Sendov and V. A. Popov, *The averaged moduli of smoothness*, John Wiley, New York, 1988.

[24] K. L. Teo, C. J. Goh, and K. H. Wong, *A unified computational approach to optimal control problems*, Longman, Harlow, 1991.

[25] F. Tröltzsch, *Optimality conditions for parabolic control problems and applications*, Teubner Verlag, Teubner-Texte zur Mathematik, Band **62**, Leipzig, 1984.

[26] V. Veliov, *Second-order discret approximations to linear differential inclusions*, SIAM J. Numer. Anal., **29** (1992), pp. 439–451.

[27] V. Veliov, *On the time discretization of control systems*, SIAM J. Control Optim., **35** (5) (1997), pp. 1470–1486.

[28] J. Zowe and S. Kurcyusz, *Regularity and stability for mathematical programming problem in Banach spaces*, Appl. Math. Optim., **5** (1979), pp. 49–62.

Numerical Optimal Control Strategies for Semi-Active Vehicle Suspension with Electrorheological Fluid Dampers

Uwe Rettig and Oskar von Stryk

Abstract. Optimal control problems for semi-active vehicle suspensions and their numerical solution are discussed in this paper. For this purpose, several models of the vehicle dynamics with different levels of details and a general formulation of different sub-criteria for rating the ride comfort and safty are presented and investigated in this paper. The benefits and drawbacks of various numerical optimal control methods such as LQR-, H^∞ and direct collocation when applied to the different optimal control problems for semi-active vehicle suspension are investigated. Furthermore, the semi-active vehicle suspension is based on a dynamic model of the recently developed prototype of a continuously controllable shock absorber with a smart, electrorheological fluid. These are smart materials and have been known for already more than 50 years. They belong to the group of colloidal suspensions which are able to change their viscosity drastically. This depends upon molecular chain formations in the fluid caused by an electric field perpendicular to the direction of flow. Very low control costs and fast response times of the ERF devices have sparked much an interest in ERFs in the last couple of years. The development of new control strategies for ERF devices integrated into complex multi body systems require a high level of knowledge of the behavior of the ERF subsystems. Dynamic models of controllable ERF devices are studied with respect to their particular dependencies, effects and requirements. An application is presented here which merges linear optimal control strategies and ERF shock absorbers within a complex model of full car dynamics. Here we give a mathematical formulation for the objectives of ride comfort and safety that takes into account various measurement possibilities. The result demonstrates the large potential of optimally controlled ERF devices.

1. Introduction

Electrorheological fluids (ERF) change their viscosity depending on an externally applied electrical field strength. An analogous phenomenon exists for the viscosity of magnetorheological fluids which depends on the applied magnetic field strength. The bandwidth of the resulting flow properties is large; the state varies between

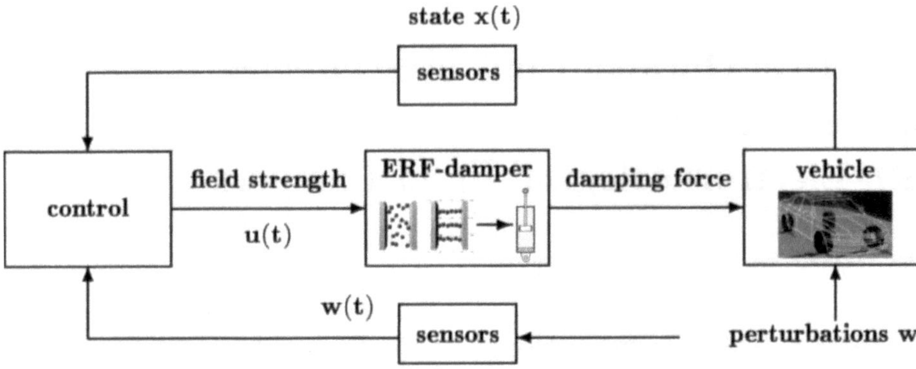

FIGURE 1. Submodels and flow chart of control, perturbation, and state variables for the problem of optimal semi-active vehicle suspension using ERF shock absorbers.

fluid and nearly solid material. The phenomenon of these fluids, caused by polarizable particles within a nonconducting carrier fluid which disturb the flow when excited, has been known since the late 1950's [2]. Hence flow, shear and squeeze processes can be controlled using easily generated electrical fields. ERF devices have additionally several advantageous control properties. The response time between one and 15 ms is extremely fast. Furthermore, ERF devices are continuously controllable and operate subject to almost no wear.

ERF devices thus represent an excellent class of interfaces between electronic control units and mechanical components which have gained increased scientific and economic interest in recent years. As a result, new generations of ERFs with optimized properties are now available. In particular, the difficulties like stability over long time periods and sedimentation of the polarizable particles have largely been resolved. The application areas for ERF devices are numerous. High frequencies and forces may be relatively easily controlled using flexible electronic units. Already many different applications have been reported [7] including a prototype of an adaptively controllable ERF-shock absorber by Schenck AG, Darmstadt.

There are two major problems which arise when investigating the simulation and optimal control of the continuously controllable ERF-shock absorbers and their application to (semi-)active vehicle damping. First there lies the question of modeling and simulating the dynamic behavior of ERF-shock absorbers which depend on the applied electrical field. Second, with regards to the computation of controls that maximize safety and comfort, suitable models for the vehicle dynamics and the safety and comfort objectives must be developed together with applicable numerical methods (Figure 1). A control framework which serves well

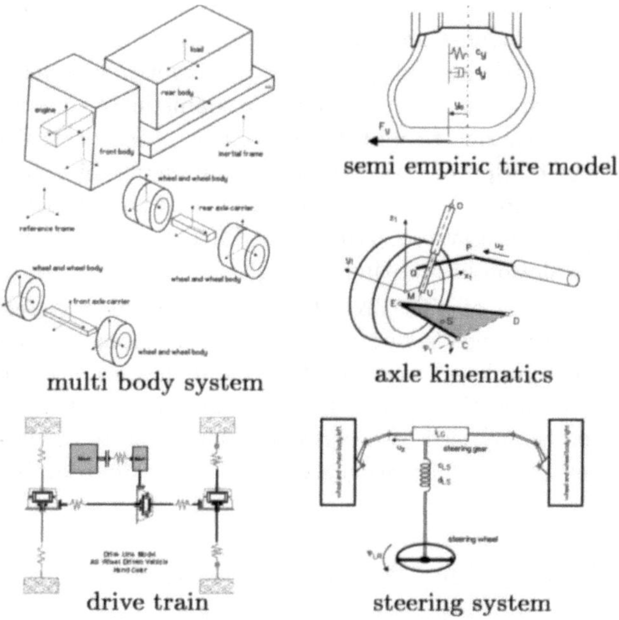

FIGURE 2. Submodels of the full motor vehicle dynamics model of VEDYNA

to evaluate the potential of semi-active damping with optimally controlled ERF-shock absorbers is that of deterministic optimal control methods which provide an open-loop optimal trajectory numerically. For its application in a semi-actively suspended vehicle, feedback controls must be computed under real-time conditions and, thus, the control will likely only be suboptimal.

2. Optimal Control Problems for Optimal Semi-Active Suspension

2.1. Models of Motor Vehicle Dynamics

Motor vehicles are very complex systems which include many significant components for its dynamical driving behavior and yield a high number of degrees of freedom. A detailed and comprehensive vehicle model is needed to represent the nonlinear kinematics of the wheel and axle and to describe the drive train, the steering mechanism, the tire dynamics and ground contact forces [12, 18]. Depending upon the specific design purpose for the dynamical model, a reduction of the system dimension can often be achieved. The influence of the neglected components of the full model are interpreted then as general perturbations of the reduced system.

2.1.1. FULL VEHICLE DYNAMICS MODEL. The numerical, real-time simulation of a full motor vehicle dynamics model that accounts for all significant effects is used in our investigations to verify optimal and suboptimal controls that have been computed using reduced and small scale models of the vehicle dynamics. Our vehicle model consists of a suitable multibody system with kinematical connections and force elements which is supplemented by a sophisticated tire model. A general purpose modeling approach to multibody systems based on the descriptor form of the equations of motion results in a large-scale system of differential-algebraic equations (DAEs) of index 3. However, we make use of an optimally tailored model description which yields a system of ordinary differential equations (ODEs) and is well suited for simulation in real-time.

The vehicle model of VEDYNA [9, 28] consists of a system of nine rigid bodies comprising the vehicle body, the axle suspensions and the wheels. Further submodels are employed to depict the characteristics of the drive train, the steering mechanism, and the tires (Figure 2). Suitable minimum coordinates and generalized velocities are used to describe the spatial state of the vehicle and its components. The equations of motion are derived from Jourdain's Principle yielding

$$\begin{aligned}
M_{BV}(y_{BV})\,\dot{z}_{BV} &= Q_{BV}(y_{BV}, z_{BV}, y_{ST}, z_{ST}, y_{DT}, z_{DT}) && (1)\\
\dot{y}_{BV} &= K_{BV}^{-1}(y_{BV})\,z_{BV} && (2)\\
M_{DT}\,\dot{z}_{DT} &= Q_{DT}(y_{DT}, z_{DT}) && (3)\\
\dot{y}_{DT} &= V_{DT}\,z_{DT} && (4)\\
M_{ST}(y_{ST}, y_{BV})\,\dot{z}_{ST} &= Q_{ST}(y_{ST}, z_{ST}) && (5)\\
\dot{y}_{ST} &= V_{ST}\,z_{ST} && (6)\\
D\,\dot{y}_T &= F_{stat} - C\,y_T. && (7)
\end{aligned}$$

Thus, the vehicle dynamics are fully characterized by the system of 24 first-order ODEs comprising the vehicle body and the axles, (1) and (2). Eight ODEs (7) describe the lateral and longitudinal deviations of the tires by means of spring and damper elements. The vertical deformations of the tires are covered by (1). The dynamic model of the drive train consists of 19 ODEs, (3) and (4), including four equations governing the angular wheel speeds. Five additional ODEs account for the dynamics of the steering system (5) and (6). Couplings between the separate systems occur via the generalized forces and torques Q_{BV}. Wind forces and moments result in additional forces applied to the multibody system of the vehicle [9, 28].

The tire forces have a significant impact on the dynamical behavior of a vehicle. The semi-empirical tire model that is used here describes the behavior of a real tire accurately [9, 28]. About 80 parameters which can be measured or estimated enter the model for each tire in VEDYNA. The model covers different driving situations, including effects at the driving limits such as sliding and spinning. The actual tire model is selected online depending on the respective road and weather conditions.

Due to the stiffness of the system (1)–(7) its numerical integration is carried out recursively with a semi-implicit one-step Euler scheme using a constant step size [9, 28]. In particular, the integration method makes efficient use of the special block structure of the ODEs. It turns out that a fast and stable solution is possible in real-time on modern PC hardware.

For a realistic implementation of virtual test-drives using the simulation, models for the driver and the road have been developed [9, 28].

2.1.2. MODELS OF QUARTER CARS AND LINEAR VEHICLE DYNAMICS. The oscillating behavior of a vehicle excited by perturbations of the ground and driving maneuvers are investigated using *vertical dynamics models*, i.e. reduced models for vertical displacements of the vehicle and wheel bodies. The unknown maneuvers such as braking, acceleration and cornering under high velocities, if considered, may be regarded as perturbations.

Vertical models of vehicles may be classified into different levels of detail. There are so-called *full-car-models* (in the context of vertical models) including two axles, which reflect both vertical deflections and inclinations. Bounce, roll and pitch motions are investigated simultaneously. Separated and decoupled investigations are possible using *half-car-models* (cf. Figure 3, right). The inclination is

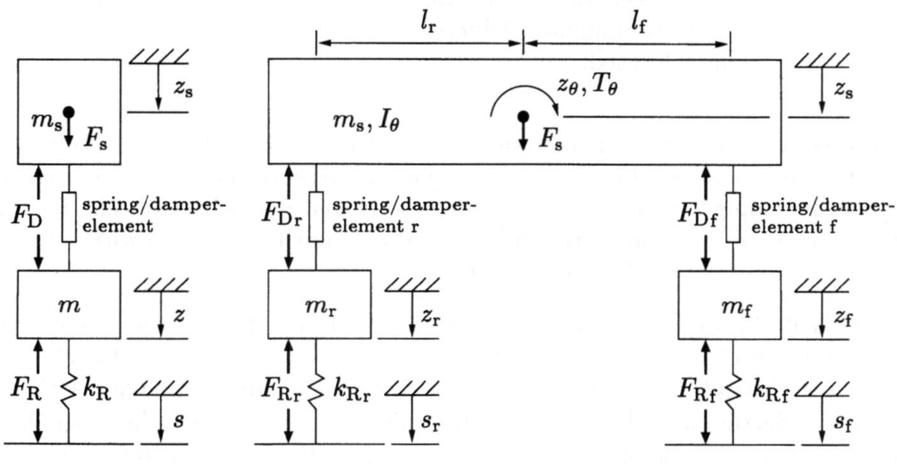

FIGURE 3. Quarter car model (left) and half car model (right).

variable	denotation
$z_s, z_\theta, z_{r/f}$	displacement and inclination of vehicle body, displacement of the wheel
m_s, I_θ	mass and moment of inertia of the vehicle body
F_D	force at the spring/damper element
F_R	wheel load (wheel model assumed to be a spring, often in addition with a parallel damping element or as a more complex model)
$s_{r/f}$	perturbation of the ground, (r - rear, f - front)
F_s, T_θ	perturbing force and moment (induced by maneuvers)

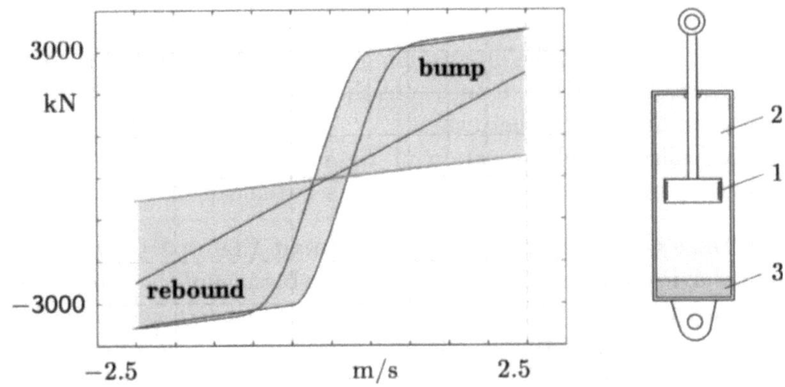

FIGURE 4. Typical ERF damper characteristics (left) under minimum and maximum excitation of the ERF, the enclosed area describes the bandwidth of adjustable damping rates compared to conventional linear damping. Sketch of a ERF damper (right), 1 - valve between the electrodes, 2 - chambers filled with ERF, 3 - gas filled accumulator. Different designs with longer electrodes are commonly used.

interpreted as roll *or* pitch motion. The most common and simple model is the *quarter-car-model* (cf. Figure 3, left), which represents the vertical motion of a system including a quarter of the vehicle body and one of the wheels. Its corresponding simplified equations will be the base of our calculations with respect to our goal of optimal semi-active suspension.

2.1.3. NONLINEAR SINGLE TRACK VEHICLE MODEL. The vehicle dynamics model should also take into account the particular properties of the ERF shock absorbers. Substitution of the commonly used linear damping behavior within the models of Section 2.1.2 by a realistic dynamic model of the ERF damping characteristics yields a highly nonlinear system of differential equations (cf. Section 2.2). This may be extended by further substitutions such as nonlinear models for tires. Such a more realistic upgrade of a half car model is commonly used for calculation of different optimal controls [28]. Specially calculated damping controls take into account pitch motions which are significant for time shifted perturbations on front and rear wheels.

2.2. Dynamic Model of a Continuously Controllable ERF Shock Absorber

Conventional damping behavior is usually described by a fixed damping characteristic defined by the force-velocity rate. In general, dampers exhibit nonlinear behavior. Depending on the actual damper design, the damping rates are different in the bump and rebound phases while within each phase the system is best

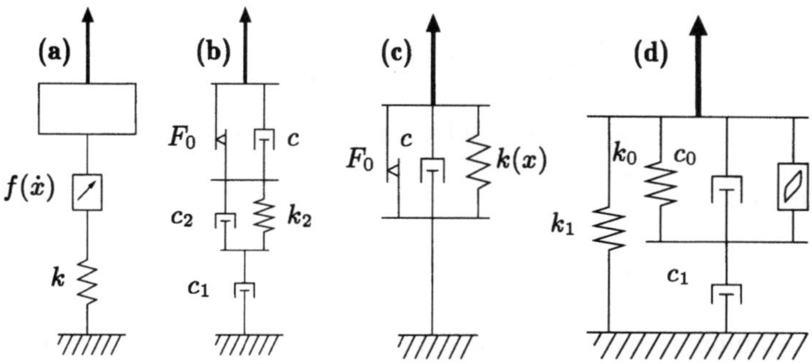

FIGURE 5. Proposed phenomenological models for ERF devices. (a) Peel, Stanway, Bullough [24] (b) Powell [19]; (c) Ehrgott, Masri [11] resp. Kamath Wereley [14]; (d) Spencer et.al. [23].

described by progressive damping characteristics. For a more realistic simulation, piecewise linear or higher order polynomial approximations of realistic tabular data are used.

Measurements of ERF dampers indicate strong nonlinear behavior and characteristics with hysteresis. The dynamics of the characteristics depend upon the damping velocity *and* on the exciting electrical field.

Poiseuille flow within the valve has already been calculated under highly simplifying assumptions (steady state, laminar, incompressible flow neglecting mass inertia, cf. [1, 20]). The results demonstrate that the *Bingham* property of the ERF, defining the ratio between shear stress and strain rate, dominates the force-velocity characteristics of the damper. A comparison of the calculated and measured damping rates also suggests the necessity for more complex simulations, including the effects of dynamic flow, particular geometry, vortices, etc.

Another approach which avoids these complex and expensive calculations are parametric models. A sufficiently precise prediction of the dynamic behavior of an ERF shock absorber using parametric models depends upon an optimal set of parameters that minimize a particular distance between model prediction and provided measurements. Of course, the type of dynamic model dominates the error between prediction and reality. General approximation techniques, for instance Čebyšev polynomials (cf. [11]) and multilayer neural networks (cf. [6, 15]) have previously been investigated. More common models are phenomenological models. Here the approximation of the ERF effects is usually induced by friction elements or nonlinear spring or damper elements (cf. Figure 5). However, only a few of them take into account the dependency on a variable electrical field. An overview is given in [7, 25].

The *augmented Bouc-Wen* model (cf. [23], see also [7, 13]), Figure 5(d)) is the most flexible one and is used in the following sections. It describes a dynamic system depending on $\dot{z}_{\rm D}$, the velocity of the piston rod.

$$\begin{aligned}\dot{s}_1 &= \frac{1}{c_0 + c_1}\left(c_1 \dot{z}_{\rm D} - \alpha s_2 - k_0 s_1\right) \\ \dot{s}_2 &= \left(A - \beta(1 + \operatorname{sgn}(\dot{s}_1 s_2)) s_2^2\right) \dot{s}_1\end{aligned} \tag{8}$$

The output function

$$F = c_1(\dot{z}_{\rm D} - \dot{s}_1) + k_1(z_{\rm D} - z_0) \tag{9}$$

denotes the damping force and depends on $\dot{z}_{\rm D}$ and $z_{\rm D}$, the relative displacement of the damper. The system (8) describes a hysteresis operator, and its properties are parameterized with respect to the applied electrical field. This gives

$$\dot{s}_3 = \eta(u_{\rm D} - s_3) \tag{10}$$

with control $u_{\rm D}$ as the applied field strength and

$$\begin{aligned}c_0 &= c_{01} + s_3 c_{02} \\ c_1 &= c_{11} + s_3 c_{12} \\ \alpha &= \alpha_1 + s_3 \alpha_2 \,.\end{aligned} \tag{11}$$

The states $\mathbf{s} = (s_1, s_2, s_3)^{\rm T}$ denote the inner variables of the virtual model of each ERF damper. For a more realistic ERF damper model, the 11-dimensional parameter vector $\mathbf{p} = (c_{11}, c_{01}, \alpha_1, A, \beta, c_{11}, c_{02}, \alpha_2, k_0, k_1, z_0, \eta)^{\rm T}$ must be estimated numerically from measurements of a prototype damper.

2.3. Objectives for Comfort and Safety

2.3.1. Parameterized Cost Functional. The two primary objectives for a vehicle ride with semi-active suspension are ride safety and ride comfort. For both, mathematical models must be provided.

The characterization of safety in the vehicle dynamics depends primarily on wheel loads. High loads have greater longitudinal and lateral transmission forces between wheels and ground. On the contrary, low wheel loads can cause the loss of controllability of the vehicle. Larger magnitudes for the roll and pitch angles will indirectly influence ride safety as the wheels' contact force magnitudes might reach zero causing lift-off.

The comfort of a ride can be of almost equal importance to passengers than safety. Comfort is mainly characterized by the accelerations of the vehicle body, often called the *sprung mass* contrary to the wheel body, which in this sense is called the *unsprung mass*. With respect to the vertical models (cf. Section 2.1.2), vertical accelerations are treated here. For models with a higher level of detail, angle accelerations of pitch and roll motions may also be considered.

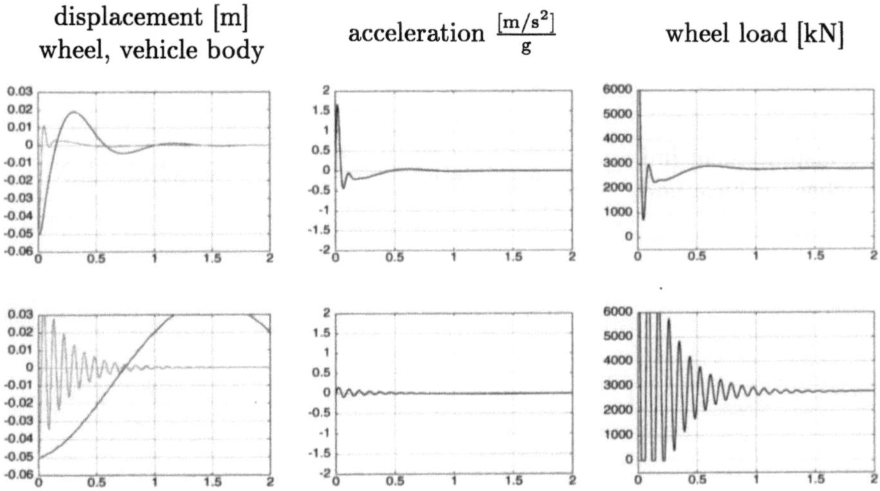

FIGURE 6. Comparison of optimal controls and trajectories resulting from the two extremes of cost functional weights. The curves in the upper row are obtained for the optimal control with respect to safety only, the lower ones for optimal control with respect to comfort only. All curves are depicted over time in [s] and are based on a quarter car model and an initial step disturbance. (wheel displacement: bright, vehicle body displacement: darc line)

Altogether a performance index consisting of a weighted sum of various criteria of safety and comfort may be used

$$
\begin{aligned}
L(\mathbf{x},\mathbf{u}) = \mu_{\text{safety}} & \left[\sum_{i=f,r} \left[\frac{F_{\text{dyn.load},i}}{F_{\text{stat.load},i}} \right]^2 + \left[\frac{z_\theta}{z_{\theta,\max}} \right]^2 \right] \\
& + \mu_{\text{comfort}} \left[\left[\frac{\ddot{z}_s}{\ddot{z}_{s,\max}} \right]^2 + \left[\frac{\ddot{z}_\theta}{\ddot{z}_{\theta,\max}} \right]^2 \right] .
\end{aligned} \tag{12}
$$

For an example of the state and control variables $\mathbf{x}$ and $\mathbf{u}$, we refer to the half-car-model of Figure 3. For a full-car-model (in the sense of vertical models) one can extend the functional by terms of the second angular magnitude. The variables $F_{\text{dyn.load},i}, \ddot{z}_s, z_\theta, \ddot{z}_\theta$ denote the deviation from the corresponding stationary value.

2.3.2. THE WEIGHTS. The weights $\mu_\diamond$ have to be chosen properly, i.e. depending on the purpose of investigation. For example, their actual value may depend on the actual driving situation where either comfort or safety may be more desirable. Figure 6 shows suspension behavior under different optimal semi-active damping with respect to either only safety or comfort. Here, the purpose is to demonstrate the antagonistic character of the two goals. Optimized safety leads to fast regulation of the oscillations under high accelerations of the vehicle body, whereas optimized

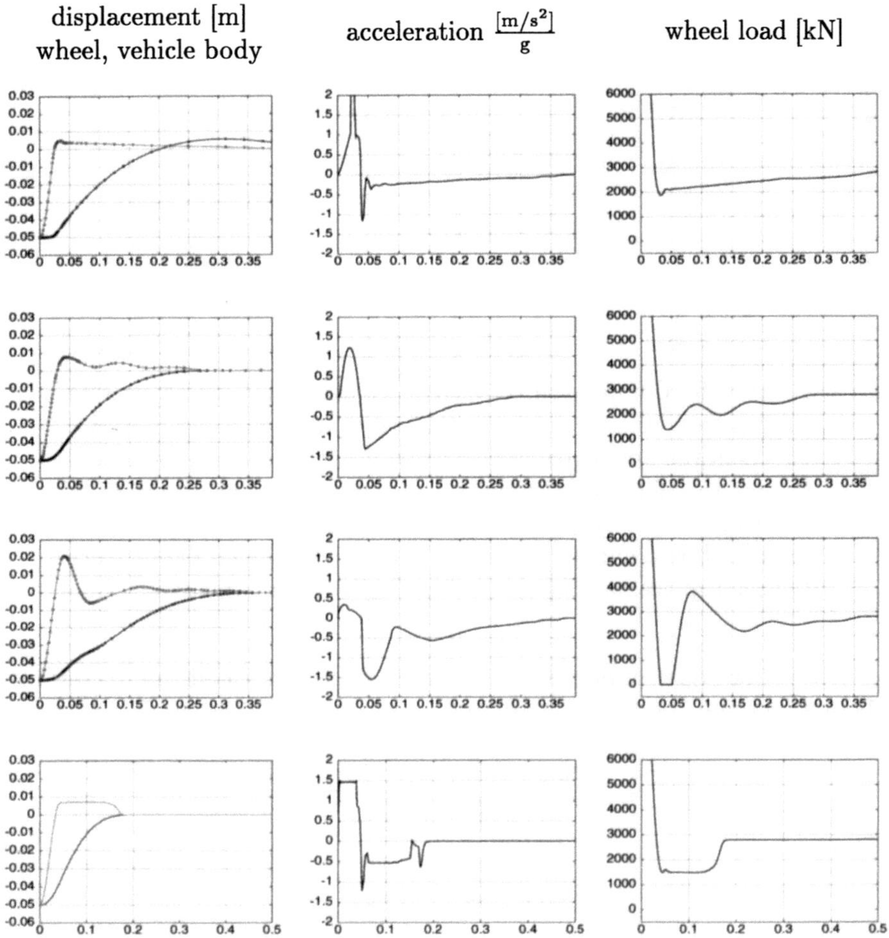

FIGURE 7. Comparison of curves by different optimization configurations (all curves over time in [s], wheel displacement: bright line, vehicle body displacement: darc line).
first row: optimal safety, comfort ignored
second row: optimal comfort with constrained wheel load
third row: optimal comfort, safety ignored
fourth row: version with a slack variable.

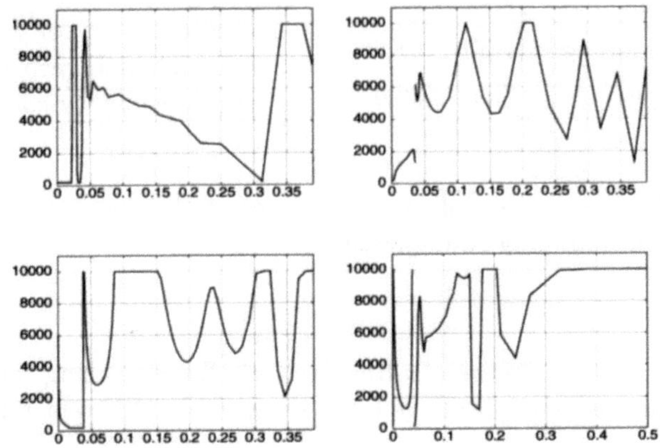

FIGURE 8. Controls corresponding to the four rows depicted in Figure 7 (damping rates over time in [s]: upper row, left - one; upper row, right - two; lower row, left - three; lower row, right - four).

comfort yields almost decoupled motions of vehicle and wheel body under very low accelerations of the sprung mass and high frequencies of the unsprung mass. Please note the periods of time intervals of contact losses to the ground.

2.3.3. DIFFERENT VERSIONS OF OBJECTIVES. Evidently it is not possible to simultaneously ensure a maximum value for the respective cost functionals corresponding to both safety and comfort by choosing particular weights of the cost functional

$$L = \mu_{\text{safety}} L_{\text{safety}} \quad + \quad \mu_{\text{comfort}} L_{\text{comfort}}.$$

Another approach to handle this problem is to optimize one of the objectives and to restrict the other one to suitable bounds.

$$\min L_1 \qquad \text{under} \qquad L_2 \leq L_{2,\max}.$$

This approach usually yields satisfactory results. A third approach is to maximize a positive slack variable σ which is subtracted from both objective values of safety and comfort

$$\max \sigma \qquad \text{under} \qquad L_i \leq L_{i,\max} - \sigma, \quad i = 1, 2 \quad \sigma > 0\,.$$

Each difference is then constrained to a proper maximum value. Figure 7 shows the behavior of the state variables of a quarter car model for the resulting optimal controlled semi-active suspension. Although the displacement state variable shows similar behavior for the first three examples, the optimal damping rates are very different (see Figure 8). The last optimization configuration using a slack variable yields high gain damping.

Optimization with respect to safety		
constant parameters	18.85	
semi-active damping	11.57	(61,4%)
active damping	7.05	(37,4%)

Values derived by general nonlinear calculation of optimal damping forces for a quarter car model with respect to occurring inequality constraints.

TABLE 1. Comparison of optimal values of the cost functionals for different damping configurations.

However, weighting remains a subjective task. Sporty racing cars, ambulance vehicles or ordinary automobiles have to be designed by different requirements.

3. Numerically Calculated Controls and Results

The goal of the control problem for controllable ERF-shock absorbers integrated into a semi-active vehicle suspension is a combination of optimal safety and optimal comfort of the ride. Nowadays available standard techniques for damping control are rather simple: The driver or a control unit selects a fixed damping characteristic from several optional settings. The new technology of ERF-shock absorbers permits continuous control. This supposes the calculation of suitable (if possible optimal) controls taking into account the dynamics of the vehicle and of the damper (Section 2.1). The resulting dynamical system may be highly nonlinear and further constraints have to be taken into account.

3.1. State Feedback Controller based on LQR and H^∞ Techniques

In order to test the capability of semi-active damping with ERF-shock absorbers, state feedback controls for various linearized, reduced vehicle models (cf. Section 2.1.2) were computed and implemented. The general problem formulation for a linear vehicle dynamics model is written as

$$\dot{\mathbf{x}} = \underbrace{\begin{pmatrix} \mathbf{0}_{n_x,n_x} & \mathbf{I}_{n_x,n_x} \\ \mathbf{M}^{-1}\mathbf{A}_\mathrm{k} & \mathbf{M}^{-1}\mathbf{A}_\mathrm{c} \end{pmatrix}}_{\mathbf{A}} \mathbf{x} + \underbrace{\begin{pmatrix} \mathbf{0}_{n_x,n_u} \\ \mathbf{M}^{-1}\mathbf{B}_\mathrm{m} \end{pmatrix}}_{\mathbf{B}} \mathbf{u} + \underbrace{\begin{pmatrix} \mathbf{0}_{n_x,n_g} \\ \mathbf{M}^{-1}\mathbf{D}_\mathrm{m} \end{pmatrix}}_{\mathbf{D}} \mathbf{g}, \tag{13}$$

with zero matrix $\mathbf{0}$, unit matrix $\mathbf{I}$, mass matrix $\mathbf{M}$, stiffness matrix $\mathbf{A}_\mathrm{k}$, damping matrix $\mathbf{A}_\mathrm{c}$, control matrix $\mathbf{B}_\mathrm{m}$ and perturbation matrix $\mathbf{D}_\mathrm{m}$. The state $\mathbf{x}$ is controlled by $\mathbf{u}$ and perturbed by $\mathbf{g}$, $n_\diamond$ denote their dimensions, $\diamond = x, u$, or g. Perturbations include shocks caused by an uneven ground as well as forces and moments acting on the vehicle body caused by driving maneuvers. The linear quadratic cost functional

$$J[\mathbf{u}] = \int_0^\infty \mathbf{x}^\mathrm{T}\mathbf{Q}\,\mathbf{x} + 2\mathbf{u}^\mathrm{T}\mathbf{S}\,\mathbf{x} + \mathbf{u}^\mathrm{T}\mathbf{R}\,\mathbf{u}\;\mathrm{d}t \tag{14}$$

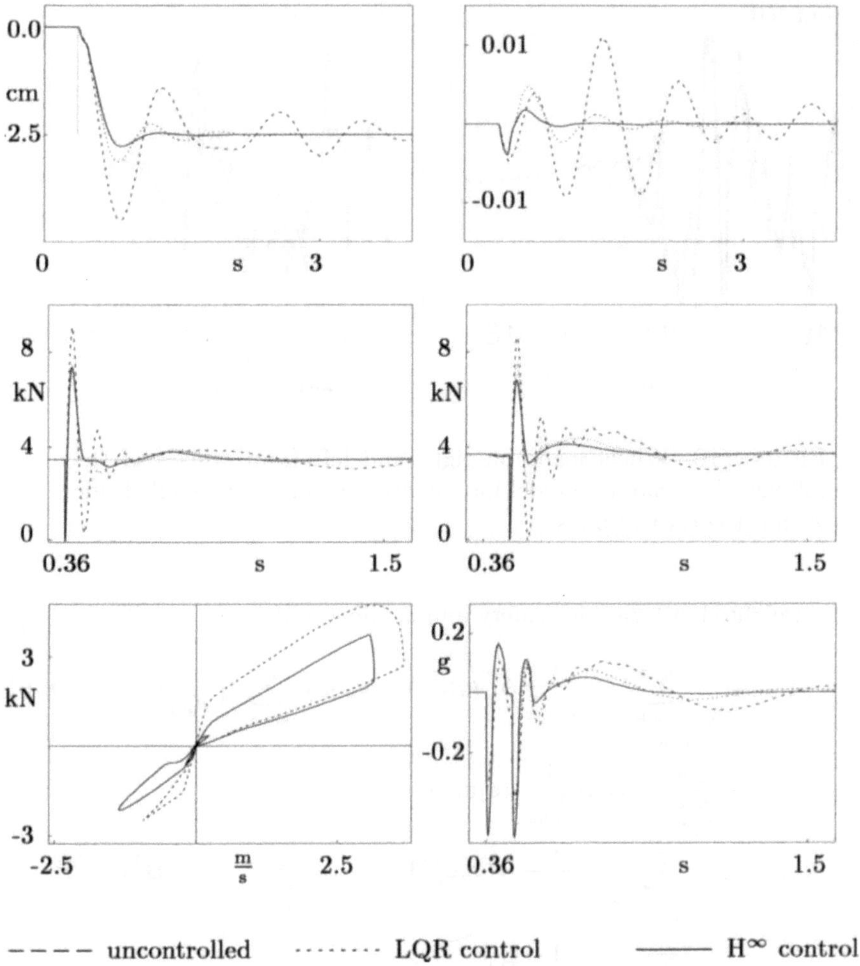

FIGURE 9. Comparison of some of the state variables of a single track model for a ride over a step of 2.5 cm height at a speed of 100 km/h for different controls.

upper left	motion of center of gravity of the sprung mass	z_s
upper right	curve of the pitch angle	z_θ
middle left	wheel load of the front wheel	$F_{\mathrm{R}f}$
middle right	wheel load of the rear wheel	$F_{\mathrm{R}r}$
lower left	front(-·-) and rear (–) characteristics	$F_{\mathrm{D}r/f}$
lower right	sprung mass accelerations	$\ddot{z}_s$

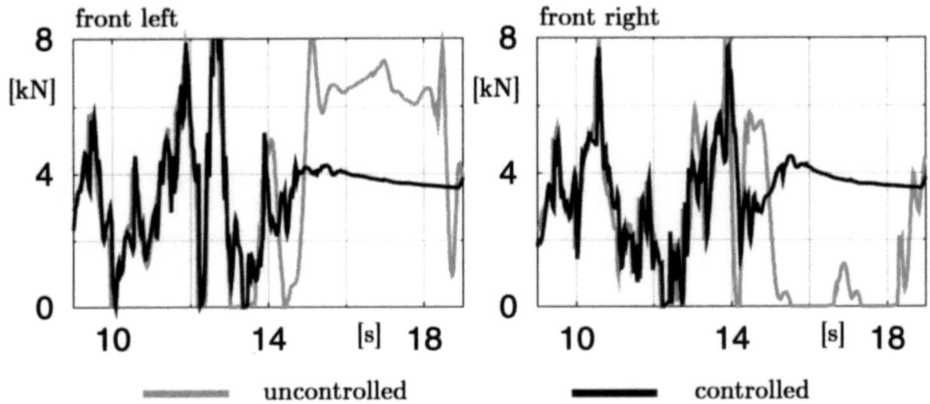

FIGURE 10. Wheel loads on right and left front wheels of the full vehicle dynamics model for a controlled and uncontrolled ride within the critical phase.

again is a weighted criterion for safety and comfort with

$$\begin{aligned}
\mathbf{Q} &= \mu_{\text{comfort}} \sum_{\substack{\text{vehicle body}\\\text{velocities}}} \frac{1}{\ddot{z}^2_{i,\max}} \mathbf{A}^{\mathrm{T}} \mathbf{e}_i \mathbf{e}_i^{\mathrm{T}} \mathbf{A} + \mu_{\text{safety}} \sum_{\text{wheels}} \left[\frac{k_{\mathrm{R}i}}{F_{\text{stat.load},i}} \right]^2 (\mathbf{e}_i \mathbf{e}_i^{\mathrm{T}}) \\
&\quad \ldots + \sum_{\text{states}} \mu_{i,\text{state}} \frac{1}{x^2_{i,\max}} (\mathbf{e}_i {\mathbf{e}_i}^{\mathrm{T}}) \\
\mathbf{R} &= \mu_{\text{comfort}} \sum_{\substack{\text{vehicle body}\\\text{velocities}}} \frac{1}{\ddot{z}^2_{i,\max}} \mathbf{B}^{\mathrm{T}} \mathbf{e}_i \mathbf{e}_i^{\mathrm{T}} \mathbf{B} + \mu_{\text{costs}} \sum_{\text{wheels}} \frac{1}{u_{i,\max}} \mathbf{B}^{\mathrm{T}} \mathbf{e}_i \mathbf{e}_i^{\mathrm{T}} \mathbf{B} \\
\mathbf{S} &= \mu_{\text{comfort}} \sum_{\substack{\text{vehicle body}\\\text{velocities}}} \frac{1}{\ddot{z}^2_{i,\max}} \mathbf{B}^{\mathrm{T}} \mathbf{e}_i \mathbf{e}_i^{\mathrm{T}} \mathbf{A}.
\end{aligned}$$

The optimal active damping force can now be obtained by

$$\mathbf{u}^* = -\mathbf{K}\mathbf{x} = \mathbf{R}^{-1}(\mathbf{S} + \mathbf{B}^{\mathrm{T}}\mathbf{P})\mathbf{x} \tag{15}$$

with the solution $\mathbf{P}$ of the algebraic Riccati equation corresponding to the cost functional of Equation (14). Expanding the approach of conventional optimal control (LQR, cf. [5, 10, 16]), H^∞ control techniques include the worst influence of occurring perturbations (cf. [3, 10, 13]). Here the extreme (optimal resp. pessimistic) controls are saddle-point solutions of the dynamic game with the parameterized cost functional

$$J_\gamma\,[\mathbf{u}, \mathbf{g}] = \int_0^\infty \mathbf{x}^{\mathrm{T}} \mathbf{Q}\, \mathbf{x} + 2\mathbf{u}^{\mathrm{T}} \mathbf{S}\, \mathbf{x} + \mathbf{u}^{\mathrm{T}} \mathbf{R}\, \mathbf{u} - \gamma^2 \mathbf{g}^{\mathrm{T}} \mathbf{g}\; dt\,, \tag{16}$$

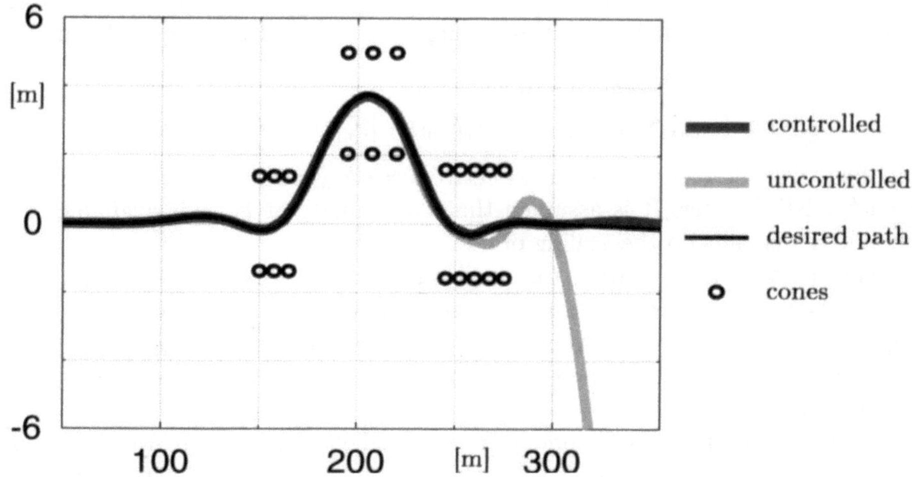

FIGURE 11. Double lane change maneuver (path through the cones) on a very uneven street simulated with the full vehicle dynamics model. The uncontrolled car, i.e. with a conventional passive suspension, swerves off the road whereas the controlled car, i.e. with an (sub-)optimally controlled semi-active suspension, follows the desired time optimal path.

where $\mathbf{u}$ is the minimizing and $\mathbf{g}$ is the maximizing player. For $\gamma > 0$ and optimal $\mathbf{u}^0 = \mathbf{u}^0(\mathbf{g})$ we have $J\left[\mathbf{u}^0; \mathbf{g}\right] / \|\mathbf{g}\|^2 \leq \gamma^2 \quad \forall \mathbf{g}$, i.e. the disturbance of the system is bounded. The optimal disturbance attenuation is related to a minimal bound γ^*, which is calculated iteratively. Since the existence of an optimal controller for the "infimum" is not guaranteed, suboptimal solutions for $\gamma = \gamma^* + \varepsilon$ may be sought for (see [13]). Both the approaches LQR and H^∞ are feasible and yield useful damping performances (cf. Figure 10).

3.2. Verification within a Dynamic Full Car Simulation

The linearity of the problem formulation merely allows the calculation of the optimal damping force but not the optimal damping rate or field strength at the valve of the ERF-shock absorber as in the general nonlinear case. Hence the application of ordinary algorithms for *linear quadratic regulators* causes active damping. Technically speaking this requires energy input into the vehicle suspension system which is not supported by the technique of ERF-shock absorbers. Table 1 shows the difference between optimal cost functionals under various damping configurations. Please note that obtaining the best performance with fully active damping must in practice be payed with a substantially more expensive technical effort than for semi-active damping.

Based on the calculated optimal damping force (Section 3.1), a prediction of a capable semi-active damping rate is computed by a heuristic compensation regulator. A "clipped optimal" algorithm is suggested in [22]. Here a related approach

$$u = \begin{cases} u_{\min} \text{ for } F_{\text{opt}} F_{\text{appl}} < 0 \\ \min(u_{\max}, \max(u_{\min}, u + (|F_{\text{opt}}| - |F_{\text{appl}}|) * K)) \text{ otherwise} \end{cases} \tag{17}$$

is applied with the requested optimal active force F_{opt} and the current force F_{appl} for each ERF damper. It is assumed that an estimate of F_{appl} is available. The constant K depends on the scaling of u.

Generally in this context, the measurability and observability of the state variables is important. In addition to an estimate of F_{appl}, the damping velocities of the shock absorber and the vehicle body accelerations are assumed to be given, e.g., by measurements. A *Kalman state estimator* is used to provide state estimates for the system (13), cf. [10].

The calculated feedback controllers are verified within the real-time full car simulation VEDYNA (Section 2.1.1), which includes 56 dynamic state variables in its basic version and accounts for the most relevant effects of a complex vehicle model. A typical test maneuver for vehicles, the *double lane change*, cf. Figure 11, simulated on a very uneven street demonstrates the improvement in safety. A car equipped with an ERF-shock absorber remains on a commanded minimum time path through the cones (cf. [28]), whereas the conventional car begins to slide. Higher wheel loads are provided by controlled vehicle suspensions as shown in Figure 10.

3.3. Nonlinear deterministic optimization

3.3.1. THE DIRECT COLLOCATION METHOD. as implemented in DIRCOL, cf. [26], computes the numerical solution for general, nonlinear optimal control problems with the following general problem formulation

$$\min J[u] = \Phi(x(t_f), t_f)$$

$$\text{under} \quad \begin{aligned} \dot{x}(t) &= f(x(t), u(t), t) \\ 0 &= r(x(0), x(t_f), t_f) \\ 0 &\leq g(x(t), u(t), t), \quad 0 \leq t \leq t_f. \end{aligned} \tag{18}$$

The method belongs to the class of direct transcription methods [4]. The optimal control $u^*(t)$ and trajectory $x^*(t)$ minimize the cost functional J subject to the given inequality and boundary constraints and subject to the differential equations. Tasks with cost functionals of different types, namely with an integral part, have to be transformed into a problem with a Mayer-type objective, cf. Equation (18). The discretization process approximates the controls $u(t)$ on a time grid $\{t_k : t_0 \leq t_k < t_{k+1} \leq t_n = t_f\}$ by a continuous, piecewise linear function; the states $x(t)$ are interpolated by continuously differentiable cubic functions on the same time grid. DIRCOL further permits the modeling of multiple phases used for switching dynamics or controls. In this case, the time interval is decomposed within each

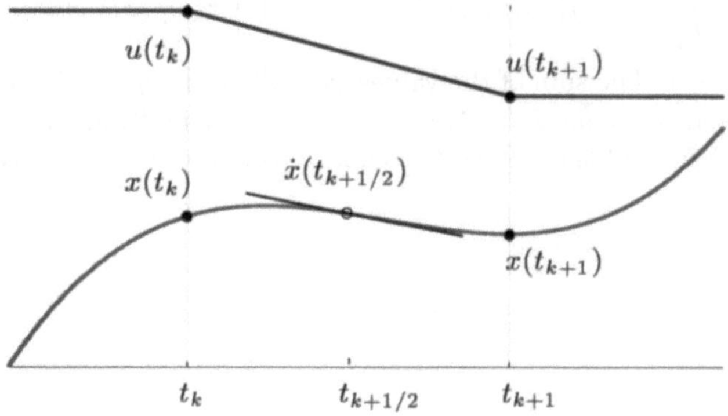

FIGURE 12. Principle of the direct collocation method

phase. This yields additional conditions on the phase transitions. In order to save expensive writing effort a one phase problem is described here. The discretization scheme leads to

$$\min J[u] = \Phi(x(t_f), t_f)$$

$$\begin{aligned} f(t, x(t), u(t)) &= \dot{x}(t) & \forall t = t_{k+1/2} \\ r(x(t_0), x(t_f)) &= 0 & \\ g(t, x(t), u(t)) &\geq 0 & \forall t = t_k . \end{aligned} \tag{19}$$

For $Y = \big(x(\tau_0), u(\tau_0), \ldots, x(\tau_n), u(\tau_n), t_f\big)$ a *nonlinearly constrained minimization problem* results in

$$\min \Phi(Y)$$

$$\begin{aligned} a(Y) &= 0 \\ b(Y) &\geq 0 . \end{aligned} \tag{20}$$

Solutions of (20) are currently obtained by *SQP-methods*. If structure and sparsity of the NLP functions and their gradients are exploited and the large-scale SQP method SNOPT [21] is applied, a computational speed-up and storage savings by two orders of magnitude can be obtained [27].

3.3.2. RESULTS. Several local error estimators are investigated to monitor the accuracy of the computed approximation of the solution and the need for a local grid refinement. The estimates of the adjoint variables $\tilde{\lambda}$ and the multipliers $\tilde{\mu}$ of the constraints resulting from the optimal control problem and calculated from the Lagrangian multipliers of the discretized minimization problem make the following calculation possible

$$\tilde{\omega}_k = \int_{t_k}^{t_{k+1}} \tilde{\lambda}(t)\Big(f(\tilde{x}(t),\tilde{u}(t),t) - \dot{\tilde{x}}(t)\Big) + \tilde{\eta}(t)g(\tilde{x}(t),\tilde{u}(t),t)\mathrm{d}t \qquad (21)$$

for $k = 1, \ldots, n-1$. The sum of the values $\tilde{\omega}_k$, called the *optimality error*, vanishes for exact optimal solutions $\tilde{x} = x^*$, $\tilde{u} = u^*$ and exact multipliers $\tilde{\lambda} = \lambda^*$, $\tilde{\mu} = \mu^*$. Furthermore, *defects* of the differential equations and of the nonlinear constraints

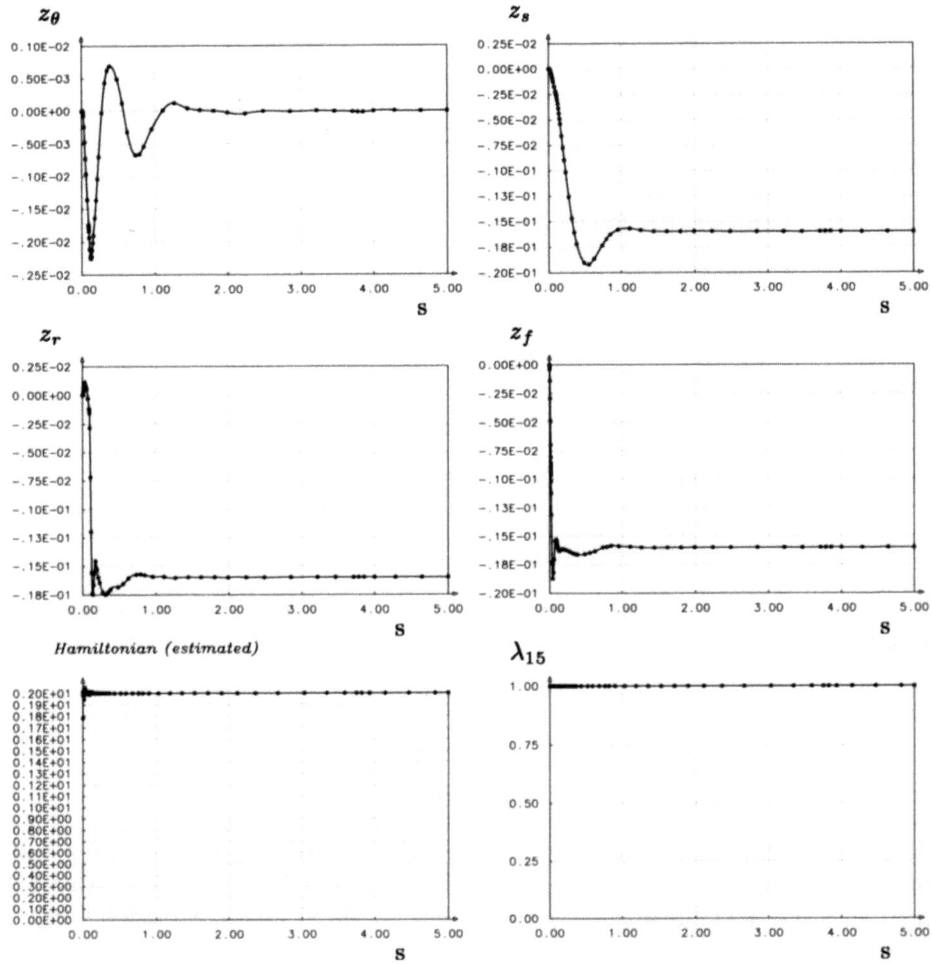

FIGURE 13. Calculated results of an optimal semi-active damping control for a single track vehicle model equipped with ERF shock absorber during a ride over a step (speed 100 km/h). The solutions shows the 68 marked grid points.

are evaluated at the points $t = t_{k+l/4}$, $l = 0, 1, 2, 3$ (cf. [26, 27]). These error estimators are used for an automatic iterative grid refinement which can produce more exact and robust solutions. This yields problems with a high number of grid points; nevertheless, they can be calculated very fast with the sparse version of DIRCOL.

Figure 13 depicts a calculated example. Here the lower figures show the estimated Hamiltonian and the adjoint variable for the scaled time variable. These trajectories demonstrate the high accuracy of the solutions although adjoint differential equations have neither been formulated nor solved numerically. The location of the marked grid points are obtained by evaluation of the error estimators in the sequence of solutions.

4. Conclusions

The problem of optimal semi-active suspension of vehicles using the new technology of electrorheological fluid dampers has been investigated. For the formulation of the corresponding optimal control problems, several models of the vehicle dynamics with different levels of detail for the ERF shock absorber dynamics and the safety and comfort objectives are presented and investigated.

Reduced, linear vehicle dynamic models permit the application of LQR and H^∞ control techniques which provide an optimal and real-time capable feedback control for the reduced model. The resulting controls are only suboptimal with respect to the full scale vehicle dynamics model but provide remarkable improvements over passive suspensions as has been demonstrated in numerical experiments. In spite of the improvements (in particular for the ride safety), significant nonlinearities in the dynamics of the ERF-suspensions continue to counteract the calculated control, whereby the potential of the ERF technology can not fully be exploited by this approach. On the other hand, direct transcription methods can deal with general, nonlinear dynamic models and constraints, but they only provide optimal open-loop state and control trajectories.

Furthermore, it has been demonstrated that comfort and safety have antagonistic properties. Several ways to deal with comfort and safety in a single objective have been investigated and discussed here.

Future work will focus on nonlinear H^∞ control and the approximation of optimal feedback controls in the case of general nonlinear dynamics.

References

[1] R. J. Atkin, Xiao Shi, and W. A. Bullough, *Solutions of the constitutive equations for the flow of an electrorheological fluid in radial configurations*, Journal of Rheology, **35(7)** (1991), pp. 1441–1461.

[2] H. T. Banks, R. C. Smith, and Y. Wang, *Smart Material Structures — Modeling, Estimation and Control*, J. Wiley & Sons, 1996.

[3] T. Basar and P. Bernhard, *H^∞-Optimal Control and Related Minimax Design Problems Problems, A Dynamic Game Approach*, Berlin, Birkhäuser, 1991.

[4] J. T. Betts, *Survey of numerical methods for trajectory optimization*, AIAA J. Guidance, Control, and Dynamics, **21**, 2 (1998), pp. 193–207.

[5] A. E. Bryson and Y. C. Ho, *Applied Optimal Control*, Ginn and Company, 1969; Rev. printing, Hemisphere, 1975.

[6] S. A. Burton, N. Makris, I. Konstantopoulos, and P. J. Antsaklis, *Modeling the response of ER damper: phenomenology and emulation*, Journal of Engineering Mechanics, **122** (1996), pp. 897–906.

[7] T. Butz and O. von Stryk, *Modelling and simulation of electro- and magnetorheological fluid dampers*, Z. Angew. Math. Mech., 2001, to appear.

[8] J. D. Carlson and B. F. Spencer Jr., *Magneto-rheological fluid dampers: scalability and design issues for application to dynamic hazard mitigation*, in: Proc. Intern. Workshop on Structural Control, Hong Kong, Dec. 18-20, 1996, to appear.

[9] C. Chucholowski, M. Vögel, O. von Stryk, and T.-M. Wolter, *Real time simulation and online control for virtual test drives of cars*, in: H.-J. Bungartz, F. Durst, Chr. Zenger, eds., High Performance Scientific and Engineering Computing, Lecture Notes in Computational Science and Engineering, **8**, Springer-Verlag, 1999, pp. 157–166.

[10] P. Dorato, C. Abdallah, and V. Cerone, *Linear-Quadratic Control — An Introduction*, Englewood Cliffs, N.J., Prentice-Hall, 1995.

[11] R. C. Ehrgott and S. F. Masri, *Modelling the oscillatory dynamic behavior of electrorheological materials in shear*, Smart Material Structures, **1** (1992), pp. 275–285.

[12] G. Genta, *Motor Vehicle Dynamics, Modelling and Simulation*, Series on Advances in Mathematics for Applied Sciences, **43**, London, World Scientific, 1997.

[13] R. H. W. Hoppe, G. Mazurkevitch, U. Rettig, and O. von Stryk, *Modeling, simulation, and control of electrorheological fluid devices*, in: H.-J. Bungartz et al., eds., Lectures on Applied Mathematics, Springer-Verlag, 2000, pp. 251–276.

[14] G. M. Kamath and N. M. Wereley, *A nonlinear viscoelastic-plastic model for electrorheological fluids*, Smart Material Structures, **6** (1997), pp. 351–359.

[15] B. Kim and P. N. Roschke, *Linearization of Magnetorheological Behavior Using a Neural Network*, Proc. of the American Control Conf., San Diego, June 1999, pp. 4501–4505.

[16] W. Kortüm and P. Lugner, *Systemdynamik und Regelung von Fahrzeugen*, Springer, 1994.

[17] B. Koslik, G. Rill, O. von Stryk, and D. E. Zampieri, *Active suspension design for a tractor by optimal control methods*, Preprint SFB-438-9801, Sonderforschungsbereich 438, Technische Universität München – Universität Augsburg, 1998.

[18] M. Mitschke, *Dynamik der Kraftfahrzeuge*, Springer-Verlag, 1994.

[19] J. A. Powell, *Modelling the oscillatory response of an electrorheological fluid*, Smart Material Structures, **3** (1994), pp. 416–438.

[20] K. R. Rajapogal and A. S. Wineman, *Flow of electrorheological materials*, Acta Mechanica, **91** (1992), pp. 57–75.

[21] P. E. Gill, W. Murray, and M. A. Saunders, *User's Guide for SNOPT 5.3: A Fortran Package for Large-Scale Nonlinear Programming*, Draft, Department of Mathematics, University of California, San Diego, December 1998, Software Version 5.3-5, June 1999.

[22] B. F. Spencer Jr., S. J. Dyke, M. K. Sain, and J. D. Carlson, *Modeling and control of magnetorheological dampers for seismic response reduction*, Smart Materials and Structures, **5** (1996), pp. 565–575.

[23] B. F. Spencer Jr., S. J. Dyke, M. K. Sain, and J. D. Carlson, *Phenomenological model of a magnetorheological damper*, ASCE Journal of Eng. Mech., **123(3)** (1996), pp. 1–9.

[24] R. Stanway, D. J. Peel, and W. A. Bullough, *Dynamic modelling of an ER vibration damper for vehicle suspension applications*, Smart Material Structures, **5** (1996), pp. 591–606.

[25] R. Stanway, J. L. Spronston, and A. K. El-Wahed, *Applications of electrorheological fluids in vibration control: a survey*, Smart Material Structures, **5** (1995), pp. 464–482.

[26] O. von Stryk, *User's Guide for DIRCOL Version 2.1: a direct collocation method for the numerical solution of optimal control problems*, Lehrstuhl M2 Höhere Mathematik und Numerische Mathematik, Technische Universität München, 1999.

[27] O. von Stryk, *Numerical Hybrid Optimal Control and Related Topics*, Habilitationsschrift, Technische Universität München, 2000.

[28] M. Vögel, O. von Stryk, R. Bulirsch, T.-M. Wolter, and C. Chucholowski, *An optimal control approach to real-time vehicle guidance*, submitted for publication.

Adaptive Approximation Algorithm for Relaxed Optimization Problems

Tomáš Roubíček and Martin Kružík

Abstract. Nonconvex optimization problems need a relaxation to handle effectively fast oscillation (and possibly also concentration) effects. This uses Young measures or their generalizations. Approximation of the relaxed problem can then be made by various ways, but computationally the most effective way appears to use adaptively a maximum principle (if it forms also a sufficient optimality condition) with the Hamiltonian guessed approximately from a previous iteration, e.g., from a coarser mesh if a multilevel technique is exploited. A survey and a concrete optimal-control example are presented.

1. Introduction

To keep a certain generality, let us consider the following fairly abstract optimization problem:

$$(\mathsf{P}) \qquad \left\{ \begin{array}{ll} \text{minimize} & \Phi(u) \\ \text{subject to} & u(x) \in S \quad \text{for } x \in \Omega \text{ (a.a.)} \\ & u \in L^p(\Omega; \mathbb{R}^m), \end{array} \right.$$

where $S \subset \mathbb{R}^m$ is closed but not necessarily bounded, $\Omega \subset \mathbb{R}^n$, $L^p(\Omega; \mathbb{R}^m)$ denotes the Lebesgue space of measurable functions $\Omega \to \mathbb{R}^m$ whose p-power is integrable, and $\Phi : L^p(\Omega; \mathbb{R}^m) \to \mathbb{R}$ is a continuous (but not weakly continuous) functional.

It is known that (P) does not have any solution in general which is related with the phenomena that the minimizing sequences *oscillate* faster and faster or may *concentrate* "energy". This leads to a necessity to extend the problem continuously on a suitable convex (locally) compact envelope of $U_{\rm ad} := \{u \in L^p(\Omega; \mathbb{R}^m);\ u(x) \in S \text{ a.e.}\}$, the extended problem being addressed as a *relaxed problem*.

To treat the problem numerically, we approximate not the original problem (P) but directly the relaxed one, which often gives a convex (or sometimes even linear/quadratic) mathematical-programming problem but with very many variables, we refer to [9, Section 3.5] for a fairly general theory. Typically, most of the variables are, however, zeros in (at least some) optimal solutions. One is naturally tempted to try to estimate a-priori which variables will be really active, which might be understood as a certain *preconditioning*. Often, the approximate relaxed problem is only one of a sequence created by an iterative procedure and then one can naturally exploit some an information from the previous iteration. The general options are as follows:

1. Successive refinement of a discretization: a multigrid technique; cf.[1, 4].
2. Evolutionary problems: recursive formulae arising by an implicit time discretization.
3. Optimization of some parameter entering (P) by using an iterative optimization procedure.
4. Other iterative schemes: e.g., sequential quadratic programming.
5. Combination of previous cases.

Here, focusing on the 1st option, we want to give a general exposition of relaxation and approximation of (P) and present a concrete nontrivial example of optimal control problems where both oscillations and concentrations may appear.

2. Relaxation of the problem

Let us briefly present a construction of convex locally compact envelopes of $U_{\rm ad}$ sufficiently fine so that the problem (P) can be continuously extended onto it.

We define a normed linear space $\mathrm{Car}^p(\Omega;\mathbb{R}^m) := \{h : \Omega\times\mathbb{R}^m \to \mathbb{R};\ h(\cdot,s)$ measurable, $h(x,\cdot)$ continuous, $\|h\| < +\infty\}$ equipped with the norm

$$\|h\| := \inf_{|h(x,s)|\le a(x)+b|s|^p} \|a\|_{L^1(\Omega)} + b. \tag{1}$$

Depending on a concrete problem, we then choose a sufficiently large (but preferably still separable) linear subspace $H \subset \mathrm{Car}^p(\Omega;\mathbb{R}^m)$, and define the embedding:

$$i : U_{\rm ad} \to H^* : u \mapsto \left(h \mapsto \int_\Omega h(x,u(x))\mathrm{d}x\right). \tag{2}$$

Then we denote by $Y^p_H(\Omega;S)$ the weak* closure of $i(U_{\rm ad})$. The set $Y^p_H(\Omega;S)$ is always convex in H^*. If H contains an integrand having a minorant $|s|^p$, then $Y^p_H(\Omega;S)$ makes a *convex locally compact hull* of the original set $U_{\rm ad}$. The embedding i is always (norm,weak*)-continuous and, if H is rich enough, even homeomorphical; cf. [9]. This makes the hull $Y^p_H(\Omega;S)$ a very natural extension of $U_{\rm ad}$.

In concrete situations, H can be chosen so large (i.e. the convex local compactification $Y^p_H(\Omega;S)$ of $U_{\rm ad}$ will be so fine) that Φ admits a continuous extension $\bar\Phi : Y^p_H(\Omega;S) \to \mathbb{R}$, i.e. $\bar\Phi \circ i = \Phi$. This naturally suggests to think about the continuously extended problem P, called a *relaxed problem*:

$$(\mathsf{RP}) \qquad \begin{cases} \text{minimize} & \bar\Phi(\eta) \\ \text{subject to} & \eta \in Y^p_H(\Omega;S), \end{cases}$$

The relaxed problem (RP), being defined on a locally compact set, has better properties concerning existence, stability, and numerically approximation of a solution than the original problem (P), especially if Φ is *coercive* so that

$$\exists\varrho: \quad \langle\eta,|s|^p\rangle \le \varrho \quad \Rightarrow \quad \bar\Phi(\eta) > \min(\mathsf{RP}) = \min\bar\Phi\left(Y^p_H(\Omega;S)\right). \tag{3}$$

Both problems are naturally closely related with each other, e.g., inf(P) = min(RP), cluster points of minimizing sequence of (P) (embedded via i) are solutions to (RP), and conversely every sequence in $U_{\rm ad}$ whose image via i converges to some solution of (RP) is minimizing for (P).

Moreover, in concrete cases, $\bar{\Phi}$ is often *smooth* with the Gâteaux derivative $\bar{\Phi}'(\eta)$ living not only in H^{**} but even in H. Then the first-order necessary optimality condition, i.e. $h = -\bar{\Phi}'(\eta)$ belongs to the normal cone to $Y^p_H(\Omega; S)$, can be equivalently written in the form of a *maximum principle*:

$$\langle \eta, h\rangle = \sup_{u\in U_{\rm ad}} \int_\Omega h(x, u(x))\, {\rm d}x, \tag{4}$$

$h \in H$ being called a *Hamiltonian*.

Often, $\bar{\Phi}$ is *convex* at least for those η's such that

$$\langle \eta, |s|^p\rangle \le \varrho\ , \tag{5}$$

with ϱ referring to (3). Then (4) completed with (5) forms even a sufficient optimality condition.

3. Approximation of (RP) and adaptive numerical implementation

The direct numerical approximation of relaxed problem is treated rather rarely in literature; we refer to [7, 8] for problems without concentration and to [6] for problems with concentrations. Here, we will briefly present a rather special construction following [6, 9]. We consider γS a metrizable compactification of S; later, in Section 4, we will use $S = \mathbb{R}^n = \mathbb{R}^1$ and γS the usual two-point compactification $\mathbb{R} \cup \{-\infty, +\infty\}$ for calculations. Also, we denote $[\Upsilon_\alpha v](s) := v(s)/(1 + |s|^\alpha)$ and consider a discretization parameter $d = (d_1, d_2)$ with $d_1 > 0$ a mesh parameter responsible for a finite-element discretization of Ω and $d_2 > 0$ a parameter responsible for a discretization of γS.

We will construct a suitable projector $P_d : H \to H$ so that $P_d H$ is finite-dimensional and thus directly implementable on computers. Without only a slight restriction of generality, we consider a special case $H = G \otimes V$ with some separable linear space $G \subset L^\infty(\Omega)$ and $V \subset C(\mathbb{R}^m)$ a linear space of some functions v for which $\Upsilon_\alpha v$ has a continuous extension on γS, denoted then by $\overline{\Upsilon_\alpha v}$. A typical case is $\alpha = 0$ if S is bounded (then $S = \gamma S$) or $\alpha = p$ in the opposite case; note that in the latter, unbounded case, the coercivity needs H to contain some integrand of the p-polynomial growth. Then we compose $P_d := P_{d_1} P_{d_2} = P_{d_2} P_{d_1}$.

The discretization of Ω is made by a triangulation T_{d_1} with a mesh parameter $d_1 > 0$, and the projector P_{d_1} makes averages over particular simplexes $\triangle \in \mathsf{T}_{d_1}$, i.e.

$$[P_{d_1} h](x, s) := \frac{1}{{\rm meas}(\triangle)} \int_\triangle h(\xi, s) {\rm d}\xi \qquad \text{if } x \in \triangle \in \mathsf{T}_{d_1}. \tag{6}$$

or, in one formula, $[P_{d_1} h](x, s) = \sum_{\triangle\in\mathsf{T}_{d_1}} {\rm meas}(\triangle)^{-1} \int_\Omega h(\xi, s)\chi_\triangle(\xi){\rm d}\xi \chi_\triangle(x)$ where $\chi_\triangle : \Omega \to \{0, 1\}$ is a characteristic function of $\triangle$.

For the discretization of γS (=a compactification of S) with a parameter $d_2 > 0$ we consider a finite set $\gamma S_{d_2} \subset \gamma S$ and some "base functions" $w_s \in V$ with $s \in \gamma S_{d_2}$, and then construct the projector $P_{d_2} : H \to H$ by a *Lagrange-type interpolation*:

$$[P_{d_2}h](x,s) := \sum_{\sigma\in\gamma S_{d_2}} [\overline{\Upsilon_\alpha h(x,\cdot)}](\sigma)w_\sigma(s). \tag{7}$$

Assuming $[\overline{\Upsilon_\alpha w_\sigma}](s)$ equal to 1 or 0 for $s = \sigma$ or $s \neq \sigma$ ($s, \sigma \in \gamma S_{d_2}$), respectively, we have indeed $P_{d_2}P_{d_2} = P_{d_2}$. Moreover, we will need to assume that the collection $\{\Upsilon_\alpha w_\sigma\}_{\sigma\in\gamma S_{d_2}}$ is the partition of the unity on S, i.e.

$$\Upsilon_\alpha w_\sigma \geq 0 \quad \& \quad \sum_{\sigma\in\gamma S_{d_2}} \Upsilon_\alpha w_\sigma = 1. \tag{8}$$

As $P_d = P_{d_1}P_{d_2}$ in the concrete case (6)–(7), one has $P_d^* Y_H^p(\Omega; S) \subset Y_H^p(\Omega; S)$, see [9, Section 3.5], and one can even calculate more explicitly:

$$[P_d h](x,s) = [P_{d_1}P_{d_2}h](x,s) = \sum_{\triangle\in\mathsf{T}_{d_1}} \sum_{\sigma\in\gamma S_{d_2}} a_{\triangle,\sigma}(h)\chi_\triangle(x)w_\sigma(s) \quad \text{with}$$

$$a_{\triangle,s}\in H^* \quad \text{defined by} \quad a_{\triangle,s}(h) := \int_\triangle \frac{[\overline{\Upsilon_\alpha h(\xi,\cdot)}](s)}{\text{meas}(\triangle)}\mathrm{d}\xi. \tag{9}$$

Thus we get a representation of $\eta \in P_d^* Y_H^p(\Omega; S)$ in the form

$$\langle \eta, h\rangle = \langle P_d^*\eta, h\rangle = \langle \eta, P_d h\rangle = \sum_{\triangle\in\mathsf{T}_{d_1}} \sum_{s\in\gamma S_{d_2}} a_{\triangle,s}(h)\left\langle \eta, \chi_\triangle \otimes w_s \right\rangle. \tag{10}$$

This leads us naturally to the definition of the *approximate relaxed problem* as a restriction of (RP) to the finite-dimensional convex set $P_d^* Y_H^p(\Omega; S)$, i.e. we define

$$(\mathsf{RP})_d \qquad \begin{cases} \text{minimize} & \bar{\Phi}(\eta) \\ \text{subject to} & \eta\in P_d^* Y_H^p(\Omega; S), \end{cases}$$

Under the natural assumptions that

$$\sup_{\triangle\in\mathsf{T}_{d_1}} \text{diam}(\triangle) \leq d_1 \quad \& \quad \max_{\substack{s,\sigma\in\gamma S_{d_2}\\ s\neq\sigma}} d_{\gamma S}(s,\sigma) \leq d_2 \tag{11}$$

with $d_{\gamma S}$ denoting a metric on γS, we have the convergence $\lim_{d\to 0} \|h - P_d h\| = 0$ guaranteed. This implies w*-$\lim_{d\to 0} P_d^*\eta = \eta$, and then the convergence of $(\mathsf{RP})_d$ to (RP) in the sense

$$\lim_{d\to 0} \min(\mathsf{RP})_d = \min(\mathsf{RP}) = \inf(\mathsf{P}), \tag{12a}$$

$$\operatorname*{Limsup}_{d\to 0}\ \text{Argmin}(\mathsf{RP})_d = \text{Argmin}(\mathsf{P}), \tag{12b}$$

where "Limsup" denotes the set of all cluster points of all selected sequences. The convexity of the set $P_d^* Y_H^p(\Omega; S)$ allows us to pose the first-order necessary and also sufficient *optimality conditions* similarly as for (RP), i.e.

$$\bar{\Phi}'(\eta) \in -N_{P_d^* Y_H^p(\Omega;S)}(\eta) \quad \& \quad \langle \eta, |s|^p \rangle \leq \varrho, \tag{13}$$

where $N_K(\eta)$ denotes the normal cone to K at η. The former part of (13) is equivalent to

$$\langle \eta, h_{d,\eta} \rangle = \max_{\tilde{\eta} \in P_d^* Y_H^p(\Omega;S)} \langle \tilde{\eta}, h_{d,\eta} \rangle \tag{14}$$

where $h_{d,\eta} = P_d \bar{\Phi}'(\eta) \in H$ can be called a *discrete Hamiltonian*, or in a pointwise form

$$\forall x \in \triangle \in \mathsf{T}_{d_1} : \qquad [h_{d,\eta} \bullet \eta](x) = \max_{\tilde{\eta} \in P_d^* Y_H^p(\triangle;S)} \langle \tilde{\eta}, h_{d,\eta} \rangle =: M_{d,\eta}^{\triangle}, \tag{15}$$

where $[h \bullet \eta](x) = |\triangle|^{-1} \langle \eta, \chi_\triangle h \rangle$ for $x \in \triangle \in \mathsf{T}_{d_1}$.

The advantage of the above approximation method is that it creates a convex problem – at least on η's satisfying (5). On the other hand, a disadvantage is certainly a *great dimension* – effectively we face a discretization of $n \times m$-dimensional domain $\Omega \times \gamma S$. Thus, we are ultimately challenged to perform real calculations on a smaller problem. The desired information can be extracted from (15):

Proposition 3.1. *Let* η *be an optimal solution to* $(\mathsf{RP})_d$. *Then* $\overline{\Upsilon_\alpha h_{d,\eta}}|_{\gamma S_{d_2} \setminus S} \leq 0$ *and* $M_{d,\eta}^{\triangle} = \max_{s \in \gamma S_{d_2} \cap S} h_{d,\eta}(x,s)$ *for some (or, equivalently, any)* $x \in \triangle$. *Moreover,* η *is in the form (10) with* $\langle \eta, \chi_\triangle \otimes w_s \rangle = 0$ *for* $s \in \gamma S_{d_2}$ *such that*

$$\left.\begin{array}{ll} h_{d,\eta}(x,s) < M_{d,\eta}^{\triangle} & \text{if } s \in S \,, \\ [\overline{\Upsilon_\alpha h_{d,\eta}}](x,s) < 0 & \text{if } s \notin S \,. \end{array}\right\} \tag{16}$$

Sketch of the proof. As $\tilde{\eta} \in P_d^* Y_H^p(\triangle; S)$ can bear arbitrary energy on the remainder $\gamma S \setminus S$ (cf. [9] or also the formula (20) below), $\overline{\Upsilon_\alpha h_{d,\eta}}$ cannot be positive in any point of $\gamma S_{d_2} \setminus S$ otherwise $M_{d,\eta}^{\triangle} = +\infty$ and (15) cannot be fulfilled on $\triangle$ in question. As always $i(u)$ with $u(x) = s \in \gamma S_{d_2} \cap S$ belongs to $P_d^* Y_H^p(\triangle; S)$, we always get $M_{d,\eta}^{\triangle} \geq \max_{s \in \gamma S_{d_2} \cap S} h_{d,\eta}(x,s)$. Let us consider the p-nonconcentrating modification of η (cf. [9, Sect.3.4]) having a Young-measure representation ν of the form $\nu_x = \sum_{s \in \gamma S_{d_2} \setminus S} c_{\triangle,s} \delta_s$ with $x \in \triangle$, δ_s denoting the Dirac measure supported at s, and some coefficients satisfying $\sum_{s \in \gamma S_{d_2} \setminus S} c_{\triangle,s} = 1$ and $c_{\triangle,s} \geq 0$. We then call $\{c_{\triangle,s}\}_{s \in \gamma S_{d_2} \setminus S}$ admissible. As the contribution to $M_{d,\eta}^{\triangle}$ coming from $\gamma S_{d_2} \setminus S$ must vanish, we get that, in fact, $M_{d,\eta}^{\triangle} = \max_{\{c_{\triangle,s}\}\text{admissible}} \sum_{s \in \gamma S_{d_2} \setminus S} c_{\triangle,s} h_{d,\eta}(x,s) = \max_{s \in \gamma S_{d_2} \cap S} h_{d,\eta}(x,s)$ for some (or equivalently each) $x \in \triangle$ and the relations (16) then follows. □

The condition (16) motivates us to construct a smaller problem by a guessed Hamiltonian, let us denote it by $\mathbf{h} \in P_d H$, and by a tolerance $\varepsilon > 0$: we denote

$$\gamma S_{d_2}^{\mathbf{h},\varepsilon}(\triangle) := \Big\{ s \in \gamma S_{d_2} \cap S;\ \mathbf{h}(x,s) \geq M_{d,\eta}^{\triangle} - \varepsilon \Big\} \\ \bigcup \Big\{ s \in \gamma S_{d_2} \setminus S;\quad [\overline{\Upsilon_\alpha \mathbf{h}}](x,s) \geq -\varepsilon \Big\} \tag{17}$$

and consider the problem

$$(\mathsf{RP})_d^{\mathbf{h},\varepsilon} \quad \begin{cases} \text{minimize} & \bar{\Phi}(\eta) \\ \text{subject to} & \eta \in P_d^* Y_H^p(\Omega; S) \ \text{ has the representation :} \\ & \langle \eta, h \rangle = \displaystyle\sum_{\triangle \in \mathsf{T}_{d_1}} \sum_{s \in \gamma S_{d_2}^{\mathbf{h},\varepsilon}(\triangle)} a_{\triangle,s}(h) \, \langle \eta, \chi_\triangle \otimes w_s \rangle. \end{cases}$$

One can see that obviously $\mathbf{h} = h_{d,\eta}$ implies that $\mathrm{Argmin}(\mathsf{RP})_d^{\mathbf{h},\varepsilon} = \mathrm{Argmin}(\mathsf{RP})_d$ for any $\varepsilon > 0$. In general, if $\mathbf{h} \neq h_{d,\eta}$, one must take the tolerance ε sufficiently large; indeed, one has

$$\varepsilon \geq 2 \max \Big(\max_{\substack{x \in \Omega \\ s \in \gamma S_{d_2} \cap S}} |\mathbf{h}(x,s) - h_{d,\eta}(x,s)|, \ \max_{\substack{x \in \Omega \\ s \in \gamma S_{d_2} \setminus S}} |[\overline{\Upsilon_\alpha \mathbf{h}}](x,s) - [\overline{\Upsilon_\alpha h_{d,\eta}}](x,s)| \Big) \\ \Longrightarrow \quad \mathrm{Argmin}(\mathsf{RP})_d^{\mathbf{h},\varepsilon} = \mathrm{Argmin}(\mathsf{RP})_d. \tag{18}$$

However, taking ε unnecessarily large makes the problem $(\mathsf{RP})_d^{\mathbf{h},\varepsilon}$ unnecessarily large, hence ineffective. So, the general strategy is to guess the Hamiltonian $h_{d,\eta}$ as accurately as possible, and then to take ε small but still satisfying the premisse in (18). Yet, the last point cannot be directly verified without knowledge of $h_{d,\eta}$. Fortunately, in concrete cases the maximum principle for the full problem $(\mathsf{RP})_d$ can be easily (and computationally "cheaply") verified. As we assume sufficiency of this maximum principle, we have an effective criterion to decide whether the chosen tolerance ε was sufficiently large and, if not, to increase it (say twice) and thus calculate $(\mathsf{RP})_d^{\mathbf{h},\varepsilon}$ for this (possibly repeatedly) increased ε until the maximum principle is satisfied. It opens obviously a great freedom for set up all the parameters in the iteration procedure and, fortunately, it is usually not difficult to reach much greater effectivity in comparison with the non-adaptive scheme solving the full problem $(\mathsf{RP})_d$. In general, our adaptive algorithm consists in the following steps:

1. Take an initial (very coarse) discretization d, activate all points of γS_{d_2} (e.g. by putting $\mathbf{h} := 0$), choose some $\varepsilon > 0$.
2. Solve $(\mathsf{RP})_d^{\mathbf{h},\varepsilon}$, obtaining η.
3. If the maximum principle is not satisfied, go to Step 5.
4. If the discretization level is the final one, then end, otherwise put $\mathbf{h} := h_{d,\eta}$, refine the discretization $d := d/2$, put $\varepsilon := \varepsilon/2$, and goto Step 2.
5. Increase the tolerance (e.g. $\varepsilon := 2\varepsilon$) and go to Step 2.

4. Example: optimal control with oscillations/concentrations

We will illustrate the above general considerations on a Lagrange optimal-control problem where minimizing sequences of control necessarily exhibit fast oscillations together with concentrations of their "energy", i.e. the p-power. In particular, we want to illustrate the following steps:

1. Relaxation by continuous extension on a suitable convex (local) compactification.
2. Analysis of convexity of the relaxed problem.
3. Approximation (with keeping the convexity).
4. Maximum principle for the approximate relaxed problem.

Possibly, Steps 2 and 3 can be re-ordered. Such problem (namely Steps 1 and 3) together with a non-adaptive numerical algorithm was presented in [6] where some more detailed assumptions can be found. Here, however, we additionally assume smoothness of coefficients with respect to y-variable to investigate the increment formula (24) below.

We consider $n = 1$, $\Omega := (0,T)$ with writing t instead of x, and, following DiPerna and Majda [2], $G = C([0,T])$, $V = \{v \in C(S);\ \Upsilon_p v$ has a continuous extension on $\gamma S\}$. Of course, we consider $H = G \otimes V$. It was essentially proved in [5] that, after a certain rearrangement, $\eta \in Y_H^p(\Omega;\mathbb{R})$ has a representation $\mu \in \mathrm{rca}([0,T] \times \gamma S)$ in the form

$$\langle \eta, h\rangle = \int_{[0,T]\times\gamma S} \frac{h(x,s)}{1+|s|^p}\mu(\mathrm{d}x\mathrm{d}t) \tag{19}$$

where, for $s \in \gamma S \setminus S$, $h(x,s)/(1+|s|^p)$ is considered extended by continuity and where the measure μ satisfies

$$\mu \geq 0 \quad \& \quad \int_{\gamma S} \frac{1}{1+|s|^p}\mu(\cdot,\mathrm{d}s) = \text{const.} = 1\ , \tag{20a}$$

$$\mathrm{meas}\Big\{t\in[0,T];\ \mu(\{t\}\times S) = 0\Big\} = 0\ , \tag{20b}$$

where "meas" refers to the Lebesgue measure on $[0,T]$. Let us emphasize that (20) is a precise characterization, which is also important for the numerical implementation.

As a concrete optimization problem, we will consider the additively coupled *Lagrange optimal-control problem*:

$$(\mathsf{P_{Lag}}) \left\{ \begin{array}{ll} \text{Minimize} & J(y,u) := \displaystyle\int_0^T a(t,y) + b(t,u)\,\mathrm{d}t \\ \text{subject to} & \dfrac{\mathrm{d}y}{\mathrm{d}t} = c(t,y) + e(t,u), \qquad y(0) = y_0\ , \\ & y\in W^{1,1}(0,T;\mathbb{R}^l)\ , \quad u\in L^p(0,T;\mathbb{R}^m)\ , \quad u(t)\in S, \end{array}\right.$$

with $a \in C([0,T]\times\mathbb{R}^l)$, $b \in C([0,T]\times\mathbb{R}^m)$, $c \in C([0,T]\times\mathbb{R}^l;\mathbb{R}^l)$, $e \in C([0,T]\times\mathbb{R}^m;\mathbb{R}^l)$ and, moreover, $\partial a/\partial y$ and $\partial^2 c/\partial y^2$ continuous. If $b, e_i \in \mathrm{cl}(G \otimes V)$, $i = 1,...,l$, the closure referring to the topology induced from $C([0,T]\times\gamma S)$ via Υ_p, the problem ($\mathsf{P_{Lag}}$) can be continuously extended to the above mentioned measures, which gives the following relaxed problem:

$$(\mathsf{RP_{Lag}})\qquad \begin{cases} \text{Minimize} & \bar{J}(y,\mu) := \int_0^T a(t,y)\,\mathrm{d}t + \int_{[0,T]\times\gamma S} \frac{b(t,s)}{1+|s|^p}\mu(\mathrm{d}t\mathrm{d}x) \\ \text{subject to} & \dot{y} = c(t,y) + \int_{\gamma S} \frac{e(t,s)}{1+|s|^p}\mu(\mathrm{d}x\,\cdot), \qquad (y,\dot{y})|_{t=0} = y_0\ , \\ & (y,\dot{y}) \in W^{1,\mathrm{M}}([0,T];\mathrm{I\!R}^l), \\ & \mu \in \mathrm{rca}([0,T]\times\gamma S) \ \ \text{satisfying (20)}, \end{cases}$$

where we denoted by $W^{1,\mathrm{M}}([0,T];\mathrm{I\!R}^l) := \{(y,\dot{y}) \in L^\infty(0,T;\mathrm{I\!R}^l) \times \mathrm{rca}([0,T];\mathrm{I\!R}^l);\ \exists y_k \in C^1([0,T]) : y_k \to y \ \&\ \frac{\mathrm{d}}{\mathrm{d}t}y_k \to \dot{y}$ weakly* in $\mathrm{rca}([0,T];\mathrm{I\!R}^l)\}$. This space, in a more generality, has been introduced by J.Souček [11] who also proved that the traces of $(y,\dot{y})$ for $t=0$ and $t=T$, denoted respectively by $(y,\dot{y})|_{t=0}$ and by $(y,\dot{y})|_{t=T}$, can be defined uniquely to satisfy the usual by-parts integration formula.

Furthermore, we will also assume coercivity of ($\mathsf{RP_{Lag}}$), which can be guaranteed, e.g., if $e(t,s) \geq |s|^p$ and $a \geq 0$. This implies that, for some $\exists\varrho$, every optimal relaxed control μ fulfills $\mu([0,T]\times\gamma S) \leq T+\varrho$. For such μ, one can also assume the a-priori estimates

$$|y(t)| \leq M_1(t)\ , \quad |\lambda(t)| \leq M_2(t) = \left[M_2(\frac{\partial a}{\partial y}, \frac{\partial c}{\partial y})\right](t) \tag{21}$$

where the *adjoint state* $\lambda \in W^{1,1}(0,T;\mathrm{I\!R}^n)$ solves the linear terminal-value problem, called the adjoint equation:

$$\frac{\mathrm{d}\lambda}{\mathrm{d}t} + \left[\frac{\partial c}{\partial y}\right]^{\mathrm{T}} \lambda = \frac{\partial a}{\partial y}\ , \qquad \lambda(T) = 0. \tag{22}$$

We moreover assume that the controlled system is only "*slightly nonlinear*" (i.e. $\partial^2 c/\partial y^2$ small) and the cost functional is "*enough uniformly convex*" (i.e. $\partial a/\partial y$ uniformly monotone), namely

$$\left(\frac{\partial a}{\partial y}(t,r_1) - \frac{\partial a}{\partial y}(t,r_2)\right)(r_1 - r_2) \geq K(t)|r_1 - r_2|^2\ , \tag{23a}$$

$$K(t) \geq M_2(t) \sup_{|r|\leq M_1(t)} \left|\frac{\partial^2 c}{\partial y^2}(t,r)\right|\ . \tag{23a}$$

By analyzing the following *increment formula*

$$\begin{aligned} \bar{\Phi}(\tilde{\mu}) - \bar{\Phi}(\mu) \ &= \int_{[0,T]\times\gamma S} \frac{\lambda(t)e(t,s) + b(t,s)}{1+|s|^p}[\tilde{\mu} - \mu](\mathrm{d}t\mathrm{d}s) \\ &\qquad + \int_0^T \Delta_a(t) + \lambda(t)\Delta_c(t)\mathrm{d}t \end{aligned} \tag{24}$$

with $\bar{\Phi}(\mu) := \bar{J}(y(\mu),\mu)$ and the second-order correcting terms $\Delta_a(t)$ and $\Delta_c(t)$ defined by

$$\Delta_a(t) := a(t,\tilde{y}(t)) - a(t,y(t)) - \frac{\partial a}{\partial y}(t,y(t))(\tilde{y}(t)-y(t)), \tag{25a}$$

$$\Delta_c(t) := c(t,\tilde{y}(t)) - c(t,y(t)) - \frac{\partial c}{\partial y}(t,y(t))(\tilde{y}(t)-y(t)), \tag{25b}$$

(see, e.g., Gabasov and Kirillova [3], or also [10, Lemma 2]) one can show that $\bar{\Phi}$ is convex on the set $\{\mu;\ (20) \text{ holds},\ \mu([0,T]\times\gamma S) \le T+\varrho\}$.

To make a discretization, let us consider a (say, equidistant) partition T_{d_1} of the interval $[0,T]$ with a time-step $d_1 > 0$ and a finite set $\gamma S_{d_2} \subset \gamma S$, assuming $\gamma S_{d_2} \subset \gamma S_{d_2'}$ for $d_2 \ge d_2' > 0$ and $\bigcup_{d_2>0} \gamma S_{d_2}$ dense in γS. The projectors P_{d_1} and P_{d_2} are then as in Section 3 with $\alpha = p$. Then, it results [6] that the discretization of $(\mathsf{RP}_{\mathrm{Lag}})$ can be made by restricting it on μ intervalwise constant on T_{d_1} with $\mathrm{supp}(\mu) = [0,T]\times\gamma S_{d_2}$; let us denote the restricted problem by $(\mathsf{RP}_{\mathrm{Lag}})_d$.

To implement the adaptive procedure described in Section 3, we need to have sufficient optimality conditions for $(\mathsf{RP}_{\mathrm{Lag}})_d$ at our disposal. Let us note that the Gâteaux differential $\bar{\Phi}'(\mu)$ equals $\Upsilon_p(\lambda e - b)$ with $\lambda = \lambda(y)$ solving (22) and $y = y(\eta)$ solving, together with $\dot{y}$, the state equation in $(\mathsf{RP}_{\mathrm{Lag}})$. Then the "elementwise constant" maximum principle (15) reads, for a.a. $t \in [0,T]$, as

$$\int_{\gamma S} h_{d,\lambda}(t,s)\mu(\cdot,\mathrm{d}s) = \max_{\substack{\int_{\gamma S}\frac{\tilde{\mu}(\mathrm{d}s)}{1+|s|^p}=1\\ \tilde{\mu}\ge 0,\ \mathrm{supp}(\tilde{\mu})\subset\gamma S_{d_2}}} \int_{\gamma S} h_{d,\lambda}(t,s)\tilde{\mu}(\mathrm{d}s) \tag{26}$$

with the *discrete Hamiltonian* given by

$$h_{d,\lambda} := \Upsilon_p P_d(\lambda\cdot e - b). \tag{27}$$

Due to the convexity under the condition (5), this maximum principle is also sufficient for the optimality of μ if, in addition, $\mu([0,T]\times\gamma S) \le T+\varrho$ (which just means (5)) is fulfilled.

It is important that, on each interval, the relation (26) can be checked effectively by verifying whether both $\max_{\gamma S_{d_2}\cap S} P_d(\lambda\cdot e - b)$ does not exceed the left-hand side of (26) and $h_{d,\lambda}|_{\gamma S_{d_2}\setminus S} \le 0$, cf. also (16). Alternatively, (26) can be checked by verifying the Karush-Kuhn-Tucker necessary (and here also sufficient) optimality conditions. Indeed, the equality constraint $\int_{\gamma S_{d_2}}(1+|s|^p)^{-1}\mu(\mathrm{d}s) = 1$ can be put off via augmenting the Hamiltonian (27), which results to $\Upsilon_p P_d(\lambda\cdot e - b + \ell\otimes 1)$ with ℓ being a scalar-valued intervalwise constant multiplier produced by usual linear-quadratic routines as a by-product. The remaining condition has a pointwise character and its verification over the whole γS_{d_2} is then computationally cheap.

To demonstrate computational efficiency of the algorithm, we will consider the following model problem, using $p = 1$ and $S = \mathbb{R}$:

$$
(\mathsf{P}_{\mathbf{Example}}) \quad \left\{ \begin{array}{lll} \text{Minimize} & J(y,u) := \displaystyle\int_0^1 g_\beta(u(t)) + (y_1(t)-t)^2 + y_2(t)^2 \, \mathrm{d}t & \\ \text{subject to} & \dfrac{\mathrm{d}y_1}{\mathrm{d}t} = |u|, & y_1(0) = 0 \,, \\ & \dfrac{\mathrm{d}y_2}{\mathrm{d}t} = u \,, & y_2(0) = 0 \,, \\ & y \in W^{1,1}(0,1;\mathbb{R}^2) \,, \quad u \in L^1(0,1) \,, & \end{array} \right.
$$

where $g_\beta(s) = s^2/(1+s^4)+\beta^2|s|$ for some $\beta \geq 0$. Note that the problem is coercive for $\beta > 0$ and thus fits with the previous theory. Also note that $c(t,\cdot)$ is affine, which simplifies considerably the preceding considerations, allowing us to consider $K \equiv 0$ in (23) and to set simply $\varrho = +\infty$. A similar type of problems was already considered in [6]. The relaxed problem, denoted by $(\mathsf{RP}_{\mathbf{Example}})$, now uses γS the usual two-point compactification of $\mathbb{R} \cup \{-\infty, +\infty\}$. In the following proposition we find the precise solution to $(\mathsf{RP}_{\mathbf{Example}})$.

Proposition 4.1. *Let* $\beta \in [0,1]$. *Then* $\min(\mathsf{RP}_{\mathbf{Example}}) = \beta^2(1-2\beta/3)$ *and the optimal solution is*

$$
\mu(\mathrm{d}t\mathrm{d}s) = \left\{ \begin{array}{ll} \mathrm{d}t \otimes \frac{1}{2}(\delta_{-\infty} + \delta_{\infty} + 2\delta_0) & \textit{if } 0 \leq t \leq 1-\beta \\ \mathrm{d}t \otimes \delta_0 & \textit{if } 1-\beta < t \leq 1 \,. \end{array} \right. \tag{28}
$$

The corresponding states are

$$
y_1(t) = \left\{ \begin{array}{ll} t & \textit{if } 0 \leq t \leq 1-\beta \\ 1-\beta & \textit{if } 1-\beta \leq t \leq 1 \end{array} \right. , \qquad y_2(t) = 0 \,. \tag{29}
$$

Proof. First, notice that (20) is indeed satisfied for μ from (28). The formulae for the response y, i.e. (29), and for the value $\bar{J}(y,\mu)$ can then be found by straightforward calculations. Now our task in only to show that (μ, y_1, y_2) defined by (28)–(29) satisfies the optimality conditions (26)-(27) in the continuous case. Having in mind the adjoint equation (22), one gets that in this case the adjoint state $\lambda = (\lambda_1, \lambda_2)$ has the components

$$
\lambda_1(t) = \left\{ \begin{array}{ll} \beta^2 & \text{if } 0 \leq t \leq 1-\beta \\ -(1-\beta-t)^2 + \beta^2 & \text{if } 1-\beta < t \leq 1 \end{array} \right. , \qquad \lambda_2 = 0 \,. \tag{30}
$$

Hence, for the Hamiltonian $h_\lambda(t,s) = (\lambda_1(t)|s| - g_\beta(s))/(1+|s|)$ the right hand side in (26) equals zero and obviously the maximum principle (26) is satisfied for μ from (29). By convexity of $(\mathsf{RP}_{\mathbf{Example}})$, μ is thus optimal. □

Remark 4.1. If $\beta \geq 1$ then neither oscillations nor concentrations appear in the analysis of the problem and $u = y_1 = y_2 = 0$ is the optimal solution.

Let us divide the interval $(0,1)$ into $1/d_1$ subintervals and consider an intervalwise constant μ in the form

$$\mu(\mathrm{d}t\mathrm{d}s) = \sum_{s\in\gamma S_{d_2}} q_{is}\delta_s \quad \text{if } (i-1)d_1 \le t \le id_1,\ 1 \le i \le 1/d_1\ , \tag{31}$$

where

$$q_{is} \ge 0 \quad \text{and} \quad \sum_{s\in\gamma S_{d_2}} \frac{q_{is}}{1+|s|} = 1 \quad \text{for any } 1 \le i \le 1/d_1\ , \tag{32}$$

which makes μ compatible with (20). Note that (20b) is fulfilled automatically for all piecewise constant μ's.

In terms of $q \equiv (q_{is})$, the discrete relaxed problem looks as follows:

$$(\mathsf{RP}_{\mathrm{Example}})_d \left\{ \begin{array}{ll} \text{Minimize} & \bar{J}(y,q) := \displaystyle\sum_{i=1}^{1/d_1} \sum_{s\in\gamma S_{d_2}} \frac{g_\beta(s)}{1+|s|} q_{is} \\ & \qquad\qquad + \displaystyle\int_0^1 (y_1(t)-t)^2 + y_2(t)^2\,\mathrm{d}t \\ \text{subject to} & \dot{y}_1 = \displaystyle\sum_{s\in\gamma S_{d_2}} \frac{|s|}{1+|s|} q_{is}\ , \qquad y_1(0) = 0\ , \\ & \dot{y}_2 = \displaystyle\sum_{s\in\gamma S_{d_2}} \frac{s}{1+|s|} q_{is}\ , \qquad y_2(0) = 0\ , \\ & \displaystyle\sum_{s\in\gamma S_{d_2}} \frac{q_{is}}{1+|s|} = 1, \quad q_{is} \ge 0\ ,\ 1 \le i \le \frac{1}{d_1},\ s\in\gamma S_{d_2}, \\ & y \in W^{1,\infty}(0,1;\mathbb{R}^2)\ , \end{array} \right.$$

This is, in fact, a linear-quadratic programming problem for $q = (q_{is})$.

In our numerical calculations, we took $\beta = 0.3$ and a fixed time step $d_1 = 1/20$. The minimum of the non-discretized problem ($\mathsf{RP}_{\mathrm{Example}}$) is then 0.072. In the table below one finds statistics verifying the efficiency of the algorithm. The sets γS_{d_2}, $d_2 > 0$, always contain $\pm\infty$ and in each next step we double the number of points of γS_{d_2} belonging to S. Intentionally, $0 \notin \gamma S_{d_2}$ to avoid coincidence of γS_{d_2} with the support of the exact solution. We initialized $\varepsilon = 0.005$ at the beginning and divided it by two in each next step.

card(γS_{d_2})	# of active variables	# of active variables/element	minimum
6	120	6	0.232
10	120	6	0.096
18	114	5.5	0.073
34	70	3.5	0.073
66	110	5.5	0.072

TABLE 4.1.

We emphasize that the maximal possible number of variables is $\mathrm{card}(\gamma S_{d_2})/d_1$ which is in our latest case 1320, meanwhile we computed with 110 variables only. Taking into account a polynomial (degree 3) complexity of usual LQ-routines, we can see that the adaptive algorithm is, in this concrete case, about 10^3-times faster than a non-adaptive algorithm working with 1320 variables. Of course, in the adaptive case, one must also count with preliminary calculations on coarser discretizations but this does not essentially change the effectivity picture.

The next figures show our solution for card $(\gamma S_{d_2}) = 10$ and 34, respectively.

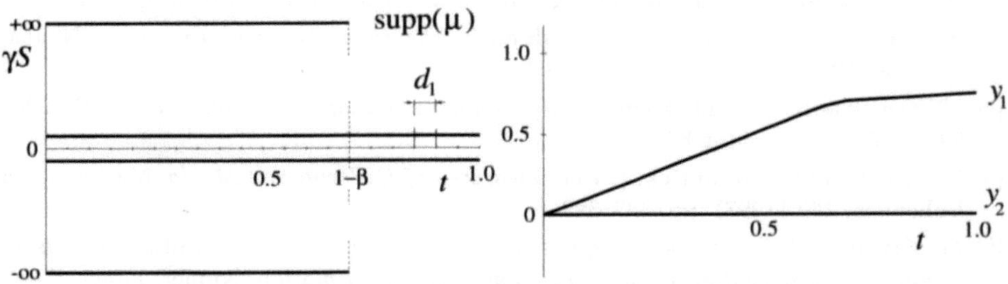

FIGURE 4.1. Solution (μ, y) to $(\mathrm{RP}_{\mathbf{Example}})_d$ for $\mathrm{card}(\gamma S_{d_2}) = 10$.

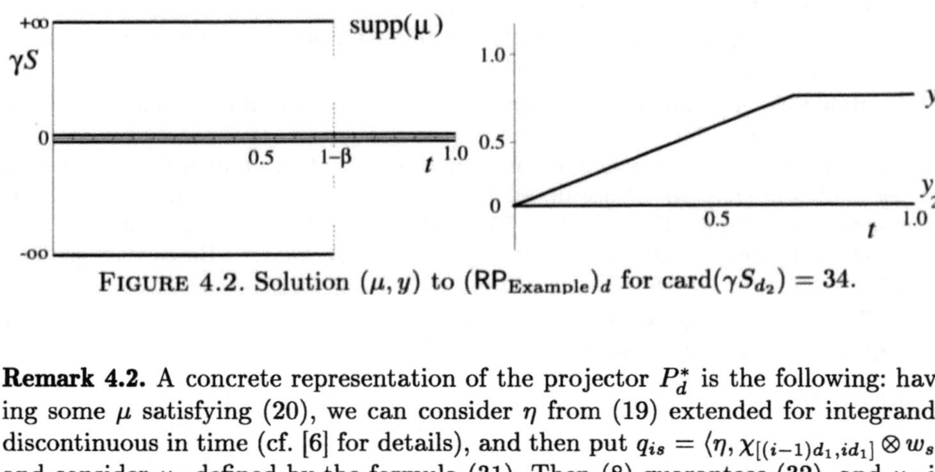

FIGURE 4.2. Solution (μ, y) to $(\mathrm{RP}_{\mathbf{Example}})_d$ for $\mathrm{card}(\gamma S_{d_2}) = 34$.

Remark 4.2. A concrete representation of the projector P_d^* is the following: having some μ satisfying (20), we can consider η from (19) extended for integrands discontinuous in time (cf. [6] for details), and then put $q_{is} = \langle \eta, \chi_{[(i-1)d_1, id_1]} \otimes w_s \rangle$ and consider μ_d defined by the formula (31). Then (8) guarantees (32), and μ_d is then related by $P_d^*\eta$ by the formula (19) written for μ_d and $P_d^*\eta$ in place of μ and η, respectively.

Remark 4.3. The above multi-level adaptive discretization of relaxed problems has been tested, but for minimization problems without concentration effects, on classical Young measures first time for a nonconvex 1D variational problem, so-called Tartar's broken extremal, and another time for a 2D problem in micromagnetics where S is a sphere in $\mathbb{R}^2$ in [1] and [4], co-authored by C. Carstensen and A. Prohl, respectively.

Acknowledgement: This research was partly supported also by the grants 201/00/0768 (GA ČR), A 107 5005 (GA AV ČR), and MSM 11320007 (MŠMT ČR).

References

[1] C. Carstensen and T. Roubíček, *Numerical approximation of Young measure in non-convex variational problems*, Numer. Math., **84** (2000), pp. 395–415.

[2] R. J. DiPerna and A. J. Majda, *Oscillations and concentrations in weak solutions of the incompressible fluid equations*, Comm. Math. Physics **108** (1987), pp. 667–689.

[3] R. Gabasov and F. Kirillova, *Qualitative Theory of Optimal Processes*, Nauka, Moscow, 1971.

[4] M. Kružík and A. Prohl, *Young measures approximation in micromagnetics*, Bericht, **00-2**, 2000, Universität Kiel.

[5] M. Kružík and T. Roubíček, *On the measures of DiPerna and Majda*, Mathematica Bohemica, **122** (1997), pp. 383–399.

[6] M. Kružík and T. Roubíček, *Optimization problems with concentration and oscillation effects: relaxation theory and numerical approximation*, Numer. Funct. Anal. Optim., **20** (1999), pp. 511-530.

[7] P. Pedregal, *Numerical approximation of parametrized measures*, Numer. Funct. Anal. Opt., **16** (1995), pp. 1049–1066.

[8] T. Roubíček, *Approximation theory for generalized Young measures*, Numer. Funct. Anal. Opt., **16** (1995), pp. 1233–1253.

[9] T. Roubíček, *Relaxation in Optimization Theory and Variational Calculus*, W. de Gruyter, Berlin, 1997.

[10] T. Roubíček, *Existence results for some nonconvex optimization problems governed by nonlinear processes*, Proceedings, 12th Conf. on Variational Calculus, Optimal Control and Applications held in Trassenheide, September 23-27, 1996, W. H. Schmidt, K. Heier, L. Bittner, and R. Bulirsch, eds., Birkhäuser, Basel, 1998, pp. 87–96.

[11] J. Souček, *Spaces of functions on domain* Ω, *whose* k*-th derivatives are measures defined on* $\bar{\Omega}$, Čas. pro Pěst. Mat., **97** (1972), pp. 10–46.

Optimization Tools for Solving Nonlinear Ill-posed Problems

Thomas Viklands and Marten Gulliksson

Abstract. Using the L- and a-curve, we consider how a nonlinear ill-posed Tikhonov regularized problem can be solved by a Gauss-Newton method. The solution to the problem is chosen from the point on the logarithmic L-curve that has maximal curvature, i.e. at the corner. The L-curve is used for analysing how the tradeoff between minimizing the solution norm versus the residual norm changes with the choice of the regularization parameter. The a-curve is used to analyse how the object function changes for different choices of the regularization parameter. In the numerical tests we solve an inverse problem in heat transfer where the L-curve solution is compared to the optimal solution and the solution given by the discrepancy principle.

1. Introduction

Consider a problem of the form

$$\min_x \frac{1}{2}\|f(x)\|_W^2 + \frac{1}{2}\lambda\|L(x - x_c)\|_V^2 \tag{1}$$

where W and V are any spaces such that $f : V \to W$, x_c the center of regularization, λ as the regularization parameter and L a transformation matrix of x.

Suppose that no information about the noise in f is known, such as the noise level or the standard deviation. Thus, the solution to (1) can not be calculated in the sense of the well known discrepancy principle, or by any other method, that results in a convergent method [12], [2]. That is when the noise $\delta \to 0$ the regularization parameter λ given by the method should also tend to zero, resulting in the nonregularized solution. A tool that can be used for analysing ill-posed problems is the controversial L-curve, that has shown to be successfull in some application areas dealing with linear problems [1]. However, it has been shown that the usage of the L-curve results in an inaccurate method when choosing regularization parameter [3], [8], [9]. Our intention is to investigate the use of the L-curve from an engineer's point of view.

2. The Nonlinear L-curve

The L-curve for nonlinear problems is defined as the curve

$$(t(x), y(x))$$

where

$$t(x) = \frac{1}{2}||f(x)||^2, \; y(x) = \frac{1}{2}||L(x - x_c)||^2$$

and

$$x = x(\lambda)$$

is the solution to (1). It can be shown that the L-curve $y(t)$, $y : \Re \to \Re$ have the basic local properties

$$\frac{dy}{dt} = -\frac{1}{\lambda} < 0 \; , \quad \frac{d^2y}{dt^2} > 0 \tag{2}$$

which defines y as a *monotonically decreasing, strictly convex function of* t [4].

The L-curve in a logarithmic scale, $(\log(t), \log(y))$, is expected to have the shape of the letter "L" if the problem is ill-posed and contains noise. The shape is a result of a radical growth of the solution norm y as the regularization parameter λ gets small, which is common for ill-posed problems. A resonable solution should lie in the vicinity of the "corner", where y is about to start growing and t almost remains fixed.

For linear problems the construction of the L-curve is pretty straightforward [1]. We just solve the linear problem for a wide range of regularization parameters λ, and locate the corner solution. Doing this for nonlinear problems can be very time consuming, since attaining every point (t, y) requires the solution of a nonlinear problem. Instead of computing the "exact" L-curve, we compute an approximation to it.

2.1. The shadow L-curve

To attain an approximation to the L-curve for (1) we can use an algorithm similar to the following:

1. While no solution found
   - 1.1. Choose regularization parameter λ_i
   - 1.2. Compute a search direction $p(\lambda_i)$
   - 1.3. Check for descent $f(x_i + p) < f(x_i)$
   - 1.4. Take $x_{i+1} = x_i + p$
   - 1.5. $t_i = \frac{1}{2}||f(x_{i+1})||^2$, $y_i = \frac{1}{2}||L(x_{i+1} - x_c)||^2$
   - 1.6. $i = i + 1$

By gathering points $\{t_i = t(x(\lambda_i)), y_i = y(x(\lambda_i))\}$ given during iteration we may pick out a subset M from the set $\{t_i, y_i\}$ defining a *monotonically decreasing convex* function. These points will always lie on or above the exact L-curve, hence approximating it. The set M defines another set of linear functions from point (t_i, y_i) to (t_{i+1}, y_{i+1}) which is called the polygon shadow L-curve, y_{sp}. Since it lacks of smoothness let $y_{sm}(t)$ be a *convex monotonically decreasing* spline function interpolating the set M. $y_{sm}(t)$ is the *smooth* shadow L-curve that contains higher order of information, such as first and second order derivatives. For further details see [10].

Iterating with a fixed λ and getting closer to the exact L-curve is not meaningful unless it is assumed that keeping it fixed would give the corner solution or a better approximation of the L-curve is needed. Our method is to choose the regularization parameters so that the algorithm converges towards the corner. Close to the corner we compute a more precise approximation of the L-curve.

2.2. The a-curve

The a-curve is defined as the curve $(\lambda, a(\lambda))$ where

$$a(\lambda) = \min_x t(x) + \lambda y(x)$$

which is just the object function itself. Clearly it describes how the optimization problem (1) behaves when changing the regularization parameter. Having a good approximation of the a-curve gives information on how the optimization problem depends of the regularization parameter. However, the a-curve contains the same information as the L-curve but it is just represented differently.

The a-curve has the local properties

$$\frac{da}{d\lambda} = y > 0 \ , \quad \frac{d^2a}{d\lambda^2} < 0$$

defining it as a *strictly concave, strictly increasing* function of λ [4].

As in the case with the L-curve, there exist piecewise linear functions $a_{sp}(\lambda) = t_i + \lambda y_i$ that represent the polygon shadow a-curve [5]. A smooth shadow a-curve can easily be constructed when using the smooth shadow L-curve.

3. Definitions of the corner

As stated before, the L-curve plotted in a logarithmic scale is expected to have a corner if the ill-posed problem contains noise. The corner may be defined as the point where the logarithmic L-curve has its maximal curvature [1], [7]. The corner describes the point where the tradeoff between minimizing the solution norm and residual norm is somehow balanced. This does not mean that the solution $x(\lambda)$ given at the corner is optimal but in any case a resonable solution (the engineering point of view).

3.1. Formulation of Reginska

The corner with maximum curvature can be found by solving

$$\min_\lambda t(x(\lambda))y(x(\lambda))^\alpha \ , \quad \alpha > 0 \qquad (3)$$

for linear functions $f(x) = Ax - b$ [7], and it is generalized to the nonlinear case in [4]. This function can be quite nonlinear and experience strong oscillations. We consider instead the minimization problem

$$\min_{\lambda}\{\log t + \log y\} \tag{4}$$

which has the same minimum as (3) under the assumption $\alpha = 1$. (4) corresponds to the minimum of the logarithmic L-curve rotated $\pi/4$ radians [4].

3.2. Definition by curvature

Another way of computing the corner is using the formula for the curvature of the logarithmic L-curve [1]. With

$$\tau = \log t \ , \ \eta = \log y$$

the curvature of $y(t)$ in a logarithmic scale becomes

$$\kappa = \frac{\frac{d^2\eta}{d\tau^2}}{(1 + (\frac{d\eta}{d\tau})^2)^{3/2}} \, .$$

Thus finding the corner corresponds to solving a maximization problem

$$\max_{\lambda} \kappa$$

which has to be solved by using the smooth approximation of the L-curve $y_{sm}(t)$.

4. Estimating the corner using Reginska formulation

Consider the set of points $\tau_i = \log t_i$, $\eta_i = \log y_i$ given during the iteration when solving (1). When locating the corner every set of three points

$$\{(\tau_{i-1}, \eta_{i-1}), (\tau_i, \eta_i), (\tau_{i+1}, \eta_{i+1})\}$$

is rotated $\pi/4$ radians. Then, if the midpoint (τ_i, η_i) is a minimum, the corner is approximated by the minimum of a quadratic spline interpolating these three points.

5. Approaching the corner

The aim is to locate an interval of the L-curve in which a corner defined according to the Reginska formulation exists. Later on we shall show that when this interval is found and that convergence towards the corner is not problematic.

Using the Gauss-Newton method the linearized problem,

$$\min_{p} \frac{1}{2}\|f(x_i) + J(x_i)p\|^2 + \frac{1}{2}\lambda_i\|x_i + p - x_c\|^2 \tag{5}$$

where

$$J(x_i) = \frac{\partial f}{\partial x}(x_i),$$

is solved giving the search direction $p = p(\lambda)$. The regularization parameter must be chosen properly so that a 'safe' step is taken. We regard the search direction to be 'safe' if x_k and $x_k + p_k$ is close to the same nonlinear L-curve. Choosing a small λ-value will give a very large steplength, that may lead to a different trajectory. Consequently, it is recommended to begin with a rather big λ and gradually decrease λ during iteration if the center of regularization is considered well chosen. Since a problem may exist of more than one trajectory, resulting in many different L-curves, we define the *global* L-curve as the convex set of all these different L-curves.

A general iteration may look as follows.

1. Calculate λ_i
2. Compute direction $p(\lambda_i)$
3. $x_{i+1} = x_i + \alpha p_i$ is the new approximation to the solution of (1).
4. Determine if x_{i+1} belongs to the convex hull of the polygon shadow L-curve.
5. If the convex set approximates a corner, steer the solutions towards it.

5.1. A simple way of choosing λ

If using a Gauss Newton method, the linear problem

$$(J^T J + \lambda L^T L)p = J^T f$$

is to be solved in each iteration. Hence it is reasonable to choose λ_0 initially in the order of the largest singular values of $J^T J$ and then gradually decrease λ. The nonlinearity and ill-posedness combined with a problem consisting of many local minima makes the choice of λ crucial in order to get convergence towards the corner solution.

An easy way to update λ is to divide it by a constant greater than one in each iteration.

$$\lambda_{i+1} = \frac{\lambda_i}{k} \quad , \quad k > 1.$$

Though a large value of the constant k might lead to loosing convergence, this method can be effective if the problem is large and not too sensitive to the choice of the regularization parameter.

5.2. Updating λ using the linear L-curve

Standing at x_i with Jacobian J_i, compute the step $p(\lambda_i)$ for a large set of regularization parameters from (5). Then use the linear L-curve

$$y_{lin} = \frac{1}{2}||x_i + p(\lambda_i) - x_c||^2 \ , \ t_{lin} = \frac{1}{2}||f(x_i) + J_i p_i||^2$$

defined by λ_i and compare with the actual residual reduction

$$y = \frac{1}{2}||x_i + p(\lambda) - x_c||^2 \ , \ t = \frac{1}{2}||f(x_i + p_i)||^2.$$

Depending on the nonlinearity of the problem, the differences between the L-curve for the linearized and nonlinear problem might vary. The safest way to choose the regularization parameter is to pick a λ where these two L-curves starts to differ. Clearly, if the L-curves do not differ much, a λ corresponding to the corner of the linear L-curve should be chosen. The cost of calculating the linear L-curve will decide if this method is prefered.

5.3. Using the a-curve

Since the a-curve is the object function as a function of λ, analysing it could yield a more robust and mathematically correct method of updating the regularization parameter. Our idea is to use a shooting algorithm, but these methods are often very inaccurate. In this case there is no information about the function $a(\lambda)$, except that it is a decreasing concave function. Another difficulty is that the points approximating the a-curve are very logarithmically distributed. The a-curve can have a very drastic change for small λ values so it seems a good idea to work in a logarithmic scale and additionally eliminate the possibility to get negative values of the regularization parameter.

Investigation of the curvature for $a(\lambda)$

$$\kappa = \frac{d^2 a}{d\lambda}/(1+y) \tag{6}$$

could maybe yield some information making it possible to perform a more accurate shooting algorithm. However, it all boils down to that, at (λ_i, a_i), the regularization parameter λ_{i+1} must be estimated on the basis of minimizing $a(\lambda)$ some amount δa. An effective and safe way of doing this is yet to be discovered.

6. Local convergence towards the corner

Assume that we have attained three points $\{(t_{i+1}, y_{i+1}), (t_i, y_i), (t_{i-1}, y_{i-1})\}$. All of them are assumed to be close to the exact L-curve and approximating a corner defined in Section 4.1. Obviously if these points are lying on the exact L-curve the inequality

$$\lambda_{i+1} \leq \lambda_{corner} \leq \lambda_{i-1} \tag{7}$$

is satisfied and it is reasonable to assume that (7) holds also for points in the vicinity to the L-curve. The general idea of the algorithm is to shrink the distances between the three points and steer them closer to the corner, which they approximate.

Consider now the solutions $x(\lambda)$ at each point and define the new solutions

$$x_{i+1,i} = \frac{x_{i+1} + x_i}{2}, \quad x_{i,i-1} = \frac{x_i + x_{i-1}}{2}, \quad x_{i+1,i-1} = \frac{x_{i+1} + x_{i-1}}{2}$$

and regularization parameters

$$\lambda_{i+1,i} = 10^{\left(\frac{1}{2}(\log(\lambda_{i+1}) + \log(\lambda_i)\right)}, \quad \lambda_{i,i-1} = 10^{\left(\frac{1}{2}(\log(\lambda_i) + \log(\lambda_{i-1})\right)},$$

$$\lambda_{i+1,i-1} = 10^{\left(\frac{1}{2}(\log(\lambda_{i+1}) + \log(\lambda_{i-1})\right)}.$$

From each point

$$\{x_{i+1,i}, x_{i,i-1}, x_{i+1,i-1}\}$$

calculate a new solution with respective regularization parameters

$$\{\lambda_{i+1,i}, \lambda_{i,i-1}, \lambda_{i+1,i-1}\}.$$

Since the L-curve is locally convex, these new solutions give rise to three new points on the L-curve with a better approximation of the corner.

7. The algorithm in total

The algorithm constructed is not to be regarded as a black box, since modifications might be needed depending on the optimization problem. The choice of the regularization parameter in each iteration may be very tedious and result in a trial and error technique. Further, the algorithm can handle situations where the approximated corner is not well defined. This might happen if the shadow L-curve is a bad approximation of the exact L-curve.

1. While no corner solution is found
2. Compute the Jacobian J_i
3. Choose a regularization parameter λ_i
4. While (t_k, y_k) not belongs to the convex set M
   (a) Compute Jacobian J_k
   (b) Iterate with $x_{k+1} = x_k + \alpha p(\lambda_i)$ where λ_i is fixed
   (c) $t_k = ||x_{k+1} - x_c||^2$, $y_k = ||f(x_{k+1})||^2$
5. If there exist three points in M approximating a corner
   (a) While approximating corner exist
      (i) Calculate $x_{i+1,i}, x_{i,i-1}, x_{i+1,i-1}$ and the new regularization parameters.
      (ii) While the three points not belong to the convex set M
         (A) Compute the Jacobians $J_{i+1,i}, J_{i,i-1}, J_{i+1,i-1}$
         (B) Iterate for all three points using fixed regularization parameters.
      (iii) If $x_{i+1,i} \approx x_{i,i-1} \approx x_{i+1,i-1}$ return $x_{i,i-1}$ as corner solution
      (iv) else update M with the new points
   (b) If the corner is lost, continue from 1 with corner searching.

8. Numerical Simulations

The inverse problem is to determine the conductivity $\sigma(x)$ from the heat transfer equation

$$-\frac{d(\sigma(x)\frac{du}{dx})}{dx} = f(x), \quad 0 < x < 1,, \tag{8}$$

$$u(0) = u_0, \quad u(1) = u_1.$$

where $f \in L^2$ and $|u_x| > 0$. The measured quantity is denoted $\tilde{u}(x)$ and then the problem can be formulated as

$$\mathcal{F}(\sigma) \approx \tilde{u},$$

where the nonlinear operator $\mathcal{F} : H_1 \to L_2$ is Fréchet-differentiable with a Lipschitz-continuous derivative.

The minimization problem is stated as

$$\min_{\sigma} \frac{1}{2}||\mathcal{F}(\sigma) - \tilde{u}||^2_{L_2} + \frac{1}{2}\lambda||\sigma - \sigma_c||^2_{H_1}. \tag{9}$$

In order to calculate $\mathcal{F}(\sigma)$ equation (8) is solved using a finite element representation with linear spline approximations according to

$$u_{fe}(x) = \sum_{j=1}^{n} \beta_j \varphi_j(x) \ , \quad \sigma(x) \approx \sum_{s=1}^{m} \theta_s \varphi_s(x) \, .$$

After discretization, the search direction $p(\lambda)$ is found by solving the discrete version of (9)

$$(J^T W_n J + \lambda W_{pm})p = b$$

where the matrix J is the Jacobian, b is a vector and W_n and W_{pm} are inner product matrices.

In the examples δ is the noise level, λ_* is the optimal regularization parameter that minimizes $\frac{||\sigma(\lambda)-\sigma_*||}{||\sigma(\lambda)||}$ where σ_* is the exact solution. λ_c and λ_d are the regularization parameters given by the L-curve and discrepancy principle method respectively.

8.1. Example 1

For this example that is the most ill-posed, the initial value

$$\sigma_0(x) = 2 + 1.428x^5 - 4.382x^4 + 1.04x^3 + 3.63x^2$$

was used. The exact solution is assumed to be $\sigma_*(x) = 1$ and the exact data $u(x) = e^x$.

The regularization parameter corresponding to the corner of the L-curve is for this example too small compared to the optimal λ_*. And when the noise increases, the estimated regularization parameter λ_c differs even more from the optimal λ_*. The regularization parameter given by the discrepancy principle is in this case a very good approximation of the optimal one.

$\delta(\%)$	λ_*	λ_c	λ_d
0.6	$2.0 \cdot 10^{-3}$	$6.0 \cdot 10^{-4}$	$2.0 \cdot 10^{-3}$
0.05	$8.2 \cdot 10^{-4}$	$1.7 \cdot 10^{-4}$	$1.1 \cdot 10^{-3}$
0.003	$2.5 \cdot 10^{-4}$	$8.1 \cdot 10^{-6}$	$3.3 \cdot 10^{-4}$

8.2. Example 2

Here the initial value used was

$$\sigma_0(x) = 1 + \frac{1}{10\sinh(1)}(9 - 4x + 4x^2 - 4(\cosh(x) - \cosh(x-1)))$$

and with assumptions that $\sigma_*(x) = 1$ and $u(s) = s(1-s)$. For this example the corner solution lies very close to the optimal one. The solution given by the discrepancy principle is too big and not as good as the corner solution.

$\delta(\%)$	λ_*	λ_c	λ_d
4	$1.1 \cdot 10^{-3}$	$2.1 \cdot 10^{-3}$	$2.7 \cdot 10^{-3}$
0.5	$7.4 \cdot 10^{-5}$	$2.2 \cdot 10^{-4}$	$1.1 \cdot 10^{-3}$
0.05	$1.7 \cdot 10^{-5}$	$1.6 \cdot 10^{-5}$	$1.4 \cdot 10^{-4}$
0.005	$6.7 \cdot 10^{-6}$	$1.6 \cdot 10^{-6}$	$4.1 \cdot 10^{-5}$

8.3. Example 3

In this example the initial guess was $\sigma_0 = 2$, $u(s) = s(1-s)$ and the exact solution was assumed to be $\sigma_*(s) = 1 + 0.1 sin(2\pi s)$. The result for this example is nearly the same as for the previous one. The corner solution is again quite close to the optimal solution, while the discrepancy principle results in a bit too large regularization parameter.

$\delta(\%)$	λ_*	λ_c	λ_d
5	$6.1 \cdot 10^{-4}$	$2.1 \cdot 10^{-3}$	$1.5 \cdot 10^{-3}$
0.5	$1.0 \cdot 10^{-4}$	$1.6 \cdot 10^{-4}$	$6.1 \cdot 10^{-4}$
0.05	$6.7 \cdot 10^{-6}$	$1.3 \cdot 10^{-5}$	$2.5 \cdot 10^{-4}$
0.005	$1.4 \cdot 10^{-7}$	$1.4 \cdot 10^{-6}$	$3.3 \cdot 10^{-4}$

8.4. Example 4

Here the exact solution has a discontinuity and is assumed to be 1 on the interval $0 \le x \le 0.5$ and 2 on $0.5 < x \le 1$. Further the data was $u(s) = s(1-s)$ and the initial guess used was $\sigma_0 = 2$.

This problem is not so ill-posed, regularization was only needed at the beginning of the iteration if the noise level was small. However, in this case λ_c and λ_d differ quite much from the optimal λ_*.

$\delta(\%)$	λ_*	λ_c	λ_d
0.5	$6.7 \cdot 10^{-6}$	$1.3 \cdot 10^{-3}$	$2.5 \cdot 10^{-4}$
0.05	$5.0 \cdot 10^{-9}$	$1.4 \cdot 10^{-5}$	$4.1 \cdot 10^{-5}$
0.005	$1.0 \cdot 10^{-10}$	$1.0 \cdot 10^{-6}$	$1.6 \cdot 10^{-5}$

9. Conclusions

We have come to the conclusion that there indeed exist cases where the "heuristic" L-curve method is capable to approximate a reasonable solution to a nonlinear ill-posed problem. Theoretically the solution given by the corner of the L-curve x_{Lc}, might even be a worse approximation of the exact x_* than the center of regularization x_c, due to the fact that the method is nonconvergent. However, when dealing with nonartificial problems, the solution $x(\lambda_c)$ is to be considered as the solution in the vicinity of x_c that results in a small residual, not as an optimal or best possible solution.

References

[1] P. C Hansen, *Rank-Deficient and Discrete Ill-posed Problems. Numerical aspects on linear inversion*, SIAM, Philadelphia, 1997.

[2] A. B Bakushinskii and A. V Goncharskii, *Ill-posed Problems: Theory and Applications,* Kluwer, Dordrecht, 1994.

[3] A. S Leonov and A. G Yagola, *The L-curve method always introduces a nonremovable systematic error*, Moscow University, Physics Bulletin, **52**, No. 6 (1997), pp. 20–23.

[4] M. Gulliksson and P. A Wedin, *Analyzing the nonlinear L-curve,* Dept. of Comp. Science, Umea University, Sweden, 1998.

[5] M. Gulliksson and P. A Wedin, *Algorithms for using the nonlinear L-curve,* Dept. of Comp. Science, Umea University, Sweden, 1998.

[6] T. Viklands, *Routines for constructing approximating L- and a-curves*, UMNAD 24.99 (Master Thesis), Dept. of Comp. Science, Umea University, Sweden, 1999.

[7] T. Reginska, *A regularization parameter in discrite ill-posed problems*, SIAM J. Sci. Comput., **17**, No. 3 (1996), pp. 223–228.

[8] M. Hanke, *Limitations of the L-curve in ill-posed problems*, BIT, **36**, No. 2 (1996), pp. 287–301.

[9] C. R Vogel, *Non-convergence of the L-curve regularization parameter selection method*, Inverse Problems, **12** (1996), pp. 535–547.

[10] T. Elfving and L.-E. Andersson, *An Algorithm for computing constrained smoothing spline functions*, Numer. Math., **52** (1988), pp. 583–595.

[11] O. Scherzer and M. Gulliksson, *Adaptive Strategy for Damping Parameters in an Iteratively Regularized Gauss-Newton Method,* Institut für Industriemathematik, Universität Linz, Austria; Department of Computing Science, Umea University, Umea, Sweden.

[12] A. S. Leonov and A. G. Yagola, *Can an ill-posed problem be solved if the data error is unknown?*, Moscow University, Physics Bulletin, **50**, No. 1 (1995), pp. 25–28.

Material Optimization with a Penalty Barrier Multiplier Method

Ralf Werner

Abstract. Free material design deals with the question of finding the stiffest elastic structure with respect to a given load. Starting with the well known formulation as a control problem, we motivate the need for a reformulation of the problem. We do not use the standard saddle point approach for a further treatment, but use methods from Lagrange duality theory. We obtain a constrained convex programming problem, which is open to standard discretization schemes. After using a finite element ansatz we get a sparse convex quadratic system which needs due to its high dimension a suitable choice of the optimization routine. We propose the use of a penalty barrier multiplier method which takes care of the sparsity structure. For this reason the Newton system is solved by a very efficient sparse Cholesky factorization. Some examples computed with the design tool MOPED illustrate the effectiveness of our approach.

1. Problem formulation

We only give a brief introduction into the material problem; for a more detailed discussion we refer the reader to [3, 6]. For simplicity, we restrict ourselves to problems in $\mathbb{R}^2$; the extension of the results to the spatial case is straightforward.

1.1. Mechanical background

We motivate the free material optimization approach by the cantilever problem shown in Figure 1. We also use this example to introduce the basic notions of

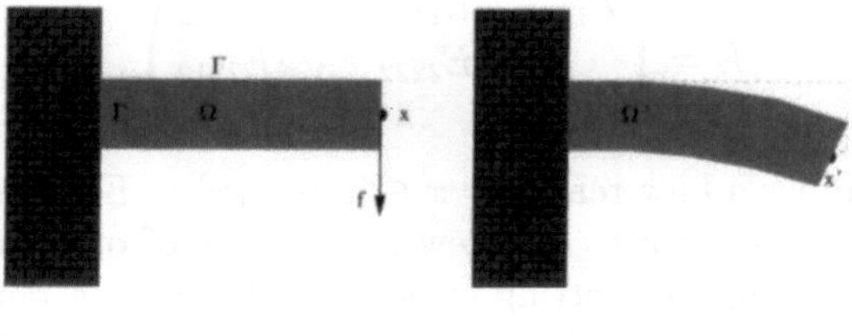

FIGURE 1. Deformation of a cantilever beam under load

the theory of linear elasticity. Let us denote by Ω the undeformed body (i.e. the cantilever beam) and by $\tilde{\Omega}$ the deformed one. Each point $x \in \Omega$ moves to a point $\tilde{x} \in \tilde{\Omega}$ due to the deformation caused by the load f. These *displacements* can be described by $u(x) := \tilde{x} - x$, i.e. u is a mapping from Ω to $\mathbb{R}^2$. To get a local characterization of the displacements u one works with the *(small) strain tensor* $\varepsilon(u) := \frac{1}{2}(u' + u'^T)$ where u' denotes the Jacobian of the mapping u. The entries of this matrix can be seen as relative extensions and shearing. The deformation will be such that an equilibrium between inner forces (measured by the stress tensor σ) and outer forces (the load f) is reached. The resulting displacements u and stresses σ are the solution of the classical formulation of the basic boundary value problem of elasticity:

$$\begin{aligned} &\operatorname{div}(\sigma) = 0 \text{ in } \Omega && \text{equilibrium condition} \\ &\sigma \cdot n = f \text{ in } \Gamma && \text{Neumann boundary condition} \\ &u|_{\Gamma_0} = 0 && \text{Dirichlet boundary condition.} \end{aligned}$$

The Dirichlet condition expresses the fact that the beam is fixed at the left side Γ_0 to the wall, the 0 on the right-hand-side of the equilibrium conditions results from neglecting inner forces like gravity (we only work with surface forces). The missing link between the inner and outer forces is (for a linear model) Hooke's Law: $\sigma(x) = E(x) \cdot \varepsilon(x)$. Here $E(x)$ is a 4-th order tensor, called *elasticity tensor*, which describes the material behaviour at a point x. So the mechanical problem of elasticity in displacements is given by:

$$\begin{aligned} &\operatorname{div}(E\varepsilon(u)) = 0 \text{ in } \Omega && \text{equilibrium condition} \\ &E\varepsilon(u) \cdot n = f \text{ in } \Gamma && \text{Neumann boundary condition} \\ &u|_{\Gamma_0} = 0 && \text{Dirichlet boundary condition.} \end{aligned} \tag{CF}$$

1.1.1. CONTRACTED NOTATION. Using the symmetries of the tensors ε and E we can reduce the second order tensor ε pointwise to a vector and the material tensor E to a matrix by the following contracted notation (let us neglect x for a moment in the notation):

$$e = (\varepsilon_{11}, \varepsilon_{22}, \sqrt{2}\varepsilon_{12})^T \in \mathbb{R}^3, \qquad \sigma = (\sigma_{11}, \sigma_{22}, \sqrt{2}\sigma_{12})^T \in \mathbb{R}^3.$$

Correspondingly, the 4−tensor E can be written as a symmetric 3×3 matrix

$$E = \begin{pmatrix} E_{1111} & E_{1122} & \sqrt{2}E_{1112} \\ & E_{2222} & \sqrt{2}E_{2212} \\ \text{sym.} & & 2E_{1212} \end{pmatrix}.$$

In this notation, Hooke's Law reads for $x \in \Omega$ as $\sigma(x) = E(x)e(u)(x)$. As we have seen for $d = 2$ we get $d' = 3$ as the new dimensions of our variables. The same trick also works for the spatial setting where we get $d' = 6$. A detailed explanation of this contracted notation can be found in the textbook of Pedersen, which is provided in the internet[1].

[1]see `http://www.fam.dtu.dk/html/pp.html`

1.1.2. THE WEAK FORMULATION. In the first step we have switched from tensor notation to contracted notation. In the second step we replace the partial differential equation by an optimization problem which is equivalent to the weak formulation of the classical problem (CF):

$$\inf_{u\in U} \frac{1}{2} \int_{\Omega} e(u)(x)^T \cdot E(x) \cdot e(u)(x)\,\mathrm{dx} - \int_{\Gamma} f(x)^T u(x)\,\mathrm{dx} \tag{WF}$$

with $U = \{u \mid u = 0 \text{ on } \Gamma_0\}$.

This is often called *principle of minimum potential energy.* One can show that each solution of the classical formulation is a solution of the weak formulation. Under certain smoothness assumptions on the solution of the weak problem the reversion is valid, too. For a deeper discussion of the equivalence see [16] or [21].

1.1.3. FREE MATERIAL DESIGN. So far we have only described the mechanical background, so let us now briefly describe the task of the designer – for a more detailed introduction we refer to [5]. The free material design approach intends to replace the cantilever beam by a structure which is better suited for the given load f. There are two main aims: first, to distribute the material only in places where material is really needed (which saves material, cf. Figure 2) and second, to take the best material available (e.g., wood, steel or glass fibres). Therefore we allow

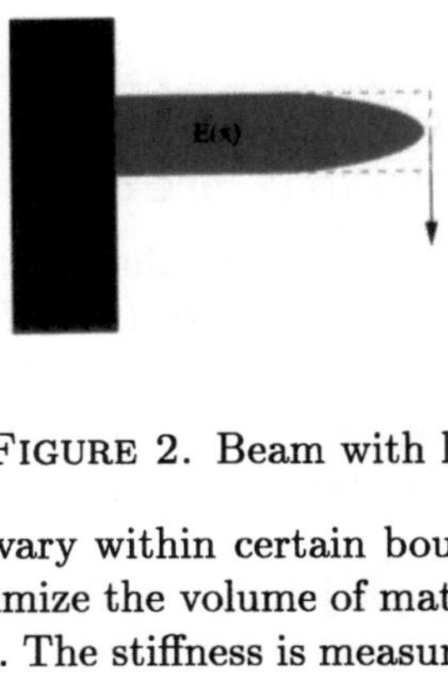

FIGURE 2. Beam with less material

the material tensor to vary within certain bounds as a function of $x \in \Omega$. As an objective we could minimize the volume of material needed to guarantee a certain stiffness of the material. The stiffness is measured by the *compliance* c (depending on the material E)

$$c(E) := \frac{1}{2} \int_{\Gamma} f(x)^T u_E(x)\,\mathrm{dx}$$

where u_E is the solution of (WF). The 'amount' (or 'total stiffness') of material can be measured by

$$\mathrm{vol}(E) := \int_{\Omega} \mathrm{tr}(E(x))\,\mathrm{dx}\ .$$

An (almost) equivalent approach is to minimize the compliance of the structure when only a given amount of material is available. Let us denote the set of feasible materials by $\mathbb{E}$, i.e. $E(x)$ should be a symmetric positive semidefinite matrix for almost all $x \in \Omega$. Furthermore, $E \in \mathbb{E}$ has to satisfy the volume condition and certain stiffness restrictions must hold:

$$\mathbb{E} = \{E \mid \operatorname{vol}(E) \leq V, \rho_l \leq \operatorname{tr}(E) \leq \rho_u\}.$$

Since we want to allow void areas and for consistency reasons we require

$$0 \leq \rho_l < V \ < \rho_u < \infty.$$

With the abbreviations

$$a_E(u,v) := \int_\Omega e(u)(x)^T \cdot E(x) \cdot e(v)(x)\, \mathrm{dx} \quad \text{and } l(u) := \int_\Gamma f(x)^T u(x)\, \mathrm{dx}$$

we can replace the formulation of the standard control problem

$$\begin{aligned}&\inf_{\substack{E\in\mathbb{E}\\ u\in U}} \int_\Gamma f(x)^T u(x)\, \mathrm{dx}\\ &\text{subject to}\\ &\operatorname{div}(E\epsilon(u)) = 0 \text{ in } \Omega\\ &E\epsilon(u)\cdot n = f \text{ in } \Gamma\end{aligned}$$

by its weak formulation

$$\begin{aligned}&\inf_{\substack{E\in\mathbb{E}\\ u\in U}} \int_\Gamma f(x)^T u(x)\, \mathrm{dx}\\ &\text{subject to}\\ &a_E(u,v) = l(v) \quad \forall v \in U.\end{aligned}$$

Although it is well known for several years now, that this formulation has several major drawbacks, it is often used for numerical treatment (see, e.g., the very recent article of Schulz [14]). Let us just mention the three most important problems:

1. It is not obvious that the above problem possesses an optimal solution (E^*, u^*) at all – please note, that in contrast to $\mathbb{E}$ the set U is not a compact set.
2. The constraint mapping is not jointly weakly continuous in (E, u) as a bilinear expression – therefore we do not even get closedness of the feasible set. This also leads to problems when applying a finite element ansatz.
3. The bilinear constraint mapping leads to a nonconvex problem, therefore it is possible to end up only with local solutions.

Having in mind these problems, it is in our opinion senseless to try to solve the above control formulation. However, we can get rid of all these problems using some ideas from duality theory. Before we do that let us have a closer look at the objective to discover that the compliance equals the optimal value of the following optimization problem

$$c(E) = \sup_{u\in U} \Big(-\frac{1}{2}a_E(u,u) + l(u)\Big).$$

Then the free material optimization problem can be written as

$$\inf_{E\in\mathbb{E}} c(E) = \inf_{E\in\mathbb{E}} \sup_{u\in U} \left(-\frac{1}{2}a_E(u,u) + l(u)\right) \tag{MO}$$

which is nothing else than minimization of the compliance. The optimal u^* of the inner maximization solves the state equation in the original formulation, thus we can neglect it. Still, this problem as a saddle point problem is hard to solve, however, seen as a problem in E it is convex and admits the computation of the global solution.

1.1.4. Mathematical proper problem setting. Let us summarize the standard assumptions from a mathematical point of view throughout this contribution:

- $\Omega \subset \mathbb{R}^d$ is a domain with Lipschitz boundary $\partial\Omega$ and $\text{meas}(\Omega) = 1$
- $\partial\Omega = \Gamma \uplus \Gamma_0 \uplus N$ (N set of measure zero),
  Γ and Γ_0 open with positive measure
- $f \in [L^2(\Gamma)]^d \subset [H^{-\frac{1}{2}}(\Gamma)]^d$
- $E \in [L^\infty(\Omega)]^{d'\times d'}$
- $u \in [H^1(\Omega)]^d \quad (H^1(\Omega) = W^{1,2}(\Omega))$
- $U \subset [H^1(\Omega)]^d$ linear subspace with $[H_0^1(\Omega)]^d \subset U$
- $\mathbb{E} \subset [L^\infty(\Omega)]^{d'\times d'}$ convex, weak-* compact

From these assumptions we can deduce

- $u \in [H^1(\Omega)]^d \Rightarrow e(u) \in [L^2(\Omega)]^{d'}$
- $u \in [H^1(\Omega)]^d \Rightarrow u \in [H^{\frac{1}{2}}(\Gamma)]^d \subset [L^2(\Gamma)]^d$, (see, e.g., [19], p. 138)

Indeed, we can show that our previously defined set of admissible materials fits into that context, i.e. it is a convex, weak-* compact set.

1.1.5. The variable thickness sheet problem. For later purposes we introduce the variable thickness sheet problem here. Let us assume for a moment that we only work in two dimensions to get a better impression of the problem. If we fix the material throughout the body to be a given material E_0 then the problem reduces to finding the optimal thickness distribution. For simplicity we assume

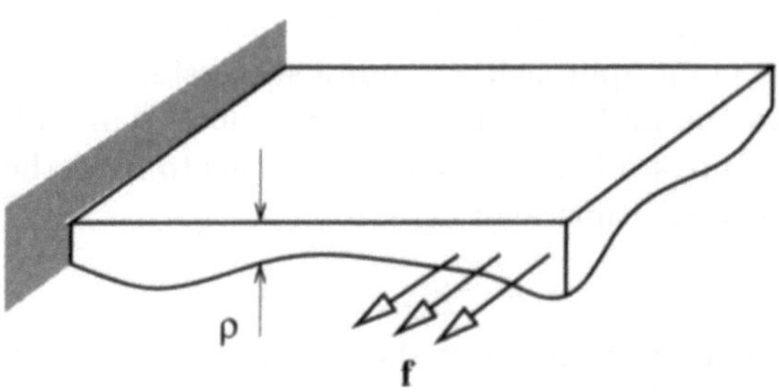

FIGURE 3. The variable thickness sheet problem

$E_0 = I_{d'}$. Then the new objective is

$$\inf_{\rho\in\mathbb{T}} \sup_{u\in U} -\frac{1}{2}\tilde{a}_\rho(u,u) + l(u) \tag{VTS}$$

$$\text{with } \mathbb{T} := \{\rho \in L^\infty(\Omega) \mid \rho_l \le \rho \le \rho_u, \int_\Omega \rho(x)\,\mathrm{dx} \le V\}$$

$$\text{and } \tilde{a}_\rho(u,v) := \int_\Omega \rho(x)\cdot e(u)(x)^T \cdot I_{d'} \cdot e(v)(x)\,\mathrm{dx}\ .$$

At first sight this looks like problem (MO) if we would put there $E(x) := \rho(x)\cdot I$. A closer look, however, shows the difference, hidden in the volume constraint: $\int_\Omega \mathrm{tr}(E(x))\,\mathrm{dx} = d' \int_\Omega \rho(x)\,\mathrm{dx}$. A different possibility of interpretation is to see $\rho(x)$ as the trace of $E(x)$, i.e. we split $E(x)$ in $E(x) = \mathrm{tr}(E(x))\cdot\tilde{E}(x)$ with $\mathrm{tr}(\tilde{E}(x)) = 1$. Now it is obvious that we cannot simply put $\tilde{E}(x) = I_{d'}$. Let us postpone the answer to this connection to the next section. In contrast to (MO) problem (VTS) is well known. Similar problems have already been treated in [7], and recently it was explored in great detail in [18]. Thus one of our aims is to reduce (MO) to (VTS) to obtain similar results. We will prove the equivalence of (MO) and (VTS) by the following steps: We first reformulate both problems to equivalent saddle point problems (MO') and (VTS') in Theorem 2.3. For these new problems we can derive primal problems (MO_P) and (VTS_P) (the meaning of *primal* will be made clear by Theorem 2.6 (2)). As a last step we see in Theorem 2.6 (1) the equivalence of these two problems, showing the required equivalence of the original ones. In Theorem 2.10 we obtain a more detailed explanation of the equivalence of (MO) and (VTS), giving a constructive connection between the saddle points of both problems

2. Lagrange Duality Approach

The problem of free material optimization as posed in (MO) has one major drawback: The formulation as a saddle point problem is not well suited for a numerical treatment. However, Zowe et. al. achieved in [6] by means of saddle point theory the derivation of a convex quadratic finite dimensional optimization problem. Though this saddle point approach was quite successful in the single load case, it is not applicable in the multiple load case any more. This difficulty lead to the question whether it is possible to use different means to treat the material problem both in the single and multiple load case.

2.1. Preliminaries

As we want to use standard Lagrange theory for the treatment of (MO), we first collect some well known facts and moreover, we introduce some notations we will need later. In our presentation we mainly follow [10]. Let us start with some basic assumptions:

(CP 1): $(X, ||.||_X)$, $(Y, ||.||_Y)$ are real separable Banach spaces,
(CP 2): $A \subset X$ is a non empty, weakly closed and convex set,
(CP 3): $f : A \to \mathbb{R}$ is continuous and convex,
(CP 4): $f : A \to \mathbb{R}$ is weakly lower semicontinuous,
(CP 5): Y is partially ordered with an order defining closed convex cone C,
(CP 6): $g : A \to Y$ is convex with respect to C,
(CP 7): $g : A \to Y$ is continuous (with respect to the norms),
(CP 8): $g : A \to Y$ is weakly lower semicontinuous with respect to C,
(CP 9): $g : A \to Y$ is bounded:

$$\sup_{||x||_X \leq r} ||g(x)||_Y \leq R_r < \infty \quad \forall r \geq 0.$$

(CP 10): $S := \{x \in A \,|\, g(x) \in -C\}$ is non empty.

Please note that it holds (see, e.g., [9], Corollary I.2.5 or [20], Theorem 6.8) for a proper convex functional in a separable Banach space that continuity is equivalent to weak lower semicontinuity. For the notion of weak lower semicontinuity with respect to a cone we refer to the following definition:

Definition 2.1. *Let X and Y be real separable Banach spaces and let $g : X \to Y$ be a bounded operator. Further let Y be ordered by an ordering cone C. Then g is called* weakly lower semicontinuous with respect to the cone C *if $\mu \circ g : X \to \mathbb{R}$ is a weakly lower semicontinuous functional for all $\mu \in C^*$.*

We do not want to go into detail here, but one can find similar definitions in some lecture notes of Luc (cf. [13], Section 1.5, cone continuous functions, and Chapter 4, scalarization).

The generic optimization problem which we want to examine in this framework becomes the *primal problem*

$$\inf_{x \in A} f(x) \qquad \text{(P)}$$
$$\text{subject to}$$
$$g(x) \in -C$$

and the *dual problem* of (P) in the sense of Lagrange is

$$\sup_{\mu \in C^*} \Theta(\mu) \qquad \text{(D)}$$

with the dual objective function $\Theta(\mu) := \inf_{x \in A} L(x, \mu)$ and the *Lagrange function* associated to (P),

$$L(x, \mu) := f(x) + \mu(g(x)).$$

Each solution x^* of (P) we call *primal solution* of (P) and each solution μ^* of (D) we call *dual solution* of (P). The standard question in Lagrange duality theory is whether it is possible to extend the weak duality – $\theta(\mu) \leq f(x)$ for all feasible pairs (x, μ) – to strong duality. For that purpose we first need the notion of a *constraint qualification* (CQ). A (CQ) is a condition on the primal problem (P) (or on its solutions) which guarantees the solvability of (D). In standard Lagrange duality theory one often works with the *Slater condition* (SC) as a (CQ). The (SC) holds iff there exists $x_0 \in A$ with $g(x_0) \in -int(C)$. This x_0 is called Slater point. However this is a very strong condition since it requires that the cone C has nonempty interior. In our situation we will work with a cone in $L^1(\Omega)$ which does not fit in here. Therefore we need a different (and weaker) constraint qualification. Indeed we have to assume now that f and g are Fréchet differentiable with derivatives f' and g' resp. We call a feasible point x_0 of (P) *regular* iff the Kurcyusz-Robinson-Zowe condition (KRZ) is fulfilled in x_0, i.e.

$$\{g'(x_0)\big(\lambda_1(x-x_0)\big) + \big(c+\lambda_2 g(x_0)\big) \,|\, x \in A, \lambda_1 \geq 0, \lambda_2 \geq 0, c \in C\} = Y.$$

Now we are ready to formulate the main result of Lagrange duality theory.

Theorem 2.2 (Strong Duality Theorem). *Let (P) be solvable with optimal solution x^*. If one of the above constraint qualification (SC) or (KRZ) holds (at x^*) then (D) is also solvable. For each solution μ^* it holds: $\theta(\mu^*) = f(x^*)$.*

Proof. See [12]. □

2.2. Connection to (MO)

So far we still do not see how problem (MO) could fit into the above context. For that reason we will proceed in two steps. First, we reformulate (MO) to an equivalent saddle point problem (MO'). Second, we recognize (MO') as the Lagrange dual of a constrained convex programming problem.

2.2.1. Deriving an equivalent saddle point problem. Let us start with the original problem (MO). The main idea of this first step is to use some penalization to move constraints on the material to the objective. Thus the equivalent problem (MO') becomes

$$\inf_{\substack{E\in\mathbb{E}\\ E\succeq 0}} \sup_{\substack{u\in U\\ \alpha\in\mathbb{R},\alpha\geq 0\\ \beta_l,\beta_u\in L^1(\Omega)\\ \beta_l,\beta_u\geq 0}} -\frac{1}{2}a_E(u,u) + l(u) + \alpha\cdot\Big(\int_\Omega \mathrm{tr}(E(x))\,\mathrm{dx} \;-V\Big) +$$

$$+\int_\Omega \beta_l(x)\cdot(\rho_l - \mathrm{tr}(E(x)))\,\mathrm{dx} \;-\int_\Omega \beta_u(x)\cdot(\rho_u - \mathrm{tr}(E(x)))\,\mathrm{dx}\,. \quad \text{(MO')}$$

We can apply the same idea to the formulation of the variable thickness sheet problem and derive in an analogous way a problem (VTS'). The important and quite obvious connection between (MO) and (MO') is given by the following theorem.

Theorem 2.3. *Problems (MO) and (MO') are equivalent in the following sense:*

1. *Each saddle point (E^*, u^*) of (MO) can be extended to a saddle point $(E^*, (u^*, \alpha^*, \beta_l^*, \beta_u^*))$ of (MO').*
2. *Each saddle point $(E^*, (u^*, \alpha^*, \beta_l^*, \beta_u^*))$ of problem (MO') yields a saddle point (E^*, u^*) of (MO).*

The optimal values of the two problems are equal. Furthermore the same relationship holds for problems (VTS) and (VTS').

Remark 2.4. *If we are interested in the equivalence of (MO) and (MO') seen as optimization problems in E only, we do not need the above theorem. This equivalence is quite obvious and follows for example from* [10], *Lemma 6.5. In fact, we do not need the above result, an equivalence in E is sufficient for our purposes. In contrast to this point of view, Theorem 2.3 states more: first, we show the equivalence of both problems seen as saddle point problems and second, we prove existence of optimal* $(\alpha^*, \beta_l^*, \beta_u^*)$. *We will see later, that this existence is by far not trivial.*

Proof. See [22], Theorem 3.3.1. □

2.2.2. Recognizing the primal problem. After we have shown the equivalence of the saddle point problems we will have a closer look at problem (MO'). We reorder the terms of the objective by collecting all terms including the unknown material tensor E. Then we can write (MO') as

$$\inf_{\substack{E\in\mathbb{E}\\ E\succeq 0}} \sup_{\substack{u\in U\\ \alpha\in\mathbb{R},\alpha\geq 0\\ \beta_l,\beta_u\in L^1(\Omega)\\ \beta_l,\beta_u\geq 0}} l(u) - \alpha\cdot V - \rho_u \int_\Omega \beta_u(x)\,\mathrm{dx} + \rho_l \int_\Omega \beta_l(x)\,\mathrm{dx} + \\ + \int_\Omega \mathrm{tr}\Big(E(x)\cdot(-\frac{1}{2}e(u)(x)e(u)(x)^T + (\alpha+\beta_u(x)-\beta_l(x))\cdot I_{d'})\Big)\,\mathrm{dx}$$

For this reformulation we have used the fact that for arbitrary $E\in\mathbb{R}^{n\times n}, E = E^T, e\in\mathbb{R}^n$ we have $\mathrm{tr}(Eee^T) = e^TEe$. If we have a closer look at the structure of (MO') we see that $E(x)$ plays the role of a Lagrange multiplier and that this problem (after multiplying by -1) can be seen as the Lagrange dual of

$$\inf_{\substack{u\in U\\ \alpha\in\mathbb{R}\\ \beta_l,\beta_u\in L^1(\Omega)\\ \beta_l,\beta_u\geq 0}} -l(u) + \alpha\cdot V + \rho_u\cdot\int_\Omega \beta_u(x)\,\mathrm{dx} - \rho_l\cdot\int_\Omega \beta_l(x)\,\mathrm{dx} \qquad (\mathrm{MO}_P)$$

subject to

$$\frac{1}{2}e(u)e(u)^T \preceq (\alpha+\beta_u-\beta_l)\cdot I_{d'}.$$

We can use the same trick for the variable thickness sheet formulation and end up with

$$\inf_{\substack{u\in U\\ \alpha\in\mathbb{R}\\ \beta_l,\beta_u\in L^1(\Omega)\\ \beta_l,\beta_u\geq 0}} \quad -l(u)+\alpha\cdot V+\rho_u\cdot\int_\Omega \beta_u(x)\,\mathrm{dx} \;-\rho_l\cdot\int_\Omega \beta_l(x)\,\mathrm{dx} \qquad (\mathrm{VTS}_P)$$

subject to

$$\frac{1}{2}e(u)^T e(u)\leq \alpha+\beta_u-\beta_l.$$

Let us explain again what we have done. For simplicity we will start with (VTS_P). If one looks at (VTS_P) as a standard convex optimization problem, it is possible to write down the associated Lagrange dual problem as previously defined. Therefore it is necessary to calculate the dual space of the space in which the constraints take values. In our case the constraint obviously takes values in $L^1(\Omega)$. The associated cone in this space is the cone of all $L^1(\Omega)$ functions which are positive almost everywhere. The dual space to $L^1(\Omega)$ is $L^\infty(\Omega)$ with the dual cone consisting of all $L^\infty(\Omega)$ functions which are positive almost everywhere. The dual problem coincides exactly with problem (VTS'). In the free material setting the problem is just a bit more complicated. Instead of taking values in $L^1(\Omega)$, the constraint takes values in $\mathbb{L}=\{A\in L^1(\Omega)^{d'\times d'}\mid A=A^T\}$. Some calculations show that the dual space to $\mathbb{L}$ is exactly $\mathbb{E}$. The order defining cone $\mathbb{L}^+$ in $\mathbb{L}$ is extended from the standard cone in $L^1(\Omega)$ via the well known Löwner ordering (see, e.g., [15]) for symmetric matrices to $\mathbb{L}^+$. We can easily deduce that $(\mathbb{L}^+)^*=\mathbb{E}^+$, which shows that (MO') is indeed the Lagrange dual of (MO_P).

Let us conclude these paragraphs with a summary of the above results. We first guarantee that problems (MO_P) and (VTS_P) fit into the general context.

Proposition 2.5. *Problems (MO_P) and (VTS_P) fulfill the assumptions (CP 1) to (CP 10).*

Proof. See [22], Proposition 3.3.3. □

Theorem 2.6. *The following statements hold:*

1. *Problems (VTS_P) and (MO_P) are equivalent.*
2. *Problem (VTS) and (MO) are (equivalent to) the Lagrange dual problems of (VTS_P) and (MO_P).*

Proof.

1. Problems (VTS_P) and (MO_P) can only be distinguished by their different constraint. However, if we have a closer look at the semidefinite constraint of (MO_P) we see that it holds:

$$\frac{1}{2}e(u)e(u)^T\preceq(\alpha+\beta_u-\beta_l)I_{d'}\Leftrightarrow\frac{1}{2}e(u)^T e(u)\leq\alpha+\beta_u-\beta_l.$$

This is due to the dyadic nature of the first term. Therefore both problems have the same solutions and the same optimal value.

2. This follows directly from the definition of the Lagrange dual.

□

2.3. Primal and dual existence

So far we have only given reformulations of the original problems (MO) and (VTS) and we have shown that their corresponding primal problems are in some sense equivalent. Now we want to explore this connection in more detail. We will state primal and dual existence of both problems and we will clarify the connection of their optimal sets.

Theorem 2.7 (Primal existence). *Problems (VTS_P) and (MO_P) possess at least one optimal solution.*

The existence of a solution of the primal problem is not apparent. With the help of Korn's inequality we are able to prove that any possible optimal solution has to lie within a norm bounded ball. However, as we work in $L^1(\Omega)$ which is a non-reflexive Banach space, this is not enough to get weak compactness of the feasible set. It needs some work to overcome this problem, which in fact could be solved using a theorem from functional analysis going back to Banach-Saks and which is strongly related to Mazur's Lemma (see, e.g., [8], Theorem 2.1.2 and [2]).

Theorem 2.8 (Dual existence). *Both problem (VTS_P) and (MO_P) possess at least one optimal dual solution.*

Proof. Some technical calculations show that indeed the constraint qualification (KRZ) is fulfilled and therefore strong duality holds. □

Let us explain again the importance of the previous two theorems. The theorem on dual existence states the existence of optimal solutions E^* and ρ^* for (MO) and (VTS). The first theorem on primal solvability shows the existence of the corresponding (optimal) displacement field u^*. If we combine these two theorems we obtain the following corollary:

Corollary 2.9. *Both problems (MO) and (VTS) possess at least one saddle point (E^*, u^*) and (ρ^*, u^*).*

Given the close connection between (MO_P) and (VTS_P) which states that both problems have the same solutions u^*, we may ask whether there is a related link between the dual solutions of both problems.

Theorem 2.10 (Equivalence of (MO) and (VTS)).

1. $(\rho^*, u^*) \in \mathbb{T} \times U$ *is a saddle point of (VTS)* $\Rightarrow$ $(E^*, u^*) \in \mathbb{E} \times U$ *is a saddle point of (MO) with* $E^* := \rho^* \cdot \frac{e(u^*)e(u^*)^T}{\|e(u^*)\|_2^2}$.
2. $(E^*, u^*) \in \mathbb{E} \times U$ *is a saddle point of (MO)* $\Rightarrow$ $(\rho^*, u^*) \in \mathbb{T} \times U$ *is a saddle point of (VTS) with* $\rho^* := tr(E^*)$.

Proof. This can be seen in two different ways. A first way to obtain this result is to write down the Karush-Kuhn-Tucker system of (MO_P) and (VTS_P). A comparison of the two systems shows the statement. The second possibility is to use characterizations of saddle points by derivatives. One can write down necessary conditions on saddle points for both (MO) and (VTS) and by means of SDP one can prove the equivalence of both conditions. □

2.4. Further results

We do not want to present all the results one can derive with the help of the above theory, but let us mention the most important ones. First, we can show, that the resulting design is governed by a weak bang bang principle. This means that the stiffness ρ only attains values in between its lower and upper bound, if the local strain energy equals some mean value. On the other hand we can show, that if the upper bound is never attained, we obtain a *fully strained design*, i.e. the local strain energy equals its average everywhere. We can also prove the uniqueness of the resulting displacement field, if we assume that $\rho_l > 0$, and without this assumption it is obvious that we can only make statements on the displacements in the regions with material, i.e. where $\rho^*(x) > 0$. Results on uniqueness like Petersson derived in [18] are also attainable, using theorems on the uniqueness of Lagrange multipliers, giving necessary and sufficient conditions. However, these conditions are hard to check and rarely satisfied.

3. Numerical treatment

The whole aim of the presented treatment was to reach a formulation which is well suited for numerical purpose. We have seen that we have started with a nonconvex optimization problem, we have reformulated it as a saddle point problem and we have finally reached a convex constrained programming problem.

3.1. Discretization scheme

Both Problems (MO_P) and (VTS_P) can be discretized using a standard finite element ansatz. Having in mind Theorem 2.6 and Theorem 2.10 we only use problem (VTS_P) for the discretization. Nevertheless, this still leads to approximations of the solutions of (MO), hence we call the discretized problem (MO_D). Let us partition the body Ω into M disjoint quadrangular elements with N nodes. We use a bilinear ansatz for u in each element, the rest of the variables is approximated by piecewise constant functions. After adjusting the degree of freedom according to the boundary condition this leads to

$$\min_{\substack{u\in\mathbb{R}^N\\ \alpha\geq 0\\ \beta_l,\beta_u\in\mathbb{R}^M_+}} \quad -f^T u + \alpha V - \rho_l \cdot \omega^T \beta_l + \rho_u \cdot \omega^T \beta_u \qquad \text{(MO}_D\text{)}$$

subject to

$$\frac{1}{2}u^T K^m u - \omega_m \cdot \alpha + \omega_m \cdot \beta_{l,m} - \omega_m \cdot \beta_{u,m} \leq 0, \quad m = 1, \ldots, M,$$

where K^m is the element stiffness matrix of the m-th element. Please note that these are large sparse positive semidefinite matrices, therefore (MO_D) is a sparse quadratic convex programming problem.

3.2. Convergence Results

Very recently we have shown in [22], Chapter 4, that this is indeed an effective approach: We can show weak-* convergence of the multiplier (the stiffness distribution ρ) to an optimal material distribution, furthermore we have weak convergence of all other variables. Under the (pretty weak) assumption of uniqueness of the optimal displacement field we can even show norm convergence of the displacements in $H^1(\Omega)$. We want to mention that this is not as easy as it seems, since we have to show weak convergence of L^1 variables.

3.3. Numerical solution of (MO_D)

Problem (MO_D) is a standard quadratic convex programming problem and as such open to standard methods. We could have used either SQP methods or Interior Point Algorithms. Nevertheless our method of choice was a penalty barrier multiplier (PBM) method. We opted for that algorithm for two reasons: First, it is quite easy to use the (constant) sparsity pattern of the Hessian of the constraints and second, we hope that the PBM method is extendable to SDP. In contrast to other approaches it seems easier to use the same idea for semidefinite problems as for convex ones. The need for that extension lies in the fact that multiple load problems no longer lead to convex, but semidefinite programming problems.

3.3.1. THE PBM METHOD. We do not want to give a complete description of the PBM method, the interested reader should refer to [1, 4]. The main idea of the PBM method is to replace the generic optimization problem by a transformed one:

$$\inf_{x\in A} f(x)$$

subject to

$$p \cdot \varphi(g(x)/p) \leq 0.$$

We simply have replaced the constraint $g(x) \leq 0$ by a new one. For that purpose we need $p > 0$ and the transformation function has to meet certain requirements. The associated Lagrange function becomes

$$L_\varphi(x, u, p) := f(x) + u \cdot p \cdot \varphi(g(x)/p).$$

Instead of looking for saddle points of the original one we try to find saddle points of the augmented Lagrange function, which are the same as of the original one. The conceptual algorithm proceeds in three steps:

1. $x^{k+1} = \arg\min_x L_\varphi(x, u^k, p^k)$,
2. $u^{k+1} = u^k \varphi'(\frac{g(x^{k+1})}{p^k})$,
3. $p^{k+1} = c^k u^{k+1}$.

The unconstrained minimization is carried out by Newton's method. For this reason we have to calculate the Hessian of the augmented Lagrangian and we have to compute the Newton direction. Here we use the sparsity of (MO_D) to accelerate the computation. We have used a very efficient sparse Cholesky factorization from Ng and Peyton [17], which yields a tremendous speed-up.

3.3.2. CONCLUDING EXAMPLE. We have used the classical MBB beam example to demonstrate the power of our implementation. In addition to the implementation of the PBM method in C, we have developed a tool for material optimization, called MOPED, see [11]. This tool is also able to produce pictures of the resulting stress fields. In the first figure we have illustrated the stiffness distribution computed with two different meshes (500 and 20000 elements). In the second figure we have

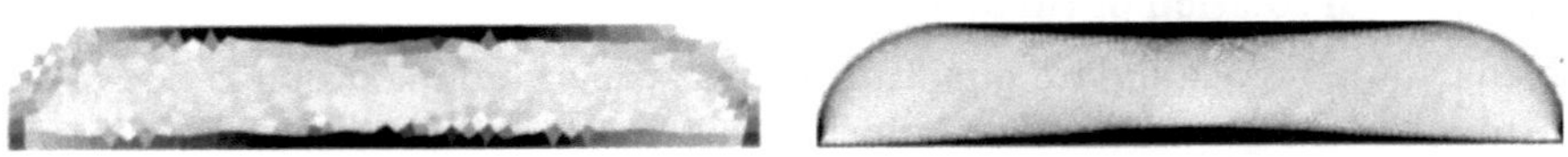

FIGURE 4. Stiffness distribution for the MBB beam

combined the plot of the von Mises stresses together with an illustration of the stress field. One can clearly distinguish between regions with high and low stress and even the main stress directions are visualized. The results of the computations

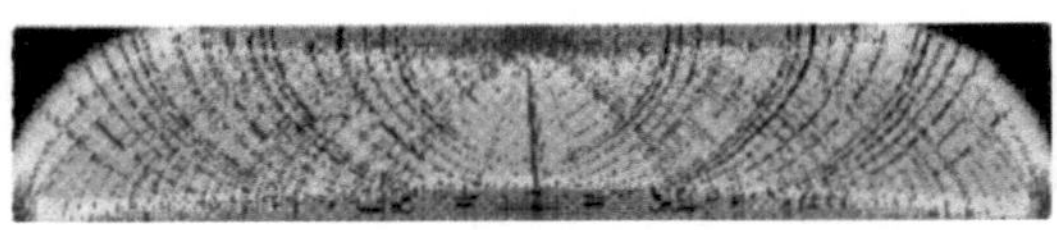

FIGURE 5. Visualization of the stress in the beam

with six different meshes are collected in Table 1. The first column contains the number of elements, followed by the number of variables and constraints. The computation time T is given in seconds, further the number of Newton steps, PBM steps and linesearch iterations. We see that our implementation of the PBM method is comparable, if not superior, to state-of-the-art interior point methods. A comparison to numerical results in [14, 18] shows the efficiency of our approach.

q	n	m	T	#N	#P	#L
453	1919	1360	10	44	21	95
4806	19641	14419	201	71	21	148 3
11073	44937	33220	835	96	47	238
16026	64913	48079	1663	90	33	223

TABLE 1. Numerical results on a standard PC

References

[1] A. Auslender, *Penalty and barrier methods: A unified framework*, SIAM J. Optimization, **10** (1999), pp. 211–230.

[2] S. Banach and S. Saks, *Sur la convergence forte dans les champs l^p*, Studia Math., **2** (1930), pp. 51–57.

[3] A. Ben-Tal, M. Kočvara, A. Nemirovski, and J. Zowe, *Free material design via semidefinite programming: The multiload case with contact conditions*, SIAM J. Optimization, **9** (1999), pp. 813–832.

[4] A. Ben-Tal and M. Zibulevsky, *Penalty/barrier multiplier methods: A new class of augmented Lagrangian algorithms for large scale convex programming problems*, SIAM J. Optimization, **7**, No. 2 (1997), pp. 347–366.

[5] M. P. Bendsøe, *Optimization of Structural Topology, Shape, and Material*, Springer, Berlin, 1995.

[6] M. P. Bendsøe, M. Kočvara, and J. Zowe, *Free material optimization via mathematical programming*, Mathematical Programming Series B, **79** (1997), pp. 445–466.

[7] J. Céa and K. Malanowski, *An example of a max-min problem in partial differential equations*, SIAM J. Control, **8** (1970), pp. 305–316.

[8] B. Dacorogna, *Direct Methods in the Calculus of Variations*, Springer, New York, 1989.

[9] I. Ekeland and R. Temam, *Convex Analysis and Variational Problems*, North Holland, Amsterdam, 1976.

[10] J. Jahn, *Introduction to the Theory of Nonlinear Optimization*, Springer, Berlin, 1994.

[11] M. Kočvara, M. Stingl, and R. Werner, *Moped user's guide – version 1.02*, Preprint 262, Lehrstuhl für Angewandte Mathematik der Friedrich-Alexander-Universität Erlangen-Nürnberg, 1999.

[12] S. Kurcyusz and J. Zowe, *Regularity and stability for the mathematical programming problem in Banach spaces*, Appl. Math. Optim., **5** (1979), pp. 49–62.

[13] D. T. Luc, *Theory of Vector Optimization*, Springer, New York, 1989.

[14] B. Maar and V. Schulz, *Interior point multigrid methods for topology optimization*, Struct. Multidisc. Optim., **19** (2000), pp. 214–224.

[15] J. E. Nesterov and A. Nemirovski, *Interior Point Polynomial Time Methods in Convex Programming: Theory and Applications*, SIAM, Philadelphia, 1996.

[16] J. Nečas and I. Hlaváček, *Mathematical Theory of Elastic and Elasto-Plastic Bodies: An Introduction*, Elsevier, Amsterdam, 1981.

[17] E. Ng and B. W. Peyton, *Block sparse Cholesky algorithms on advanced uniprocessor computers*, SIAM J. Scientific Computing, **14** (1993), pp. 1034–1056.

[18] J. Petersson, *Optimization of structures in unilateral contact*, PhD thesis, Department of Mechanical Engineering, Linköping University, Sweden, 1995.

[19] J. N. Reddy, *Applied Functional Analysis and Variational Methods in Engineering*, McGraw-Hill, New York, 1986.

[20] R. T. Rockafellar, *Conjugate Duality and Optimization*, SIAM, Soc. for Industrial and Applied Math., Philadelphia, 1974.

[21] W. Velte, *Direkte Methoden der Variationsrechnung*, Teubner, Stuttgart, 1976.

[22] R. Werner, *Free material optimization*, PhD thesis, Lehrstuhl für Angewandte Mathematik II, Friedrich-Alexander-Universität Erlangen-Nürnberg, Germany, 2000.

List of authors and co-authors

Battermann, Astrid: University of Trier, FB IV – Mathematics, D–54286 Trier, Germany, email: batt@uni-trier.de *and* Virginia Polytechnic Institute & State University, Interdisciplinary Center for Applied Mathematics, Department of Mathematics, Blacksburg, VA 24061, USA

Binder, Thomas: RWTH Aachen, Lehrstuhl für Prozesstechnik, D–52056 Aachen, Germany

Blank, Luise: RWTH Aachen, Institute for Geometrical and Practical Mathematics, D–52056 Aachen, Germany, email: blank@igpm.rwth-aachen.de

Borrvall, Thomas: University of Linköping, Department of Mechanical Engineering, Mechanical Engineering Systems, SE–581 83 Linköping, Sweden, email: thobo@ikp.liu.se

Borzi, Alfio: Karl Franzens University of Graz, Institute of Mathematics, Heinrichstrasse 36, A–8010 Graz, Austria,
email: Alfio.Borzi@kfunigraz.ac.at

Buß, Hinderk M.: University of Heidelberg, Institute of Applied Mathematics, Im Neuenheimer Feld 293, D–69120 Heidelberg, Germany,
email: hinderk.buss@iwr.uni-heidelberg.de

Dahmen, Wolfgang: RWTH Aachen, Institute for Geometrical and Practical Mathematics, D–52056 Aachen, Germany

Diwoky, Franz: Karl Franzens University of Graz, Institute of Mathematics, Heinrichstrasse 36, A–8010 Graz, Austria,
email: franz.diwoky@kfunigraz.ac.at

Dorofeev, Konstantin: Moscow State University, Physical Faculty, Department of Mathematics, Vorobyevy Gory, 199899 Moscow, Russia,
email: dorofeev@afrodita.phys.msu.su

Dostál, Zdenek: Technical University of Ostrava, Department of Applied Mathematics, 17. listopadu, CZ–70833 Ostrava–Poruba, Czech Republic,
email: zdenek.dostal@vsb.cz

Gabasov, Rafail: Belarusian State University, Faculty of Applied Mathematics and Informatics, 4, Scorina ave., 220050, Minsk, Belarus

Gulliksson, Marten: University of Umea, Department of Computing Science, S–90787 Umea, Sweden email: marten@cs.umu.se

Haase, Gundolf: Johannes Kepler University of Linz, Institute of Analysis and Computational Mathematics, Altenbergerstrasse 69, A–4040 Linz, Austria, email: ghaase@numa.uni-linz.ac.at

Hintermüller, Michael: Karl Franzens University of Graz, Department of Mathematics, Heinrichstrasse 36, A–8010 Graz, Austria,
email: michael.hintermueller@kfunigraz.ac.at

Kirillova, Faina M.: National Academy of Sciences, Institute of Mathematics, 11, Surganov str., 220072, Minsk, Belarus, email: kirill@nsys.minsk.by

Kružík, Martin: Center of Advanced European Studies And Research (caesar), Friedensplatz 16, D–53111 Bonn, Germany *and*
Academy of Sciences, Institute of Information Theory and Automation, Pod vodárenskou věží 4, CZ–182 08 Praha 8, Czech Republic,
email: kruzik@utia.cas.cz

Kunisch, Karl: Karl Franzens University of Graz, Institute of Mathematics, Heinrichstrasse 36, A–8010 Graz, Austria,
email: Karl.Kunisch@kfunigraz.ac.at

Langer, Ulrich: Johannes Kepler University of Linz, Institute of Analysis and Computational Mathematics, Altenbergerstrasse 69, A–4040 Linz, Austria, email: ulanger@numa.uni-linz.ac.at

Lindner, Ewald: Johannes Kepler University of Linz, Institute of Analysis and Computational Mathematics, Altenbergerstrasse 69, A–4040 Linz, Austria, email: lindner@numa.uni-linz.ac.at

Liu, WenBin: University of Kent, CBS CT2 7PE, UK,
email: W.B.Liu@ukc.ac.uk

Logashenko, Dmitriy: University of Heidelberg, Interdisciplinary Center for Scientific Computing (IWR), Im Neuenheimer Feld 368, D–69120 Heidelberg, Germany, email: dimitriy.logashenko@iwr.uni-heidelberg.de

Maar, Bernd: University of Heidelberg, Interdisciplinary Center for Scientific Computing (IWR), Im Neuenheimer Feld 368, D–69120 Heidelberg, Germany

Marquardt, Wolfgang: RWTH Aachen, Lehrstuhl für Prozesstechnik, D–52056 Aachen, Germany

Mittelmann, Hans D.: Arizona State University, Department of Mathematics, Tempe, AZ 85287–1804, USA, email: mittelmann@asu.edu

Mohammadi, Bijan: University of Montpellier II, Mathematiques, CC 51, 34095 Montpellier, France,
email: Bijan.Mohammadi@math.univ-montp2.fr

Mühlhuber, Wolfram: Johannes Kepler University of Linz, Spezialforschungsbereich SFB F013 "Numerical and Symbolic Scientific Computing", Freistädterstrasse 313, A–4040 Linz, Austria,
email: wmuehlhu@sfb013.uni-linz.ac.at

Nzali, Appolinaire: Humboldt University Berlin, Institute of Mathematics, D–10099 Berlin, Germany, email: a_nzali@yahoo.de

Petersson, Joakim: University of Linköping, Department of Mechanical Engineering, Mechanical Engineering Systems, SE–581 83 Linköping, Sweden, email: joape@ikp.liu.se

Rettig, Uwe: Technical University of Munich, Center for Mathematical Sciences, D–80290 München, Germany, email: rettig@ma.tum.de

Roubíček, Tomáš: Charles University, Mathematical Institute, Sokolovská 83, CZ–186 75 Praha 8, Czech Republic, email: roubicek@karlin.mff.cuni.cz *and* Institute of Information Theory and Automation, Academy of Sciences, Pod vodárenskou věží 4, CZ–182 08 Praha 8, Czech Republic

Sachs, Ekkehard W.: University of Trier, FB IV – Mathematics, D–54286 Trier, Germany *and* Virginia Polytechnic Institute & State University, Interdisciplinary Center for Applied Mathematics, Department of Mathematics, Blacksburg, VA 24061, USA, email: sachs@icam.vt.edu

Schulz, Volker: Weierstrass Institute for Applied Analysis and Stochastics, Mohrenstrasse 39, D–10117 Berlin, Germany, email: schulzv@wias-berlin.de

von Stryk, Oskar: Technical University of Darmstadt, Department of Computational Sciences, Alexanderstrasse 10, D–64283 Darmstadt, Germany, email: stryk@informatik.tu-darmstadt.de

Viklands, Thomas: University of Umea, Department of Computing Science, S–90787 Umea, Sweden, email: viklands@cs.umu.se

Volkwein, Stefan: Karl Franzens University of Graz, Institute of Mathematics, Heinrichstrasse 36, A–8010 Graz, Austria, email: stefan.volkwein@kfunigraz.ac.at

Vondrák, Vít: Technical University of Ostrava, Department of Applied Mathematics, 17. listopadu, CZ–70833 Ostrava–Poruba, Czech Republic, email: vit.vondrak@vsb.cz

Werner, Ralf: Friedrich Alexander University of Erlangen Nuremberg, Institute of Applied Mathematics, Martensstrasse 3, D–91058 Erlangen, Germany, email: werner@am.uni-erlangen.de

Wittum, Gabriel: University of Heidelberg, Interdisciplinary Center for Scientific Computing (IWR), Im Neuenheimer Feld 368, D–69120 Heidelberg, Germany

Yagola, Anatoly: Moscow State University, Physical Faculty, Department of Mathematics, Vorobyevy Gory, 119899 Moscow, Russia, email: yagola@inverse.phys.msu.su

MIX
Papier aus verantwortungsvollen Quellen
Paper from responsible sources
FSC® C105338

If you have any concerns about our products,
you can contact us on
ProductSafety@springernature.com

In case Publisher is established outside the EU,
the EU authorized representative is:
Springer Nature Customer Service Center GmbH
Europaplatz 3, 69115 Heidelberg, Germany

Printed by Libri Plureos GmbH
in Hamburg, Germany